METHODS IN MOLECULAR BIOLOGY

Series Editor
John M. Walker
School of Life and Medical Sciences
University of Hertfordshire
Hatfield, Hertfordshire, AL10 9AB, UK

For further volumes:
http://www.springer.com/series/7651

Cilia

Methods and Protocols

Edited by

Peter Satir

Albert Einstein College of Medicine, Yeshiva University, Bronx, NY, USA

Søren Tvorup Christensen

Department of Biology, University of Copenhagen, Copenhagen, Denmark

Editors
Peter Satir
Albert Einstein College of Medicine
Yeshiva University
Bronx, NY, USA

Søren Tvorup Christensen
Department of Biology
University of Copenhagen
Copenhagen, Denmark

ISSN 1064-3745 ISSN 1940-6029 (electronic)
Methods in Molecular Biology
ISBN 978-1-4939-8139-7 ISBN 978-1-4939-3789-9 (eBook)
DOI 10.1007/978-1-4939-3789-9

Printed on acid-free paper

This Humana Press imprint is published by Springer Nature
The registered company is Springer Science+Business Media LLC New York

Preface

This volume is the result of an explosion of molecular-based research on cilia which began with the discovery of the universality of intraflagellar transport (IFT) and ciliary genomics/proteomics at the turn of the millennium and attracted new interest and new investigators into the field. The cilium is a cell organelle with nanometer substructure which can be studied with techniques at the cutting edge of molecular biology. We invited expert contributors associated with these techniques, their sometimes specialized instrumentation and their implications for ciliary biology, to join us in writing chapters for the volume. Unlike previous Methods volumes, this one is broadly based, covering motile, sensory, and primary cilia and it should be attractive to anyone interested in entering the field of ciliary biology using model organisms including flagellate algae (*Chlamydomonas*), ciliates (*Paramecium*), planaria, nematodes (*C. elegans*), insects (*Drosophila*), zebrafish and *Xenopus* or mammalian or human cells.

We appreciate the time and effort our contributors have given to making this volume highly useful to both seasoned and novice investigators. Similarly, we appreciate the efforts of John Walker in keeping our schedule mainly on track and our format mainly in order. If we have sometimes been delayed in communicating with you or in keeping to schedule, it has been largely due to the welcome arrival of Ella.

Support is acknowledged where appropriate in each chapter, but some of our overall collaboration resulting in our co-editing of this volume relied on a fellowship and support from the Lundbeck foundation.

Bronx, NY, USA — *Peter Satir*
Copenhagen, OE, Denmark — *Søren T. Christensen*

Introduction

Cilia are everywhere. By this we mean that most cells of all animal as well as many plant and protistan phyla bear cilia, defined as membrane bounded organelles based on a cylindrical nine doublet microtubule axonemal cytoskeleton arising from a centriole-like basal body. Cilia fall into different classes, which are defined by their axonemal arrangement and capacity to function as motile and/or sensory units [1]. Most types of cilia are assembled onto the basal body by intraflagellar transport (IFT), which is characterized by kinesin-2 and cytoplasmic dynein 2-mediated bidirectional trafficking of proteins along the axoneme from the ciliary base towards the tip and back [2]. Further, structural and functional barriers at the proximal region of the cilium, known as the ciliary transition zone, regulate ciliary formation, membrane composition, and function via the selective passage of proteins into and out of the ciliary compartment [3–5]. In many cases, the region between the ciliary membrane and the plasma membrane folds inwards to form a ciliary pocket that comprises an active site for exocytosis and clathrin-dependent endocytosis (CDE) of ciliary proteins [6–8].

Presumably, cilia arose during the evolution of the eukaryotic cell and were present in LECA, the last eukaryotic common ancestor, probably as motile cilia with a classic 9 + 2 microtubule-based axoneme, where dynein arms attached to the microtubule doublets hydrolyze ATP to power ciliary motility. The 9 + 2 axoneme is found in cilia as divergent as the comb plates of ctenophores, the sperm cell of *Gingko*, and the epithelium of the vertebrate respiratory tract. In certain organisms, such as *Chlamydomonas* and in male gametes, cilia are sometimes called flagella, but there are no consistent differences in structure, beat, number, or molecular biology, in the context of this book, between cilia and flagella. Although the term "flagella" is best reserved for the prokaryotic organelle, it persists for some important eukaryotic cells. Cells with motile 9 + 2 cilia may bear one or more cilia, even thousands per cell. Cilia are not found on prokaryotes; bacterial flagella and "stereocilia" of vertebrate hair cells of the ear are not cilia.

In addition to motile 9 + 2 cilia, major classes include primary and sensory cilia. Primary cilia are solitary organelles that emanate from the centrosomal mother centriole at the cell surface during growth arrest in most vertebrate cell types [1], whereas sensory cilia are modified as the receptive projections of sense organs or nerve cell dendrites [9–12]. Primary and sensory cilia usually are 9 + 0, missing the two central single microtubules and often axonemal dynein arms, which make them nonmotile. Depending on type, sensory cells may be multiciliated, although with a few, not thousands of, cilia. Where dynein arms are present, as in chordotonal organs of insects [13] or at the node in vertebrates [14], a form of motility is present.

A new era for primary and sensory cilia dawned with the discovery of IFT [15] and the subsequent link to polycystic kidney disease [16, 17]. These studies paved the way for understanding that primary cilia function as sensory organelles that detect and transmit extracellular cues to the cell. We now know that primary cilia coordinate a vast array of different signaling pathways to control specified cellular processes during development and in tissue homeostasis, such as those regulated by receptor tyrosine kinases [18], extracellular matrix receptors [19], and transforming growth factor beta receptors [20] as well as by G-protein-coupled receptors of the A, B, and F classes [21], the latter including Smoothened in Hedgehog signaling [22]. Probably, the primary cilium functions as a hot spot for the balanced integration of multiple signaling pathways into higher order networks that dictate the biological output of pathway activity during development and in tissue homeostasis. Consequently, defects in formation and compartmentalization of primary cilia lead to defective cell signaling and abnormalities in cell cycle control, migration, polarization, and differentiation, often as a specific cluster of symptoms or syndromes termed ciliopathies [23–26].

Similarly, sense organs of multicellular organisms possess receptive motile or nonmotile cilia for vision, hearing, proprioception, and chemosensation leading to olfactory and mating responses. The persistence of cilia in recognizable form in most phyla suggests that the motile and sensory functions of cilia are often of evolutionary significance. Unicellular organisms rely on motile cilia to move through water, to walk on substrates, to sense favorable and unfavorable environments, to escape predation, to feed, to disperse, and to mate. Multicellular organisms use ciliated epithelia and motile sperm for similar purposes. In addition, motile cilia are used for osmoregulation and clearance in flame cells and kidneys [27], vertebrates use nodal cilia during development for left-right symmetry determination [28], and *Mytilus* uses certain tangled gill cilia as a kind of ciliary Velcro^{R} [29]. In conclusion, whether motile or not, all cilia are sensory in that they possess specific membrane receptors that respond to extracellular cues and transmit the information to control specified ciliary and cellular processes.

The variety and specialization of cilia make it imperative that for investigation of specific aspects of cilia molecular biology, preparative techniques be carefully developed and described. In this volume, a series of experts describes techniques for the study of fundamental aspects of the biology of cilia in a series of different systems. We cover methods to examine important aspects of ciliary biology in vertebrate cells and sense organs, particularly of fish, mouse, and man, in invertebrate cells and sense organs including those of *C. elegans* and *Drosophila*, and in protists such as *Chlamydomonas* and *Paramecium*.

We have asked contributors to write on advanced methodology, including super-resolution light microscopy. The techniques presented here, in combination with known ciliary genomics and proteomics, have made possible spectacular advances in our understanding. For motile cilia, advanced structural imaging and gene knockouts have brought us to the point where the basic mechanisms of motility and its cellular controls are known in considerable molecular detail and new information that holds promise in solving remaining problems is being generated at a rapid rate [30]. Nevertheless, very interesting problems concerning ciliary orientation, length control, and sensory function remain lightly explored in morphogenesis of multiciliated cells, both in protists and in metazoan (and metaphytan) organisms. Tracking techniques are being used to follow molecular paths into and along the axoneme. For sensory and primary cilia, super-resolution is defining the organization of the transition zone/ciliary necklace region responsible for molecular selection and transport within the cilium, facilitating signaling pathways and leading to an understanding of the development of ciliopathies. New trends consider integration and exchange of ciliary molecules with other cell organelles, including endosomes, autophagosomes, and the nucleus.

Many of the methods considered in the chapters that follow were necessary in developing these ideas and they can successfully contribute to continued exploration of this fast evolving field. In these chapters, detailed information is given on handling and examining ciliated systems and organisms. We invite the reader to use the methods described to join this effort.

Peter Satir
Søren T. Christensen

References

1. Satir P, Christensen ST (2007) Overview of structure and function of mammalian cilia. Annu Rev Physiol 69:377–400. doi:10.1146/annurev.physiol.69.040705.141236
2. Lechtreck KF (2015) IFT-cargo interactions and protein transport in cilia. Trends Biochem Sci. doi:10.1016/j.tibs.2015.09.003
3. Reiter JF, Blacque OE, Leroux MR (2012) The base of the cilium: roles for transition fibres and the transition zone in ciliary formation, maintenance and compartmentalization. EMBO Rep 13(7):608–618. doi:10.1038/embor.2012.73
4. Kee HL, Verhey KJ (2013) Molecular connections between nuclear and ciliary import processes. Cilia 2(1):11. doi:10.1186/2046-2530-2-11
5. Malicki J, Avidor-Reiss T (2014) From the cytoplasm into the cilium: bon voyage. Organogenesis 10(1):138–157. doi:10.4161/org.29055
6. Benmerah A (2013) The ciliary pocket. Curr Opin Cell Biol 25(1):78–84. doi:10.1016/j.ceb.2012.10.011
7. Field MC, Carrington M (2009) The trypanosome flagellar pocket. Nat Rev Microbiol 7(11):775–786. doi:10.1038/nrmicro2221
8. Bauss K, Knapp B, Jores P, Roepman R, Kremer H, Wijk EV, Marker T, Wolfrum U (2014) Phosphorylation of the Usher syndrome 1G protein SANS controls Magi2-mediated endocytosis. Hum Mol Genet 23(15):3923–3942. doi:10.1093/hmg/ddu104
9. Bae YK, Barr MM (2008) Sensory roles of neuronal cilia: cilia development, morphogenesis, and function in C. elegans. Front Biosci 13:5959–5974
10. Insinna C, Besharse JC (2008) Intraflagellar transport and the sensory outer segment of vertebrate photoreceptors. Dev Dyn 237(8):1982–1992. doi:10.1002/dvdy.21554
11. Jenkins PM, McEwen DP, Martens JR (2009) Olfactory cilia: linking sensory cilia function and human disease. Chem Senses 34(5):451–464. doi:10.1093/chemse/bjp020
12. Keil TA (2012) Sensory cilia in arthropods. Arthropod Struct Dev 41(6):515–534. doi:10.1016/j.asd.2012.07.001
13. Corbiere-Tichane G (1971) [Fine structure of the chordotonal organs of the head appendages of Speophyes lucidulus larva]. Z Zellforsch Mikrosk Anat 117(2):275–302
14. Supp DM, Witte DP, Potter SS, Brueckner M (1997) Mutation of an axonemal dynein affects left-right asymmetry in inversus viscerum mice. Nature 389(6654):963–966. doi:10.1038/40140
15. Kozminski KG, Johnson KA, Forscher P, Rosenbaum JL (1993) A motility in the eukaryotic flagellum unrelated to flagellar beating. Proc Natl Acad Sci U S A 90(12):5519–5523
16. Pazour GJ, Dickert BL, Vucica Y, Seeley ES, Rosenbaum JL, Witman GB, Cole DG (2000) Chlamydomonas IFT88 and its mouse homologue, polycystic kidney disease gene tg737, are required for assembly of cilia and flagella. J Cell Biol 151(3):709–718
17. Yoder BK, Tousson A, Millican L, Wu JH, Bugg CE, Jr., Schafer JA, Balkovetz DF (2002) Polaris, a protein disrupted in orpk mutant mice, is required for assembly of renal cilium. Am J Physiol Renal Physiol 282(3):F541–F552. doi:10.1152/ajprenal.00273.2001
18. Christensen ST, Clement CA, Satir P, Pedersen LB (2012) Primary cilia and coordination of receptor tyrosine kinase (RTK) signalling. J Pathol 226(2):172–184. doi:10.1002/path.3004
19. Seeger-Nukpezah T, Golemis EA (2012) The extracellular matrix and ciliary signaling. Curr Opin Cell Biol 24(5):652–661. doi:10.1016/j.ceb.2012.06.002
20. Clement CA, Ajbro KD, Koefoed K, Vestergaard ML, Veland IR, Henriques de Jesus MP, Pedersen LB, Benmerah A, Andersen CY, Larsen LA, Christensen ST (2013) TGF-beta

signaling is associated with endocytosis at the pocket region of the primary cilium. Cell Rep 3(6):1806–1814. doi:10.1016/j.celrep.2013.05.020
21. Schou KB, Pedersen LB, Christensen ST (2015) Ins and outs of GPCR signaling in primary cilia. EMBO Rep 16(9):1099–1113. doi:10.15252/embr.201540530
22. Pusapati GV, Rohatgi R (2014) Location, location, and location: compartmentalization of Hedgehog signaling at primary cilia. EMBO J 33(17):1852–1854. doi:10.15252/embj.201489294
23. Bettencourt-Dias M, Hildebrandt F, Pellman D, Woods G, Godinho SA (2011) Centrosomes and cilia in human disease. Trends Genet 27(8):307–315. doi:10.1016/j.tig.2011.05.004
24. Waters AM, Beales PL (2011) Ciliopathies: an expanding disease spectrum. Pediatr Nephrol 26(7):1039–1056. doi:10.1007/s00467-010-1731-7
25. Hildebrandt F, Benzing T, Katsanis N (2011) Ciliopathies. N Engl J Med 364(16):1533–1543. doi:10.1056/NEJMra1010172
26. Valente EM, Rosti RO, Gibbs E, Gleeson JG (2014) Primary cilia in neurodevelopmental disorders. Nat Rev Neurol 10(1):27–36. doi:10.1038/nrneurol.2013.247
27. McKanna JA (1968) Fine structure of the protonephridial system in Planaria. I. Flame cells. Z Zellforsch Mikrosk Anat 92(4):509–523
28. Yoshiba S, Hamada H (2014) Roles of cilia, fluid flow, and Ca2+ signaling in breaking of left-right symmetry. Trends Genet 30(1):10–17. doi:10.1016/j.tig.2013.09.001
29. Rice EL (1908) Gill development in Mytilus. BiolBull-Us14(2):61–77.doi:10.2307/1535718
30. Satir P, Heuser T, Sale WS (2014) A structural basis for how motile cilia beat. Bioscience 64(12):1073–1083. doi:10.1093/biosci/biu180

Contents

Contributors

LEA M. ALFORD • *Department of Cell Biology, Emory University, Atlanta, GA, USA*
MAUREEN M. BARR • *Human Genetics Institute of New Jersey, Rutgers, The State University of New Jersey, Piscataway, NJ, USA*
ALEXANDRE BENMERAH • *INSERM U1163, Laboratory of Inherited Kidney Diseases, Paris, France; Université Paris Descartes-Sorbonne Paris Cité, Imagine Institut, Paris, France*
NICOLAS F. BERBARI • *Department of Biology, Indiana University-Purdue University Indianapolis, Indianapolis, IN, USA*
MONICA BETTENCOURT-DIAS • *Instituto Gulbenkian de Ciência, Oeiras, Portugal*
MARTINA BRUECKNER • *Department of Pediatrics, Yale University School of Medicine, New Haven, CT, USA; Department of Genetics, Yale University School of Medicine, New Haven, CT, USA*
TAMARA CASPARY • *Department of Human Genetics, Emory University School of Medicine, Atlanta, GA, USA*
WENG MAN CHONG • *Institute of Atomic and Molecular Sciences, Academia Sinica, Taipei, Taiwan*
SØREN TVORUP CHRISTENSEN • *Department of Biology, University of Copenhagen, Copenhagen, OE, Denmark*
RANIA GHOSSOUB • *Centre de Recherche en Cancérologie de Marseille (CRCM), Inserm, U1068-CNRS UMR7258, Institut Paoli-Calmettes, Aix-Marseille Université, Marseille, France*
JUDITH L. VAN HOUTEN • *Department of Biology, The University of Vermont, Burlington, VT, USA*
EMILY L. HUNTER • *Department of Cell Biology, Emory University, Atlanta, GA, USA*
SWADHIN CHANDRA JANA • *Instituto Gulbenkian de Ciência, Oeiras, Portugal*
STEPHEN M. KING • *Department of Molecular Biology and Biophysics, University of Connecticut Health Center, Farmington, CT, USA; Institute for Systems Genomics, University of Connecticut Health Center, Farmington, CT, USA*
JOHAN KOLSTRUP • *Analytical Imaging Facility, Department of Anatomy and Structural Biology, Albert Einstein College of Medicine, Bronx, NY, USA*
KARL F. LECHTRECK • *Department of Cellular Biology, University of Georgia, Athens, GA, USA*
JUNG-CHI LIAO • *Institute of Atomic and Molecular Sciences, Academia Sinica, Taipei, Taiwan*
LOUISE LINDBÆK • *Department of Biology, University of Copenhagen, Copenhagen, OE, Denmark*
ESBEN LORENTZEN • *Department of Structural Cell Biology, Max-Planck-Institute of Biochemistry, Martinsried, Germany*
FRANK P. MACALUSO • *Analytical Imaging Facility, Department of Anatomy and Structural Biology, Albert Einstein College of Medicine, Bronx, NY, USA*
DANIELA MARAZZITI • *Institute of Cell Biology and Neurobiology, Italian National Research Council (CNR), Rome, Italy*

SUSANA MENDONÇA • *Instituto Gulbenkian de Ciência, Oeiras, Portugal*
ANAHI MOLLA-HERMAN • *Department of Genetics and Developmental Biology, Institut Curie, Paris, France; CRNS, UMR, Paris, France*
STINE K. MORTHORST • *Department of Biology, University of Copenhagen, Copenhagen, OE, Denmark*
ROBERT O'HAGAN • *Human Genetics Institute of New Jersey, Rutgers, The State University of New Jersey, Piscataway, NJ, USA*
OLATZ PAMPLIEGA • *Institut des Maladies Neurodégénératives, Université de Bordeaux, Bordeaux Cedex, France; CNRS, Institut des Maladies Neurodégénératives, UMR 5293, Bordeaux Cedex, France*
RAMILA S. PATEL-KING • *Department of Molecular Biology and Biophysics, University of Connecticut Health Center, Farmington, CT, USA*
LOTTE B. PEDERSEN • *Department of Biology, University of Copenhagen, Copenhagen, OE, Denmark*
GEOFFREY S. PERUMAL • *Analytical Imaging Facility, Department of Anatomy and Structural Biology, Albert Einstein College of Medicine, Bronx, NY, USA*
WINFIELD S. SALE • *Department of Cell Biology, Emory University, Atlanta, GA, USA*
PETER SATIR • *Analytical Imaging Facility, Department of Anatomy and Structural Biology, Albert Einstein College of Medicine, Bronx, NY, USA*
BIRGIT HEGNER SATIR • *Department of Anatomy and Structural Biology, Albert Einstein College of Medicine, Bronx, NY, USA*
ALAIN SCHMITT • *INSERM U1016, Institut Cochin, Paris, France; CNRS, UMR8104, Paris, France; Université Paris Descartes, Paris, France*
DAISUKE TAKAO • *Department of Cell and Developmental Biology, University of Michigan Medical School, Ann Arbor, MI, USA*
MICHAEL TASCHNER • *Department of Structural Cell Biology, Max-Planck-Institute of Biochemistry, Martinsried, Germany*
MEGAN SMITH VALENTINE • *Department of Biology, The University of Vermont, Burlington, VT, USA*
PATRICIA VERDIER • *Department of Biology, University of Copenhagen, Copenhagen, OE, Denmark*
KRISTEN J. VERHEY • *Department of Cell and Developmental Biology, University of Michigan Medical School, Ann Arbor, MI, USA*
SASCHA WERNER • *Instituto Gulbenkian de Ciência, Oeiras, Portugal*
UWE WOLFRUM • *Institute of Zoology, Cell and Matrix Biology, Johannes Gutenberg University of Mainz, Mainz, Germany*
KIRSTEN A. WUNDERLICH • *Institute of Zoology, Cell and Matrix Biology, Johannes Gutenberg University of Mainz, Mainz, Germany*
T. TONY YANG • *Institute of Atomic and Molecular Sciences, Academia Sinica, Taipei, Taiwan*
SHIAULOU YUAN • *Department of Pediatrics, Yale University School of Medicine, New Haven, CT, USA*

Chapter 1

Methods for Studying Ciliary Import Mechanisms

Daisuke Takao and Kristen J. Verhey

Abstract

Cilia and flagella are microtubule-based organelles that play important roles in human health by contributing to cellular motility as well as sensing and responding to environmental cues. Defects in cilia formation and function cause a broad class of human genetic diseases called ciliopathies. To carry out their specialized functions, cilia contain a unique complement of proteins that must be imported into the ciliary compartment. In this chapter, we describe methods to measure the permeability barrier of the ciliary gate by microinjection of fluorescent proteins and dextrans of different sizes into ciliated cells. We also describe a fluorescence recovery after photobleaching (FRAP) assay to measure the entry of ciliary proteins into the ciliary compartment. These assays can be used to determine the molecular mechanisms that regulate the formation and function of cilia in mammalian cells.

Key words Cilia, Nuclear import, Ciliary import, Microinjection, FRAP

1 Introduction

Cilia (and flagella) are microtubule-based organelles that project from the surface of cells and serve both motile and sensory functions. As organelles, cilia contain a unique complement of protein and lipid (reviews [1, 2]). Furthermore, the ciliary localization of signaling factors can be dynamic and change in response to ligand stimulation (e.g., in Hedgehog signaling [3, 4]). Thus, understanding the mechanisms that regulate selective entry into the ciliary compartment is important for both cell and developmental biology.

As an organelle, the cilium is not completely enclosed by a membrane barrier. Rather, the ciliary membrane is continuous with the plasma membrane, and the intraciliary space is continuous with the cytosol. Selective entry of both membrane and cytosolic proteins thus requires mechanisms that operate where the ciliary compartment meets these neighboring regions. Indeed, the base of the cilium contains several unique structures that can be observed by electron microscopy (transition fibers and Y-links) and/or are the locale for some cilia-specific proteins (the transition zone) (reviewed in [5–7]).

Peter Satir and Søren Tvorup Christensen (eds.), *Cilia: Methods and Protocols*, Methods in Molecular Biology, vol. 1454, DOI 10.1007/978-1-4939-3789-9_1,

Understanding the mechanisms that regulate entry of ciliary proteins requires assays that specifically measure the transit of proteins from an adjacent compartment (the plasma membrane or the cytosol) into the ciliary compartment. In systems that are both genetically and biochemically tractable, such as the unicellular alga *Chlamydomonas reinhardtii*, the effects of loss of transition zone protein function can be ascertained by analysis of the protein content of isolated flagella (e.g., [8, 9]). However, these assays are not optimal for measuring the dynamics of protein import/export; rather, fluorescence microscopy provides the temporal and spatial resolution for measuring protein import/export. This chapter provides detailed experimental approaches to study passive and active transport of molecules into the ciliary compartment using fluorescence microscopy in cultured mammalian cells.

A major detriment to imaging ciliary import in mammalian cells is their small size (~0.25 μm wide × 3–10 μm long) such that many details of ciliary structure and localization are at or below the resolution limit of the fluorescence microscope. Furthermore, most cilia are assembled on the top surface of cultured cells, and thus, the presence of probes within the ciliary compartment is difficult to spatially resolve from those in the cell body using epifluorescence microscopy. To date, researchers have used several approaches to get around these difficulties and directly measure the movement of fluorescent probes within the cilium or between the cilium and adjacent compartments, including (a) probes that accumulate to very high levels within cilia due to protein-protein interactions or chemical induction of such interactions [10, 11], (b) imaging of cilia that are spatially separated from the cell body due to natural variation in ciliary position and/or manipulation of cellular orientation (e.g., [12–17]), or (c) fluorescence recovery after photobleaching (FRAP) in which bleaching the fluorescent molecules in a select area gives rise to a strong signal that is readily distinguished from the background and can be monitored quantitatively in real time.

FRAP exploits the ability of laser scanning confocal microscopes to rapidly and irreversibly photobleach fluorescently tagged molecules within a specific region of the cell. Fluorescence recovers in the bleached region due to the movement of non-bleached fluorescent molecules from adjacent regions into the bleached area. Provided that the fluorescently tagged proteins are in equilibrium between the bleached and unbleached areas (*see* **Note 1**), the rate of influx of the non-bleached fluorescent proteins provides information about the dynamics and mobility of the protein population. In terms of ciliary localization, FRAP has been used to measure the dynamics of both membrane and cytosolic proteins (e.g., [15, 18–29]).

Imaging ciliary proteins in mammalian cells requires transfecting the cells with plasmids that drive the expression of fluorescently tagged protein(s) of interest. The transfection can result in transient expression (e.g., 1–2 days) of the protein from the plasmid or long-term expression if the plasmid is stably integrated into the genome. In either case, it is imperative to analyze cells with low to moderate levels of expression that, as much as possible, mimic the protein's normal expression level and localization pattern (*see* **Note 2**). In addition, it is useful to image cells in populations where the majority of the cells have a primary cilium present on the cell surface. Many cultured cell lines generate primary cilia in the G0 or G1 phases of the cell cycle, and the percentage of ciliated cells in the population can often be increased by serum starvation. Widely utilized in the field are telomerase-immortalized human retinal pigment epithelial (hTERT-RPE1) cells, NIH 3T3 cells, mouse inner medullary collecting duct (IMCD3), and canine Madin-Darby kidney (MDCK) epithelial cells.

2 Materials

2.1 Cell Culture and Transfection

1. hTERT-RPE and/or NIH 3T3 cells (ATCC).
2. 35 mm glass-bottom cell culture dishes (MatTek) (*see* **Note 3**).
3. DMEM/F12 for RPE cells or DMEM for NIH 3T3 cells.
4. DMEM/F12 phenol red-free or DMEM phenol red-free.
5. L-Glutamine [or GlutaMax (Gibco 35050)].
6. Fetal bovine serum (FBS) for RPE cells or bovine calf serum (CS) for NIH 3T3 cells (*see* **Note 4**).
7. Penicillin-streptomycin (optional).
8. Hygromycin B.
9. Trypsin/EDTA solution: 0.25 % (w/v) Trypsin + 0.01 % EDTA.
10. TransIT-LT1 (Mirus MIR2305).
11. Opti-MEM (Gibco 31985).
12. Plasmids for expression of fluorescently tagged ciliary markers as well as the ciliary protein(s) of interest (*see* **Note 5**).
13. Leibovitz L-15 medium, phenol red-free (*see* **Note 6**).

2.2 Buffers

1. Reconstitution Buffer: 25 mM Hepes/KOH pH 7.4, 115 mM KOAc, 5 mM NaOAc, 5 mM $MgCl_2$, 0.5 mM EDTA, 1 mM GTP, and 1 mM ATP.
2. Phosphate-buffered saline (PBS).

2.3 Microinjection

1. Fluorescently labeled dextrans (Molecular Probes™). Reconstitute at 10 mg/ml in reconstitution buffer. Aliquot into 10 μl portions and store at −20 °C.
2. Recombinant proteins of different sizes: nonfluorescent proteins [e.g., α-lactalbumin (Sigma L5385), BSA (Sigma A9647), protein A (Prospec PRO-774)] should be brought to

1 mg/ml with PBS and then labeled using the Alexa Fluor 488 Microscale Protein Labeling Kit following the manufacturer's protocol. Recombinant GFP can be used at 1 mg/ml without labeling.

(a) Bring recombinant proteins to 1 mg/ml in PBS if needed.

(b) Incubate protein with Alexa Fluor 488 dye according to manufacturer's directions. The labeling reaction will proceed spontaneously.

(c) Separate the labeled protein from free dye using spin filters included in the kit.

(d) Aliquot into 10 μl portions and store at −20 °C.

3. Inverted wide-field microscope. Our system consists of a Nikon TE2000-E with DIC and fluorescence optics, 40×0.75 N.A. objective, 1.5× Optivar, and Photometrics CoolSNAP ES2 camera.
4. Micromanipulator. Our system utilizes an Eppendorf InjectMan N1 2 micromanipulator, a motorized system containing a control board with joystick. Simple mechanical micromanipulators can also be used.
5. Microinjector (*see* **Note 7**). The micromanipulator is connected to a semiautomated Eppendorf FemtoJet microinjector system, which provides control over the injection motion, time, and pressure and thus allows for consistency between different injections. To ensure that medium does not flow into the micropipette through capillary suction, a continuous compensation pressure is set to allow slight flow out of liquid from the tip. The injection pressure is the pressure applied by the microinjector during the injection. The volume injected into a cell depends on the injection pressure and time. Appropriate values for injection pressure, injection time, and compensation pressure vary with micropipettes and are determined by testing using a fluorescent dye.
6. Micropipettes. We use pre-pulled micropipettes (Femtotips, Eppendorf 930000035). Micropipettes can also be fabricated from glass capillaries using a micropipette puller.

2.4 FRAP

1. Confocal microscope. We use a Nikon A1 confocal system on a Nikon Eclipse Ti microscope equipped with a live-cell temperature-controlled chamber (Tokai Hit) with CO_2 supply and a Perfect Focus System. A 60×/1.2 N.A. water immersion objective equipped with an objective heater is used for imaging. Image size is set at 512×512 pixels with a unidirectional scan speed of one frame per second. A fast imaging speed is important to minimize photobleaching of the sample while collecting pre- and post-bleach images. Imaging of mCitrine/EGFP and mCherry channels is performed sequentially to minimize

channel-to-channel signal bleed-through. Laser power for imaging is typically held at 2% for a 488 nm line (Spectra-Physics air-cooled argon ion laser, 40 mW run at 50% power) and 30% for a 543 nm line (Melles Griot HeNe laser, 5 mW). Photodetector voltage is adjusted on a cell-to-cell basis.

3 Methods

3.1 Analysis of Ciliary Import by Microinjection

1. Plate hTERT-RPE cells on 35 mm glass-bottom cell culture dishes at 1.5×10^5 cells/dish in DMEM/F12 supplemented with 10% fetal bovine serum, 1% Pen-Strep, and 0.01 mg/ml Hygromycin B. Return dish to incubator and incubate for 16 h.
2. To express a fluorescently tagged ciliary marker, transfect cells with 1 μg of plasmid/dish using 3 μl of TransIT-LT1 and 100 μl Opti-MEM, following the manufacturer's protocol.
3. At 5 h post-transfection, wash cells two times with serum-free DMEM/F12 media then leave cells in serum-free media and return to the incubator for 48 h to allow cells to arrest in G1 and generate primary cilia.
4. Thaw aliquots of fluorescent dextrans and/or proteins on ice. Spin in a microcentrifuge for 5 min at $10{,}000 \times g$ at 4 °C to remove any aggregates. Keep thawed aliquot on ice while in use.
5. Replace cell media with warmed (37 °C) phenol red-free Leibovitz L-15 medium plus L-glutamine (*see* **Note 8**), and place the dish on the stage of an inverted wide-field microscope.
6. Load 2 μl of the desired fluorescent dextran or protein into a micropipette capillary. Flick the capillary to force solution down to the tip of the capillary.
7. Check that the micromanipulator is at the middle position for all three axes and not positioned at the lowest or highest *x*-, *y*-, *z*-limit, thus allowing flexibility in the range of movement.
8. Insert the micropipette into the micropipette holder, making sure that the micropipette tip does not touch anything and break.
9. Position the micropipette at the center of the dish and slowly bring the tip down into the media.
10. Focus on the cells using DIC optics at a low magnification, preferably using a 4× or 10× objective lens. Using the micromanipulator joystick, bring the micropipette tip to the center of field of view. At this point, the tip will be blurry and out of focus. Slowly adjust the position of the micropipette along the *Z*-axis, bringing the tip closer to the cells but still slightly out

of focus. It is important to keep the focal plane in the same position throughout this step, as movement of the objective lens may inadvertently lift the plate of cells toward the micropipette and damage the injection tip.

11. Change to the 40× objective and continue to slowly lower the micropipette tip using the micromanipulator, bringing the tip closer to the cell surface but keeping it centered in the field of view. Be sure that the micropipette tip remains in a *Z* position above the cells so that the dish can be moved in the *x*–*y* axes without scraping the cells with the tip.
12. Switch to fluorescence imaging, and, using low-intensity excitation light, scan the dish for cells that are positive for the ciliary marker and have their cilium protruding off the side of the cell into empty dish space. Finding an appropriate cell for microinjection may be facilitated by capturing an image of the fluorescently tagged marker and a DIC image and overlaying these images.
13. Once a target cell has been identified, switch back to DIC imaging mode for the microinjection. Slowly lower the micropipette tip toward the cell using the micromanipulator joystick. With the FemtoJet microinjector system, a *Z* limit has to be defined before the automated injection. To do this, manually lower the micropipette so that the tip lightly presses against the cell surface, then press the "limit" key to set the *Z* limit. Manually raise the tip above the cell, and then activate the injection function by pressing the joystick button. The micropipette will begin a predetermined injection sequence in which the injector tip will enter the cell, apply an injection pressure, and then return to its original position.

    If using a nonautomated microinjector system, lower the micropipette tip close to the cell surface and move the joystick to penetrate the cell while simultaneously activating the injection by pressing the foot switch of the microinjector.
14. Following injection, capture fluorescence images in two channels, the marker and the dextran/protein, at different time points (Fig. 1).
15. Import images into ImageJ.
16. Use the signal of the ciliary marker projecting off the cell body to define a ciliary region of interest (ROI). Switch to the dextran/protein channel and measure the average fluorescence in the ciliary ROI.
17. In the dextran/protein channel, measure the average fluorescence intensity in the cytoplasmic region half the distance between the nuclear envelope and cell periphery by moving the ROI to this area.

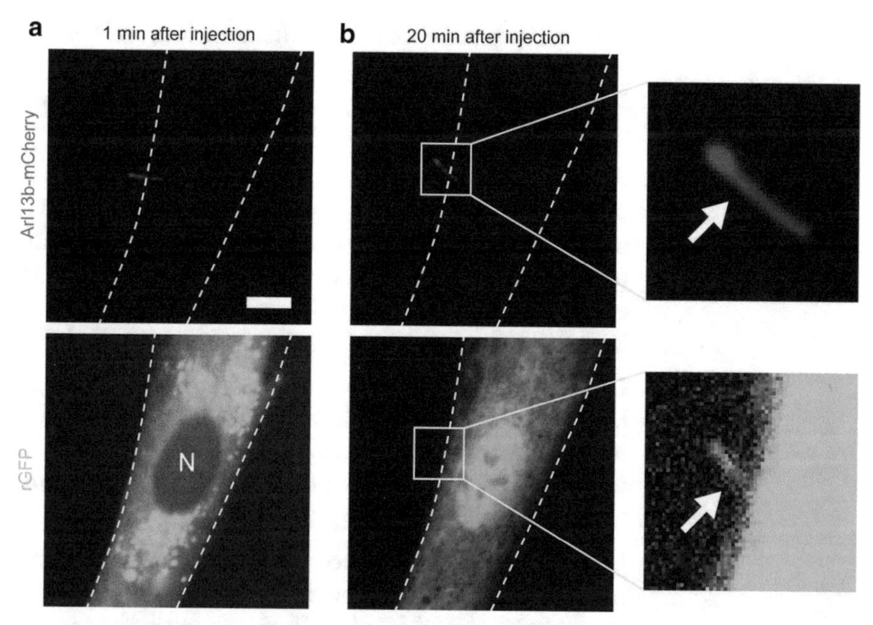

Fig. 1 Microinjection to measure protein entry into the primary cilium. hTERT-RPE cells expressing the ciliary marker Arl13b-mCherry (*red*) were scanned to identify cells with their cilium protruding off the side of the cell. The selected cell was microinjected with a fluorescent protein of interest, recombinant GFP (rGFP), and images were taken at various time points post-injection. Shown are images of Arl13b-mCherry (*red*) and rGFP (*green*) at (**a**) 1 min and (**b**) 20 min after microinjection. Far right images are higher magnification images of the cilium in the yellow boxed areas. At 1 min after microinjection, the rGFP has not entered the nuclear or ciliary compartments, but by 20 min after microinjection, the rGFP has entered both compartments. N, nucleus. White dotted line, periphery of cell. Arrow, cilium. Scale bar, 5 μm

18. In the dextran/protein channel, measure the background fluorescence by moving the ROI to a region next to the cell and measuring the average fluorescence.
19. Subtract the background fluorescence from that of the ciliary region and the cytoplasmic region.
20. Repeat for as many cells as possible and then determine the average background-corrected fluorescence for the ciliary region and the average background-corrected fluorescence for the cytoplasmic region.
21. Calculate the Diffusion Barrier Index as the ratio of mean fluorescence intensity in the ciliary region versus the cytoplasmic region.

3.2 Analysis of Ciliary Import by Fluorescence Recovery After Photobleaching (FRAP)

1. Plate NIH 3T3 cells on 35 mm glass-bottom cell culture dishes at 1.5×10^5 cells per dish in DMEM supplemented with 10 % FetalClone III, 1 % GlutaMax, and 1 % Pen-Strep.
2. After 24 h (or after ~4 h when cells have adhered to the bottom of the dish), promote ciliogenesis by replacing growth media with serum-free DMEM plus 1 % GlutaMax.

3. Immediately transfect cells with 1–1.5 μg of total expression plasmids (ciliary marker and ciliary protein of interest, *see* **Note 5**) using TransIT-LT1 in Opti-MEM media according to manufacturer's instructions.
4. After 24–48 h, turn on microscope system and set imaging chamber to 37 °C and 5% CO_2. Wash cells 2× and then leave in warmed phenol red-free and serum-free DMEM media (optional; to rid or minimize population of dead or floating cells). Transfer dish to the live-cell imaging chamber with minimal time at room temperature.
5. Configure the microscope in epifluorescence mode and use low-intensity excitation light to search for cells with moderate levels of expression of the fluorescently tagged protein of interest as well as a ciliary position that facilitates photobleaching and imaging (Fig. 2a, d).
6. Once a cell is chosen, configure the microscope in confocal imaging mode. Focus on the cilium and adjust the photodetector voltage and image offset to ensure that an appropriate dynamic range of fluorescence output is obtained.
7. Due to the tendency of cilia to move (and thus the cilium tip to drift in and out of focus), we have developed two different photobleaching protocols to maintain the cilium within the analysis region.
   (a) One protocol involves opening the confocal pinhole to a relatively wide diameter (4–7 AU which corresponds to 2.2–5.4 μm) to increase the optical section thickness. This results in a higher level of background fluorescence but allows one to image the cilium tip within a larger z area.
   (b) The other protocol allows for a thinner optical section to be used by taking multiple z-stacks for pre- and post-bleach images. While this setup minimizes background signal from the cell body, it also decreases the temporal resolution of recovery analysis and increases photobleaching of the sample.
8. Collect several pre-bleach images or z-stacks. Taking several pre-bleach images is needed to calculate an average pre-bleach intensity (Fig. 2b, e).
9. Draw a region of interest (ROI) around the desired area to be photobleached (Fig. 2b, e). Expose this region to 50% laser power (e.g., using the 488 nm line to photobleach EGFP- or mCit-tagged ciliary proteins) for a 2 s bleaching step (*see* **Note 9**).
10. Collect a series of post-bleach images or z-stacks. The time interval between images and total duration of recording will depend on the fluorescence recovery rate which varies between proteins. For a ciliary protein like the kinesin-2 motor KIF17

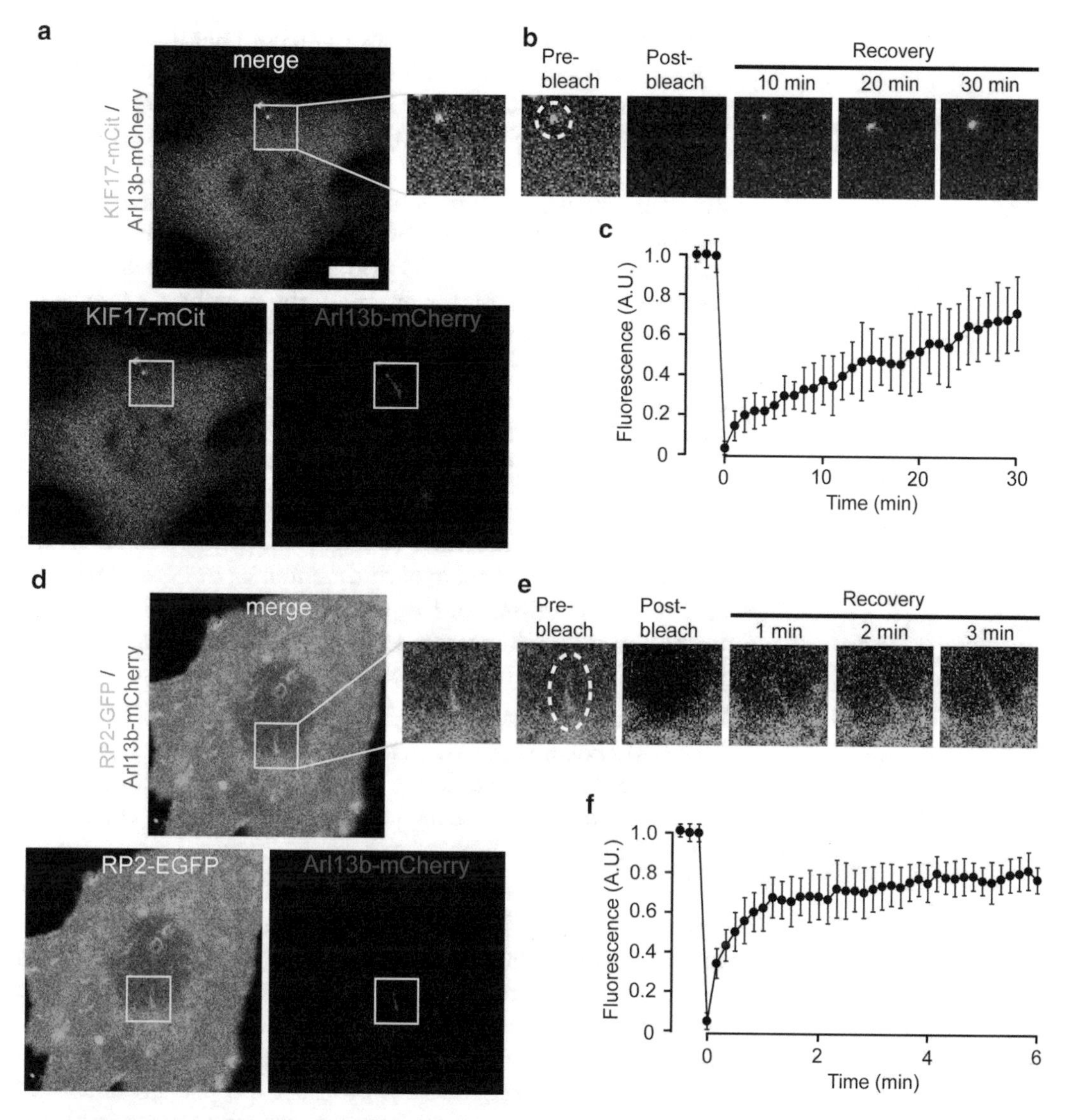

Fig. 2 FRAP to measure protein entry into the primary cilium. (**a–c**) FRAP of an NIH 3T3 cell co-expressing the ciliary marker Arl13b-mCherry and the cytosolic kinesin-2 motor KIF17-mCit. (**a**) Whole cell images showing KIF17-mCit (*lower left*) and Arl13b-mCherry (*lower right*) expression and the merged channels (*top*). (**b**) KIF17-mCit fluorescence in the ciliary compartment was photobleached with high laser power, and fluorescence recovery was measured every minute for 30 min. *Dashed white line*, ROI for photobleaching. (**c**) Quantification of the fluorescence recovery of KIF17-mCit in the distal tips of cilia (mean ± standard deviation). $N = 12$ cells. As a cytosolic protein and kinesin motor, the accumulation of new fluorescent KIF17 molecules at the tip of the cilium reflects their entry into the compartment and their movement along the doublet microtubules to the tip. (**d–f**) FRAP of an NIH 3T3 cell co-expressing the ciliary marker Arl13b-mCherry and the peripheral membrane protein RP2-EGFP. (**d**) Whole cell images of RP2-EGFP (*lower left*) and Arl13b-mCherry (*lower right*) expression and the merged channels (*top*). (**e**) RP2-EGFP in the ciliary compartment was photobleached with high laser power, and fluorescence recovery was measured every 10 s for 3 min. *Dashed white line*, ROI for photobleaching. (**e**) Quantification of the fluorescence recovery of RP2-EGFP in cilia (mean ± standard deviation). $N = 11$ cells. Scale bar, 5 μm. As a peripheral membrane protein, the recovery of RP2-EGFP fluorescence in the cilium reflects the entry of new RP2-EGFP into the compartment

with a relatively slow recovery rate [29], both methods described in **step 7a, b** can be used. When imaging using a wide diameter pinhole (**step 7a**), images are typically collected every 1 min for 30 min after photobleaching (Fig. 2b, c). When taking z-stacks (**step 7b**), images are typically collected less frequently, for example, every 5–10 min, which is still adequate for determining total levels of KIF17-mCit entering the primary cilium (data not shown). For a ciliary protein with a much faster fluorescence recovery rate, for example, the membrane-associated protein PalmPalm-mCit [29], the wide diameter pinhole imaging (**step 7a**) method should be used, and images are typically collected every 10 s for 5 min after photobleaching (Fig. 2e, f). In the following steps, we use KIF17-mCit and PalmPalm-mCit as example ciliary proteins of interest (*see* **Note 5**).

11. During the recovery imaging, images of the ciliary marker are obtained at the same time to verify that the cilium remains within the imaging area in each time frame (*see* **Note 10**). The strong beam at 488 nm for photobleaching can also partially photobleach the fluorescence signal from the ciliary marker, but the signal can be observed in most cases (*see* **Note 11**).
12. Import images into ImageJ (NIH) or MetaMorph (Molecular Devices). For analysis of data collected as z-stacks (**step 7b** above), either each stack can be flattened into a single image or measurements can be made from only the z-slice containing the greatest ciliary fluorescence.
13. Determine the average KIF17-mCit or PalmPalm-EGFP fluorescence intensity for the pre-bleach images. Draw a region of interest (ROI) around the ciliary region in the first pre-bleach image and measure the mean pixel intensity. Determine the background fluorescence by moving the ROI to a region of the image in which there are no cells or debris, and again measure the mean pixel intensity. Subtract the background from the measured fluorescence. Repeat for the next two pre-bleach images and then determine the average fluorescence for the three pre-bleach images. The intensity values are copied and pasted into an Excel sheet for further calculations such as background subtraction and normalization.
14. Repeat this process to obtain the mean pixel intensity measurements for the ciliary region post-bleach and throughout the recovery. For images where there is little or no KIF17-mCit or PalmPalm-EGFP fluorescence, the Arl13b-mCherry signal should be used to guide placement of the ROI at the cilium.
15. Normalize the resulting intensities by dividing the post-bleach and recovery values by the pre-bleach value. The pre-bleach intensity will have a normalized value of 1.
16. Average several datasets together to produce final recovery curves (Fig. 2c, f).

4 Notes

1. A key consideration for FRAP is that there is a sufficient pool of fluorescent molecules available to move into the bleached region for fluorescence recovery. For example, if FRAP is performed with a protein enriched in cilia with little to no pool of cytoplasmic or plasma membrane molecules, an absence of fluorescence recovery may be incorrectly interpreted as immobility of the protein.
2. A prerequisite for use of any epitope-tagged protein (FP or other) is that the tagged protein must be functional. In genetic organisms such as *C. elegans*, functionality can be tested by determining whether the tagged protein rescues a deletion or mutant phenotype. In mammalian cells, functionality is often tested by determining that the tagged protein displays the same subcellular localization, activity, and interactions as the native protein. Another concern is that overexpression of the protein of interest alters its subcellular trafficking, dynamics, or localization. Taking care to study cells expressing low to moderate levels of protein expression is often sufficient to alleviate such concerns.
3. There are a variety of commercially available glass slides, culture dishes, and coverslips that are suitable for live-cell imaging as alternatives to the MatTek glass-bottom culture dishes (e.g., Nunc Lab-Tek). Generally, the glass component should be #1.5 or 0.17 mm in thickness for optimal resolution with high NA objectives.
4. Some cell lines grow well with less-costly FBS alternatives. For example, we routinely culture NIH 3T3 cells in FetalClone III (HyClone SH30109).
5. Both microinjection and FRAP experiments require expression of a marker of the primary cilium in live cells. Widely utilized probes in the field include the peripheral membrane protein Arl13b [30–32] and the transmembrane proteins somatostatin receptor Sstr3 and serotonin receptor Htr6 [33–35]. FRAP experiments also require expression of the ciliary protein of interest. The dynamics of both membrane and cytosolic proteins in the ciliary compartment have measured by FRAP. Membrane proteins include Arl13b [25], Sstr3 [21, 25–27], Htr6 [21, 26], a cyclic nucleotide-gated channel [18], Smoothened [19, 21], kidney injury molecule-1 (Kim-1 [19]), polycystin 2 [22], Crumbs3 [22], retinitis pigmentosa 2 (RP2 [29]), podocalyxin/gp135 [23], and membrane-targeted (myristoylated and/or palmitoylated) fluorescent protein mCitrine [29, 36]. Cytosolic proteins include end-binding protein 1 (EB1 [19]), kinesin KIF17 [15, 20, 29], intraflagellar transport proteins IFT88

and IFT20 [21, 25, 29], transcription factor Gli2 [29], Tsga14 [29], Gtl3 [29], tubulin [24], and Bardet-Biedl syndrome 1 (BBS1 [28]).

6. Although the phenol red indicator dye is essential for routine cell culture, it should be avoided in live-cell imaging experiments as it can cause undesirable background fluorescence and its excitation can result in phototoxicity.
7. There are different micromanipulators and microinjection systems available, and the specific methods for microinjection will need to be optimized for other systems.
8. Virtually all culture media utilize sodium bicarbonate as a buffer system to regulate pH in atmosphere containing CO_2 (usually 5–7 %). For live-cell imaging on the microscope, a stage-top or other microscope chamber system is optimal to maintain temperature and CO_2 conditions for cell viability. However, such imaging chambers are not always available or may not be feasible with specific microscope accessories (e.g., the microinjection apparatus). In these cases, Leibovitz L-15 medium is favored for live-cell imaging. Leibovitz L-15 is buffered by phosphates and freebase amino acids and thus enables the culturing of cells in the absence of bicarbonate buffer (and CO_2). Although not optimal, an alternative is to supplement the culture media with 10–25 mM HEPES buffer to control pH within the physiological range in the absence of CO_2. HEPES does not eliminate the CO_2 requirement; it only slows the rate of pH changes and may thus be sufficient for short-term imaging experiments.
9. The photobleaching step will likely need to be optimized depending on the imaging system being used as well as the laser power and protein expression level. Insufficient photobleaching can result from suboptimal parameters for laser intensity and exposure timing or from too bright of a fluorescence signal (too many FP-tagged proteins). This will adversely affect the data analysis and result in an underestimation of the percent recovery and mobile fraction. On the other hand, excessive laser power and prolonged duration of photobleaching can result in phototoxicity to the cells, again causing an underestimation of the parameters revealed by quantitative analysis of the recovery phase.
10. If the cilium starts to move away from the focal plane, the microscope can be refocused during an interval of the image acquisition. Do not refocus many times to avoid photobleaching. Typically, once or twice in 30 min observation period is sufficient, and it is typically not necessary in a 5 min observation period.

11. Photobleaching can cause photodamage to other molecules and phototoxicity to the cell due to reaction of the excited fluorophore with dissolved oxygen and the release of reactive oxygen species (ROS). Although the use of oxygen scavenging systems can decrease such effects, it is best to minimize the amount of excitation light and maximize the light collection efficiency.

Acknowledgments

We thank members of the Verhey Lab for advice and discussions. We are grateful to Steve Lentz and the Morphology and Image Analysis Core of the Michigan Diabetes Research and Training Center (funded by NIDDK under NIH5P60 DK20572) for help with confocal imaging. Work in K.J. Verhey Lab is supported by NIGMS of the NIH under award numbers R01GM070862 and R01GM116204.

References

1. Emmer BT, Maric D, Engman DM (2010) Molecular mechanisms of protein and lipid targeting to ciliary membranes. J Cell Sci 123:529–536
2. Madhivanan K, Aguilar RC (2014) Ciliopathies: the trafficking connection. Traffic 15:1031–1056
3. Nozawa YI, Lin C, Chuang PT (2013) Hedgehog signaling from the primary cilium to the nucleus: an emerging picture of ciliary localization, trafficking and transduction. Curr Opin Genet Dev 23:429–437
4. Mukhopadhyay S, Rohatgi R (2014) G-Protein-coupled receptors, Hedgehog signaling and primary cilia. Semin Cell Dev Biol 33:63–72
5. Czarnecki PG, Shah JV (2012) The ciliary transition zone: from morphology and molecules to medicine. Trends Cell Biol 22:201–210
6. Garcia-Gonzalo FR, Reiter JF (2012) Scoring a backstage pass: mechanisms of ciliogenesis and ciliary access. J Cell Biol 197:697–709
7. Szymanska K, Johnson CA (2012) The transition zone: an essential functional compartment of cilia. Cilia 1:10
8. Craige B et al (2010) CEP290 tethers flagellar transition zone microtubules to the membrane and regulates flagellar protein content. J Cell Biol 190:927–940
9. Awata J et al (2014) Nephrocystin-4 controls ciliary trafficking of membrane and large soluble proteins at the transition zone. J Cell Sci 127(21):4714–4727
10. Breslow DK et al (2013) An in vitro assay for entry into cilia reveals unique properties of the soluble diffusion barrier. J Cell Biol 203:129–147
11. Lin YC et al (2013) Chemically inducible diffusion trap at cilia reveals molecular sieve-like barrier. Nat Chem Biol 9:437–443
12. Battle C et al (2015) Intracellular and extracellular forces drive primary cilia movement. Proc Natl Acad Sci U S A 112:1410–1415
13. Besschetnova TY, Roy B, Shah JV (2009) Imaging intraflagellar transport in mammalian primary cilia. Methods Cell Biol 93:331–346
14. Ishikawa H, Marshall WF (2015) Efficient live fluorescence imaging of intraflagellar transport in mammalian primary cilia. Methods Cell Biol 127:189–201
15. Kee HL et al (2012) A size-exclusion permeability barrier and nucleoporins characterize a ciliary pore complex that regulates transport into cilia. Nat Cell Biol 14:431–437
16. Mohieldin AM et al (2015) Protein composition and movements of membrane swellings associated with primary cilia. Cell Mol Life Sci 72:2415–2429
17. Williams CL et al (2014) Direct evidence for BBSome-associated intraflagellar transport reveals distinct properties of native mammalian cilia. Nat Commun 5:5813
18. Jenkins PM et al (2006) Ciliary targeting of olfactory CNG channels requires the CNGB1b subunit and the kinesin-2 motor protein, KIF17. Curr Biol 16:1211–1216

19. Boehlke C et al (2010) Differential role of Rab proteins in ciliary trafficking: Rab23 regulates smoothened levels. J Cell Sci 123:1460–1467
20. Dishinger JF et al (2010) Ciliary entry of the kinesin-2 motor KIF17 is regulated by importin-beta2 and RanGTP. Nat Cell Biol 12:703–710
21. Hu Q et al (2010) A septin diffusion barrier at the base of the primary cilium maintains ciliary membrane protein distribution. Science 329:436–439
22. Hurd TW et al (2010) The retinitis pigmentosa protein RP2 interacts with polycystin 2 and regulates cilia-mediated vertebrate development. Hum Mol Genet 19:4330–4344
23. Francis SS et al (2011) A hierarchy of signals regulates entry of membrane proteins into the ciliary membrane domain in epithelial cells. J Cell Biol 193:219–233
24. Hao L et al (2011) Intraflagellar transport delivers tubulin isotypes to sensory cilium middle and distal segments. Nat Cell Biol 13:790–798
25. Larkins CE et al (2011) Arl13b regulates ciliogenesis and the dynamic localization of Shh signaling proteins. Mol Biol Cell 22:4694–4703
26. Chih B et al (2012) A ciliopathy complex at the transition zone protects the cilia as a privileged membrane domain. Nat Cell Biol 14: 61–72
27. Ye F et al (2013) Single molecule imaging reveals a major role for diffusion in the exploration of ciliary space by signaling receptors. Elife 2, e00654
28. Liew GM et al (2014) The intraflagellar transport protein IFT27 promotes BBSome exit from cilia through the GTPase ARL6/BBS3. Dev Cell 31:265–278
29. Takao D et al (2014) An assay for clogging the ciliary pore complex distinguishes mechanisms of cytosolic and membrane protein entry. Curr Biol 24:2288–2294
30. Caspary T, Larkins CE, Anderson KV (2007) The graded response to Sonic Hedgehog depends on cilia architecture. Dev Cell 12:767–778
31. Cevik S et al (2010) Joubert syndrome Arl13b functions at ciliary membranes and stabilizes protein transport in Caenorhabditis elegans. J Cell Biol 188:953–969
32. Hori Y et al (2008) Domain architecture of the atypical Arf-family GTPase Arl13b involved in cilia formation. Biochem Biophys Res Commun 373:119–124
33. Berbari NF et al (2008) Bardet-Biedl syndrome proteins are required for the localization of G protein-coupled receptors to primary cilia. Proc Natl Acad Sci U S A 105:4242–4246
34. Brailov I et al (2000) Localization of 5-HT(6) receptors at the plasma membrane of neuronal cilia in the rat brain. Brain Res 872:271–275
35. Handel M et al (1999) Selective targeting of somatostatin receptor 3 to neuronal cilia. Neuroscience 89:909–926
36. Kee HL, Verhey KJ (2013) Molecular connections between nuclear and ciliary import processes. Cilia 2:11

Chapter 2

Targeting of ASH Domain-Containing Proteins to the Centrosome

Patricia Verdier, Stine K. Morthorst, and Lotte B. Pedersen

Abstract

A growing number of studies have used new generation technologies to characterize the protein constituents of cilia and centrosomes. This has led to the identification of a vast number of candidate ciliary or centrosomal proteins, whose subcellular localization needs to be investigated and validated. Here, we describe a simple and inexpensive method for analyzing the subcellular localization of candidate cilium- or centrosome-associated proteins, and we illustrate the utility as well as the pitfalls of this method by applying it to a group of ASH (ASPM, SPD-2, Hydin) domain-containing proteins, previously predicted to be cilia- or centrosome-associated proteins based on bioinformatic analyses. By generating plasmids coding for epitope-tagged full-length (FL) or truncated versions of the ASH domain-containing proteins TRAPPC8, TRAPPC13, NPHP4, and DLEC1, followed by expression and quantitative immunofluorescence microscopy (IFM) analysis in cultured human telomerase-immortalized retinal pigmented epithelial (hTERT-RPE1) cells, we could confirm that TRAPPC13 and NPHP4 are highly enriched at the base of primary cilia, whereas DLEC1 seems to associate specifically with motile cilia. Results for TRAPPC8 were inconclusive since epitope-tagged TRAPPC8 fusion proteins were unstable/degraded in cells, emphasizing the need for combining IFM analysis with western blotting in such studies. The method described should be applicable to other candidate ciliary or centrosomal proteins as well.

Key words ASH domain, Centrosome, Cilia, TRAPPC8, TRAPPC13, NPHP4, DLEC1

Abbreviations

ASH	ASPM, SPD-2, Hydin
ATCC	American-type culture collection
DAPI	4′ 6-Diamidino-2-phenylindole, dihydrochloride
DLEC1	Deleted in lung and esophageal cancer 1
DMEM	Dulbecco's modified Eagle's medium
DTT	Dithiothreitol
ECL	Enhanced chemiluminescence
EDTA	Ethylenediaminetetraacetic acid
EGFP	Enhanced GFP
FAP	Flagella-associated protein
FL	Full length

Peter Satir and Søren Tvorup Christensen (eds.), *Cilia: Methods and Protocols*, Methods in Molecular Biology, vol. 1454, DOI 10.1007/978-1-4939-3789-9_2,

GFP	Green fluorescent protein
hTERT-RPE1	Human telomerase-immortalized retinal pigmented epithelial
IFM	Immunofluorescence microscopy
LB	Luria-Bertani
LDS	Loading sample buffer
MFI	Mean fluorescence intensity
MSP	Major sperm protein
NPHP4	Nephronophthisis 4
OCRL	Oculocerebrorenal syndrome of Lowe
PACT	Pericentrin-AKAP450 centrosomal targeting
PBS	Phosphate-buffered saline
RT	Room temperature
siRNA	Small interfering RNA
TBS-T	Tris-buffered saline with Tween-20
TPR	Tetratricopeptide
TRAPP	Transport protein particle
TRAPPC	TRAPP complex

1 Introduction

To date, more than 50 different high-throughput studies aimed at cataloging the constituents of cilia and centrosomes from various species have been published. This has resulted in the generation of a vast amount of data that needs to be analyzed and validated at the single gene/protein level [1]. Moreover, smaller scale bioinformatics-based searches for conserved protein domains normally associated with ciliary or centrosomal functions have similarly led to identification of new candidate ciliary or centrosomal proteins, such as proteins harboring the ASH (ASPM, SPD-2, Hydin) domain [2, 3], that need to be investigated in detail.

The ASH domain was first identified by in silico analysis in proteins associated with cilia, flagella, the centrosome, and the Golgi, including ASPM, Hydin, and OCRL1, as well as in DLEC1—a protein of largely unknown function [2]. The ASH domain is evolutionarily related to the immunoglobulin (Ig)-like seven-stranded beta-sandwich fold superfamily of the nematode major sperm proteins, MSPs [2, 4], which is thought to form a protein-protein interaction interface implicated in cellular signaling and trafficking activities [4, 5]. Recently, we showed that conserved ASH domains are present in several components of the transport protein particle (TRAPP) II complex (TRAPPC9, TRAPPC10, TRAPPC11, and TRAPPC13) as well as in TRAPPC8, and we showed that endogenous TRAPPC8 and exogenously expressed ASH domains from TRAPPC10 and TRAPPC11 localized to the centrosome in human telomerase-immortalized retinal pigmented epithelial (hTERT-RPE1; hereafter, RPE1) cells [3]. Furthermore, TRAPPC3, TRAPPC9, TRAPPC10 [6], and TRAPPC8 [3] were found to

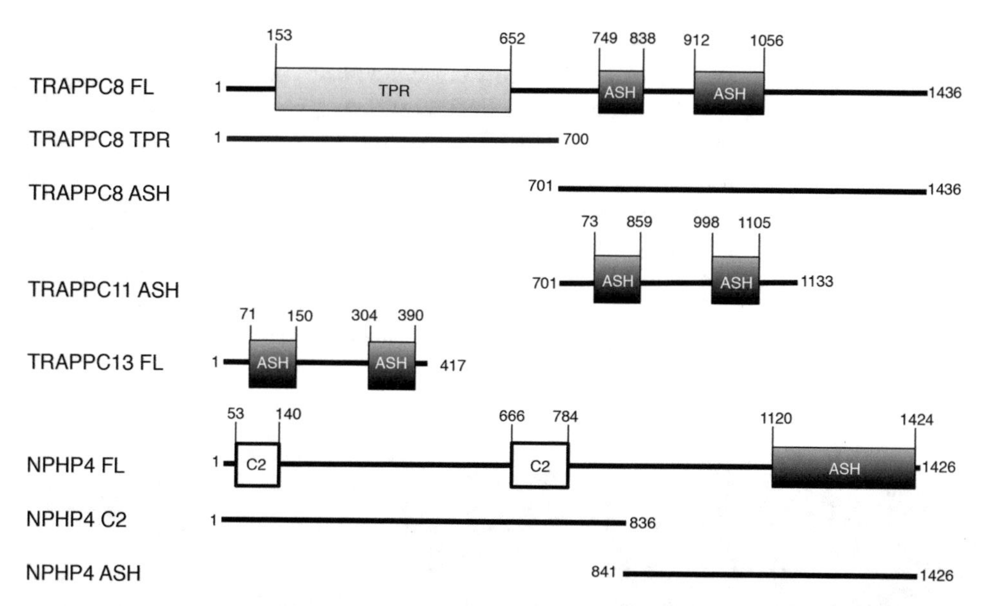

Fig. 1 Schematics of the domain structure of the ASH domain-containing proteins that were cloned and analyzed. The figure is based on data from references [3, 16,] and [17]. The corresponding NCBI Reference Sequence numbers are: TRAPPC8, NP_055754.2; TRAPPC11, NP_068761.4; TRAPPC13, NM_024941.3; NPHP4, XM_006710563.1

regulate primary cilia biogenesis by recruiting Rabin8 to the centrosome, supporting the notion that ASH domain-containing proteins may in general target to the centrosome. As such, the ASH domain itself could be sufficient to confer targeting to the centrosome or cilium. Indeed, the recruitment of specific proteins to the centrosomal-ciliary axis has been shown, in some cases, to be mediated by conserved structural domains that are enriched within centrosomal or ciliary proteomes. One well-described example of a centrosome-targeting domain is the PACT (pericentrin-AKAP450 centrosomal targeting) domain, a conserved 90 amino acid domain that was first identified within the C-terminal region of the centrosomal proteins AKAP450 and pericentrin [7]. The PACT domain is sufficient for conferring targeting of these two proteins to the centrosome [7], as well as other ectopically expressed proteins engineered to contain this domain [8, 9].

Here, we describe a method to test the hypothesis that the ASH domain confers targeting to the centrosome-cilium axis by generating plasmids coding for epitope-tagged full-length (FL) or truncated versions of the ASH domain-containing proteins TRAPPC8, TRAPPC13, NPHP4, and DLEC1, followed by expression and quantitative IFM analysis in cultured mammalian RPE1 cells. A schematic overview of the fusion proteins generated or used in this study is shown in Fig. 1. However, our method can easily be applied to other candidate ciliary or centrosomal proteins as well.

Using the described approach we could confirm that TRAPPC13 and NPHP4 are highly enriched at the base of primary cilia (Figs. 2 and 3), whereas DLEC1 could not be found at this site. This is likely because DLEC1 is specific for motile cilia and may rely on interaction with other motile cilia-specific components for its centrosomal-ciliary localization. Consistently, western blot analysis of RPE1 cells using a DLEC1-specific antibody indicated that endogenous DLEC1 is undetectable in these cells (Fig. 4a, b). Further, DLEC1 is homologous to a protein, FAP81, identified in the *Chlamydomonas* flagellar proteome [10] (BLAST E value 2e–43),

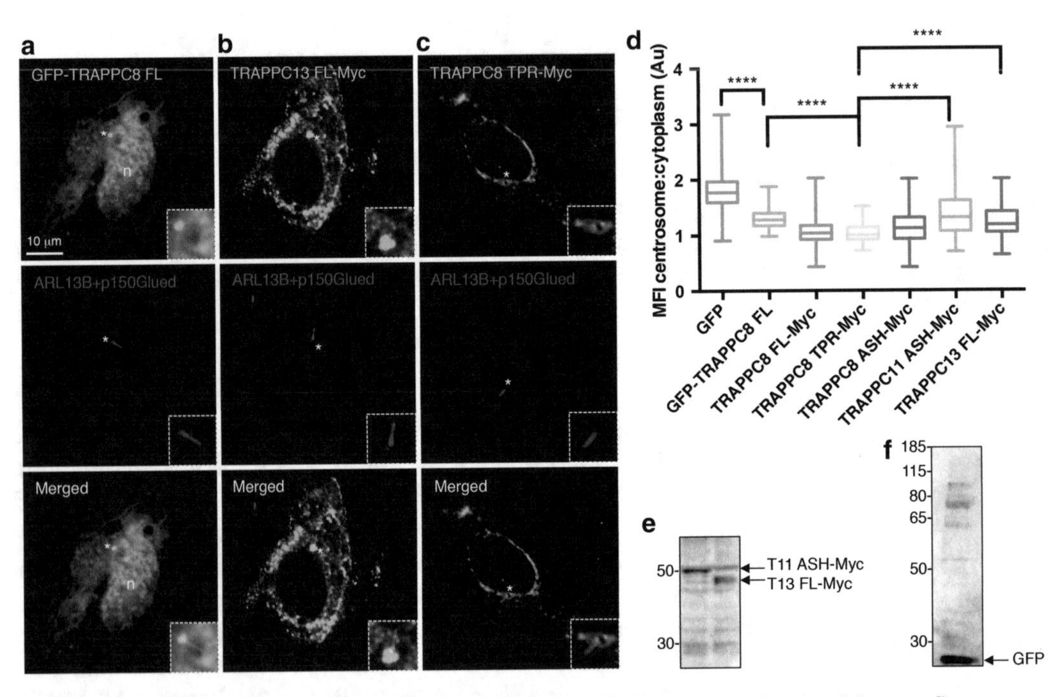

Fig. 2 Centrosomal localization of full-length and truncated TRAPP fusion proteins. (**a–c**) Immunofluorescence micrographs of transfected RPE1 cells expressing GFP-TRAPPC8 FL (**a**), TRAPPC13 FL-Myc (**b**), and TRAPPC8 TPR-Myc (**c**), respectively. Twelve hours post-transfection, cells were serum starved for 12 h, fixed with PFA (**a**) or methanol (**b, c**), and stained with antibodies against Myc (*green*), p150^{Glued} (*red*), and/or ARL13B (*red*), as indicated. Note accumulation of GFP-TRAPPC8 FL and TRAPPC13 FL-Myc around the centrosome, which is marked with an *asterisk*. *Insets* show a 2× magnification of the centrosome region. (**d**) Quantification of mean fluorescence intensity (MFI) ratio between centrosome and cytosol for the indicated fusion proteins (Au = arbitrary units). Note that TRAPPC8 FL-Myc and TRAPPC8 TPR-Myc failed to accumulate at the centrosome (likely because the epitope tag was cleaved off; see text for details), whereas GFP, GFP-TRAPPC8 FL, TRAPPC8 ASH-Myc, TRAPPC11 ASH-Myc, and TRAPPC13 FL-Myc all were found to be significantly enriched at the centrosome relative to TRAPPC8 TPR-Myc (used as negative control). In addition, the expressed GFP-TRAPPC8-FL fusion protein was detected in the nucleus (n). *p*-value: $p \leq 0.0001$. (**e, f**) western blot analysis of cells expressing TRAPPC11 ASH-Myc, TRAPPC13 FL-Myc, or GFP-TRAPPC8 FL as indicated. Note that bands of appropriate sizes were detected in TRAPPC11 ASH-Myc and TRAPPC13 FL-Myc expressing cells (**e**), whereas only a band corresponding to cleaved off GFP was seen in cells expressing GFP-TRAPPC8 FL (**f**)

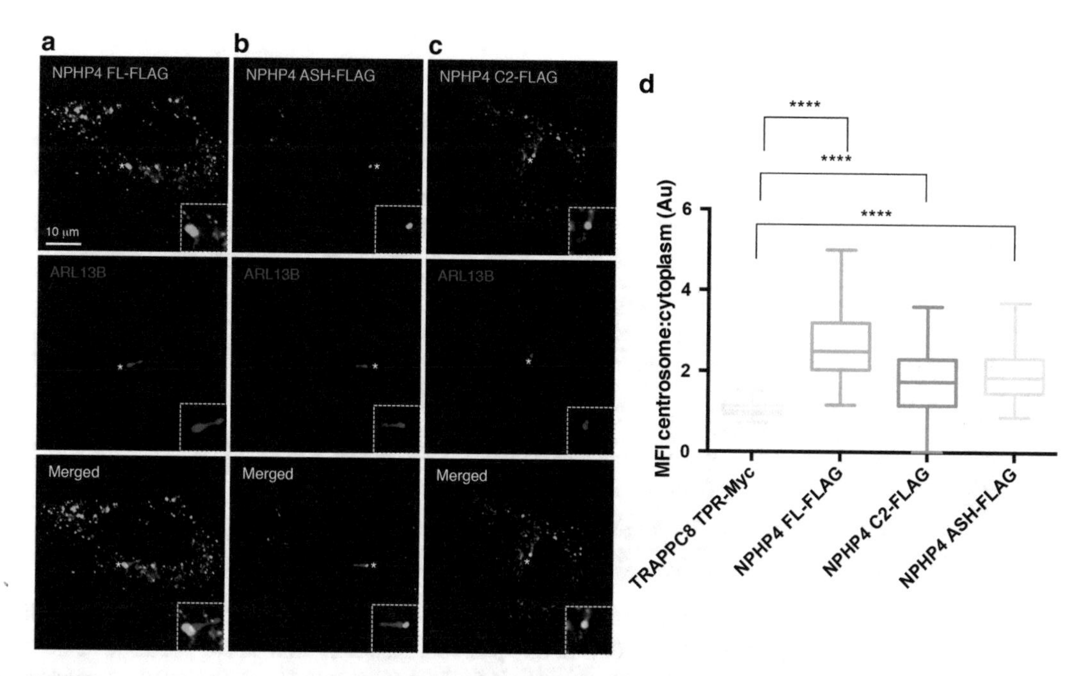

Fig. 3 Centrosomal localization of full-length and truncated NPHP4 fusion proteins. (**a–c**) Immunofluorescence micrographs of transfected RPE1 cells expressing NPHP4 FL-FLAG (**a**), NPHP4 ASH-FLAG (**b**), and NPHP4 C2-FLAG (**c**), respectively. Twelve hours post-transfection, cells were serum starved for 12 h, fixed with methanol, and stained with antibodies against FLAG (*green*) and ARL13B (*red*). Note accumulation of all three fusion proteins around the centrosome, which is marked by an asterisk. The *insets* show a 2× magnification of the centrosome region. (**d**) Quantification of mean fluorescence intensity (MFI) ratio between centrosome and cytosol for the indicated fusion proteins. Note that all NPHP4 fusion proteins display significant enrichment at the centrosome relative to TRAPPC8 TPR-Myc (p-value: ****$p \leq 0.0001$)

and data from the Human Protein Atlas (www.proteinatlas.org) [11] suggests that endogenous DLEC1 is enriched in motile cilia on, e.g., respiratory epithelial cells (Fig. 4c). Because DLEC1 also displays significant amino acid sequence homology to components of flagellar central pair-associated structures, such as *Chlamydomonas* FAP221 [12, 13] (BLAST E value 5e−10) and Hydin [14] (2e-06), DLEC1 might require association with such structures in order to localize to the cilium-centrosome axis.

Finally, it was found that epitope-tagged TRAPPC8 fusion proteins were unsuitable for IFM analysis since they seem to be unstable/degraded in cells. Nevertheless, when plasmids encoding GFP-TRAPPC8 or GFP alone were expressed in cells and analyzed by IFM, prominent accumulation of GFP at the centrosome was observed (Fig. 2), presumably owing to small size of GFP allowing it to passively diffuse into the ciliary compartment [15]. This illustrates the need for combining IFM analysis with western blotting in such studies.

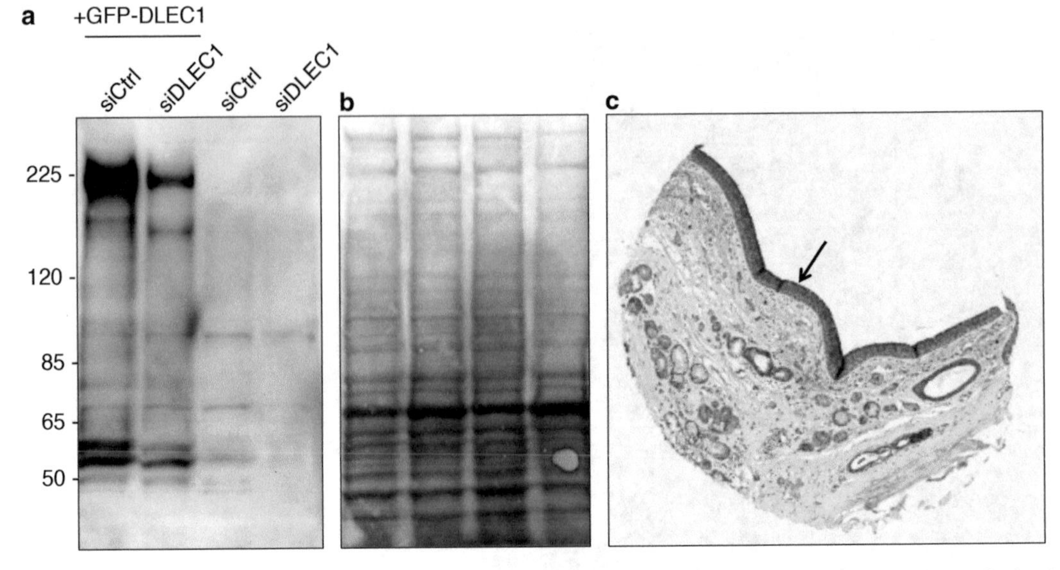

Fig. 4 Analysis of DLEC1 expression and subcellular localization in ciliated cells. (**a**) Western blot analysis of RPE1 cells transfected with plasmid encoding GFP-DLEC1 prior to transfection with mock- or DLEC1-specific esiRNA, as indicated (*lanes 1* and *2*), or RPE1 cells treated with two rounds of mock- or DLEC1-specific esiRNA, as indicated (*lanes 3* and *4*). About 18 μg of protein was loaded in each lane. The blot was probed with antibody against human DLEC1. Note the absence of a band corresponding to endogenous DLEC1 in the cells. (**b**) Ponceau S staining of the nitrocellulose membrane shown in (**a**). (**c**) Staining of a tissue section from human bronchus with the same DLEC1 antibody as that used in (**a**) (HPA019077; Sigma) showing localization of DLEC1 to motile cilia on the respiratory epithelial cells. The image is from the Human Protein Atlas (www.proteinatlas.org) [11]

2 Materials

2.1 Materials for RNA Extraction, cDNA Preparation, PCR, and Cloning

1. Kit for RNA purification (NucleoSpin®RNAII from Macherey-Nagel, includes DNase).
2. Reverse transcriptase (SuperScript®II from Invitrogen™/Thermo Fisher Scientific).
3. Specific PCR primers (*see* Table 1) and random primers (oligodeoxyribonucleotides, mostly hexamers).
4. dNTP mix, 10 mM of each dNTP.
5. RNase inhibitor (e.g., rRNasin from Promega).
6. 1 M dithiothreitol (DTT) solution.
7. PCR enzyme (e.g., Herculase®II Fusion DNA polymerase from Stratagene).
8. PCR-grade nuclease-free water.
9. Agarose.
10. Ethidium bromide stock solution (10 mg/ml).

Table 1
Primers used in this study

Insert	Vector	Primers	Restriction site
TRAPPC8 FL	pEGFP-N1	F: 5′ AA*GTCGAC***ATG**GCCCAGTGTGTACAATC 3′ R: 5′ AA*GGTACC*AA**CAC**ATTACTGATGATGATCAGGG 3′	*SalI* *Kpn*I
TRAPPC8 FL	pCMV-myc	F: 5′ AA*GTCGAC*T**ATG**GCCCAGTGTGTACAATC 3′ R: 5′ AA*GGTACC***TCA**CACATTACTGATGATGATCAG 3′	*SalI* *KpnI*
TRAPPC8 ASH	pCMV-myc	F: 5′ AA*GTCGAC*TGAATATGATTCTGAATCCTCTCAG 3′ F: 5′ AA*GGTACC***TCA**CACATTACTGATGATGATCAG 3′	*SalI* *KpnI*
TRAPPC8 TPR	pCMV-myc	R: 5′ AA*GTCGAC*T**ATG**GCCCAGTGTGTACAATC 3′ F: 5′ AA*GGTACC***TCA**TTGATCAAGATCTACATGAGTAGCT 3′	*SalI* *KpnI*
TRAPPC11 FL	pFLAG-CMV2	F: 5′ CC*ATCGAT*T**ATG**AGCCCCACACAGTGGGA 3′ R: 5′ AA*GGTACC***TCA**TGCAGCAGCAATAGAGGTATCATC 3′	*ClaI* *KpnI*
TRAPPC11 TPR	pFLAG-CMV2	F: 5′ CC*ATCGAT*T**ATG**AGCCCCACACAGTGGGA 3′ R: 5′ AAGGTACCTCAATGAAAACTTGGGGCATGAAACTTGG 3′	*ClaI* *KpnI*
TRAPPC13 FL	pCMV-myc	F: 5′ CC*GAATTC*TT**ATG**GAAGTGAATCCCCCTAAACA 3′ R: 5′ AA*GGTACC***TCA**GCTTTCCACTTTAATGGCAG 3′	*EcoRI* *KpnI*
NPHP4 FL	pFLAG-CMV2	F: 5′ AAAAA*GCGGCCGC*A**ATG**AACGACTGGCACAGG 3′ R: 5′ AAAAA*GATATC***TCA**CTGGTAGATGACCTTCAC 3′	*Not*I *EcoR*V
NPHP4 C2	pFLAG-CMV2	F: 5′ AAAAA*GCGGCCGC*A**ATG**AACGACTGGCACAGG 3′ R: 5′ AAAAA*GATATC***CTA**CAACCTCTCACTTTC 3′	*NotI* *EcoR*V
NPHP4 ASH	pFLAG-CMV2	F: 5′ AAAAA*GCGGCCGC*A**ACA**TTGCCACCGTCCAG 3′ R: 5′ AAAAA*GATATC***TCA**CTGGTAGATGACCTTCAC 3′	*Not*I *EcoR*V
DLEC1 FL	pEGFP-C1	F: 5′ AAAAA*GTCGAC***ATG**GAGACCAGGAGCTCC 3′ R: 5′ AAAAA*GGATCC***TCA**GGGCTGGTGAGG 3′	*Sal*I *BamH*I

11. Tris-acetate (TAE) buffer: (4.84 g Tris base, 1.14 ml glacial acetic acid, 2 ml 0.5 M Na_2 EDTA pH 8.0, ddH_2O up to 1 l).
12. Nucleic acid molecular weight marker (e.g., GeneRuler DNA Ladder Mix, Thermo Fisher Scientific).
13. DNA Loading Dye (e.g., 6× DNA Loading Dye, Thermo Fisher Scientific).
14. Restriction enzymes: *Kpn*I, *Cla*I, *Sal*I, *EcoR*1, *BamH*1 (available, e.g., from Roche or New England Biolabs).
15. T4 DNA ligase buffer (Invitrogen™/Thermo Fisher Scientific).
16. T4 DNA ligase (Invitrogen™/Thermo Fisher Scientific).
17. Cloning vectors: pCMV-Myc (Clontech Laboratories, Inc.), pFLAG-CMV2 (Sigma-Aldrich), pEGFP-N1 (Clontech Laboratories, Inc.), pEGFP-C1 (Clontech Laboratories, Inc.).

18. Plastic and glassware: 10 cm diameter Petri dishes, sterile test tubes and flasks for bacterial cell culture, sterile plastic tubes for PCR, microcentrifuge tubes, cryotubes for bacterial freeze culture, sterile plastic spreaders, sterile inoculation needles or toothpicks.
19. Prokaryotic culture medium: Luria-Bertani (LB) broth (Sigma).
20. Recovery medium (Invitrogen™/Thermo Fisher Scientific).
21. Bacto agar (Becton, Dickinson and company).
22. Antibiotics: kanamycin, ampicillin, tetracycline (*see* **Note 1**).
23. 96% ethanol.
24. Kit for purification of PCR products and DNA extraction from gel slices (e.g., NucleoSpin® Gel and PCR Clean-up kit from Macherey-Nagel).
25. Kit for small plasmid purification (e.g., NucleoSpin®Plasmid from Macherey-Nagel).
26. Kit for medium/large plasmid purification (e.g., NucleoBond®Xtra Midi/Maxi from Macherey-Nagel).
27. *Escherichia coli* strains for cloning, e.g., DH10α or INV110 (*see* **Note 2**).
28. Glycerol, sterile.
29. Scalpel.
30. Spectrophotometer and UV cuvettes.

2.2 Materials for Mammalian Cell Culture and Transfection

1. Plasticware: T75 flasks, 6 and 10 cm diameter Petri dishes, 6-well plates for cell culturing.
2. Phosphate-buffered saline (PBS): 8 g NaCl, 0.2 g KCL, 1.15 g $NaHPO_4$, 0.2 g KH_2PO_4, ddH_2O to 1 l. Adjust pH to 7.4.
3. Growth medium: Dulbecco's modified Eagle's medium (DMEM) with penicillin/streptomycin (p/s; 100 U/ml) plus 10% heat-inactivated fetal bovine serum (FBS).
4. Starvation medium: DMEM.
5. Transfection reagents: FuGENE®6 (Promega) and DharmaFECT Duo (Dharmacon).
6. Trypsin mix: 1% trypsin-EDTA in PBS.
7. Small interfering RNA (siRNA): mock control Stealth RNAi, Invitrogen Medium GC Duplex (Invitrogen™/Thermo Fisher Scientific); human DLEC1 esiRNA (200 ng/μl) (Sigma).
8. Mammalian RPE1 cells (*see* **Note 3**).

2.3 Materials for Immunofluorescence Microscopy Analysis

1. Microscope slides and glass coverslips (12 mm diameter).
2. Concentrated HCl.
3. Ethanol (70 and 96%).
4. 100% methanol.

5. PBS.
6. 4% paraformaldehyde (PFA) solubilized in PBS.
7. Permeabilization buffer: 0.2% (v/v) Triton X-100 and 1% (v/v) BSA in PBS.
8. Blocking buffer: 2% (w/v) bovine serum albumin (BSA) in PBS.
9. Primary antibodies (dilutions in parenthesis): goat anti-Myc (1:1000) from Abcam; rabbit anti-ARL13B (1:1000) from Protein Tech; mouse anti-p150^{Glued} (1:500) from BD Transduction Laboratories; mouse anti-FLAG (1:1000) from Sigma.
10. Secondary antibodies (all from Invitrogen™/Thermo Fisher Scientific and diluted 1:600): Alexa Fluor 488 Donkey anti-Goat; Alexa Fluor 568 Donkey anti-Mouse; Alexa Fluor 568 Donkey anti-Rabbit; Alexa Fluor 488 Donkey anti-Mouse.
11. 20 mg/ml DAPI (4′, 6-diamidino-2-phenylindole, dihydrochloride) solubilized in ddH_2O (store in aliquots at −20 °C away from light).
12. Mounting reagent (5 ml): 4.5 ml glycerol, 0.5 ml 10× PBS, 0.1 g *N*-propyl gallate (*see* **Note 4**).
13. Humidity chamber (*see* **Note 5**).
14. Nail polish.

2.4 Materials for SDS-PAGE and Western Blot Analysis

1. PBS.
2. Lysis buffer: 1.25 ml 10% SDS (0.5% final), 250 μl 1 M Tris–HCl pH 7.6 (10 mM final), 1 Complete™ Protease Inhibitor Cocktail Tablet (Santa Cruz Biotechnology), ddH_2O to 25 ml.
3. Cell scrapers/rubber policemen.
4. Kit for measurement of protein concentration (e.g., *DC* Protein Assay Reagents from Bio-Rad).
5. 1 M DTT.
6. Loading Sample Buffer (LDS): 4× Novex® NuPAGE® LDS (lithium dodecyl sulfate) sample buffer (Invitrogen™/Thermo Fisher Scientific).
7. 3–8% Tris-Acetate NuPAGE® Novex® precast gel (Invitrogen™/Thermo Fisher Scientific).
8. Nitrocellulose membrane.
9. Whatman filter paper.
10. Blotting sponge pads.
11. Gel knife.
12. XCell SureLock™ Mini-Cell Electrophoresis System (Invitrogen™/Thermo Fisher Scientific).

13. Spectra Multicolor Broad Range Protein Ladder (Thermo Fisher Scientific).
14. Running buffer: 30 ml NuPAGE® Tris-Acetate SDS Running Buffer (20×) (Invitrogen™/Thermo Fisher Scientific), 570 ml ddH_2O).
15. Transfer buffer: 30 ml NuPAGE® Transfer Buffer (20×) (Invitrogen™/Thermo Fisher Scientific), 60 ml 96% ethanol (10% final), 510 ml ddH_2O.
16. Tris-buffered saline with Tween-20 (TBS-T): 10 ml 1 M Tris–HCl pH 7.5 (10 mM final), 75 ml 2 M NaCl (150 mM final), 1 ml Tween-20 (120 mM final), add ddH_2O to 1 l.
17. Ponceau S staining solution: 0.1% (w/v) Ponceau S in 5% (v/v) acetic acid.
18. Blocking buffer: 2.5 g nonfat dry milk (5% (w/v) final) and TBS-T up to 50 ml.
19. Primary antibodies: anti-DLEC1 polyclonal rabbit antibody (Sigma, HPA019077), diluted 1:250 in blocking buffer; anti-GFP rabbit polyclonal (Santa Cruz Biotechnology, Sc-8334), diluted 1:500 in blocking buffer; anti-Myc rabbit polyclonal antibody (Abcam, Ab9106), diluted 1:1000 in blocking buffer.
20. Secondary antibody: Polyclonal Swine Anti-Rabbit Immunoglobulins/Horseradish Peroxidase (DAKO), diluted 1:4000 in blocking buffer.
21. Humidity chamber (*see* **Note 5**).
22. Amersham ECL™ Western Blotting Detection Reagents (GE Healthcare Life Sciences).

3 Methods

All experiments are carried out at room temperature unless otherwise stated.

3.1 RNA Extraction and cDNA Preparation

1. Culture RPE1 cells in growth medium in 10 cm diameter Petri dishes to 80% confluence (*see* Subheading 3.3), and prepare total RNA according to the NucleoSpin®RNAII protocol from Macherey-Nagel. This includes a DNase step for elimination of contaminating genomic DNA.
2. Calculate the total RNA concentration (C) by measuring the optical density (OD) at 260 nm in a spectrophotometer: C (μg/ml) = Absorbance at 260 nm ($A260$):0.025.
3. To prepare cDNA, mix the following in a nuclease-free microcentrifuge tube on ice: 500 ng of random primers, 2 μl dNTP mix (stock solution of 10 mM of each dNTP), 1 μg total RNA and PCR-grade, nuclease-free water to 24 μl.

4. Incubate at 65 °C for 5 min and quickly cool on ice.
5. Briefly spin tube in a microcentrifuge and add 8 µl first strand buffer, 4 µl 0.1 M DTT and 2 µl RNAse inhibitor. Incubate for 2 min at 25 °C.
6. Add 2 µl SuperScript II reverse transcriptase enzyme to the tube and incubate for 10 min at 25 °C, followed by 50 min incubation at 42 °C and 15 min at 70 °C to inactivate the enzyme.
7. Store the cDNA at −20 °C until use.

3.2 PCR and Cloning Procedures

1. For PCR, mix the following in a nuclease-free PCR tube: 10 µl 5× Herculase II Fusion DNA polymerase buffer, 2 µl forward and reverse primer (10 µM stock in nuclease-free H_2O), 1 µl dNTP mix (stock of 10 mM of each dNTP) and nuclease-free H_2O to a total volume of 46.5 µl. The primers used in this study are shown in Table 1 and a schematic overview of the fusion proteins generated is shown in Fig. 1.
2. Immediately prior to PCR add 2.5 µl cDNA from RPE1 cells (25 ng/µl) and 1 µl Herculase II Fusion DNA polymerase.
3. Run the following PCR program: 95 °C for 2 min, 10×(95 °C for 20 s, 52 °C for 20 s, 72 °C for 4 min), 20×(95 °C for 20 s, 61 °C for 20 s, 72 °C for 4 min), 72 °C for 3 min (*see* **Note 6**).
4. Analyze 5 µl of each PCR product on a 1% agarose gel in TAE buffer (include 0.5 µg/ml ethidium bromide in the gel) for 45–60 min at 130 V to verify that the PCR products have the desired sizes.
5. Purify PCR products using the NucleoSpin® Gel and PCR Clean-up kit from Macherey-Nagel.
6. Calculate the DNA concentration (C) by measuring the OD_{260} in a spectrophotometer: C (µg/ml) = A_{260}:0.020.
7. Digest the vector and the purified PCR products with appropriate restriction enzymes (e.g., *Kpn*I, *Cla*I, *Sal*I, *EcoR*1, or *BamH*1) by using buffers and protocols supplied by the manufacturer (*see* **Note 7**).
8. Run a 1% agarose gel in TAE buffer (include 0.5 µg/ml ethidium bromide in the gel) to check for correct digestion of both vector and PCR product.
9. Visualize the bands using long-wave UV light (365 nm) in order not to damage the DNA, and excise the desired bands from the gel using a scalpel.
10. Extract DNA from the gel slices by using the NucleoSpin® Gel and PCR Clean-up kit from Macherey-Nagel. Calculate the DNA concentration (*see* **step 6**).

11. Ligate the insert and vector with T4 DNA ligase over night at 16 °C (*see* **Note 7**).
12. Prepare agar plates: 10.25 g LB broth, 7.5 g agar and ddH_2O up to 500 ml. Autoclave. Add 500 μl of each relevant 1000× stock solution of antibiotics (*see* **Note 1**). Pour the agar into 10 cm diameter Petri dishes and let dry. Keep in refrigerator (bottom up) until use.
13. Transform the ligated DNA into competent *E. coli* cells (*see* **Note 2**). Thaw the competent cells on ice and add 1–5 μl of ligation mix or an equivalent amount of plasmid without insert (positive control) or nuclease-free H_2O (negative control). Incubate the cells on ice for 30 min.
14. Heat shock for 1–2 min in a 42 °C water bath.
15. Rest on ice for minimum of 1 min.
16. Add 250 μl recovery medium and incubate on shaker at 37 °C for 1 h.
17. Plate 150 μl of each cell suspension on LB agar plates with relevant antibiotics using a sterile spreader. Incubate plates bottom up at 37 °C overnight.
18. Pick selected colonies using a sterile inoculation needle or toothpick, and transfer colonies to a sterile test tube containing 3 ml LB medium and relevant antibiotics. Incubate at 37 °C overnight with vigorous shaking.
19. Purify plasmids using the NucleoSpin®Plasmid kit from Macherey-Nagel and check for correct insert by performing a control digest with relevant restriction enzymes, followed by agarose gel electrophoresis (*see* **step 8**).
20. Prepare a larger amount of relevant plasmids using NucleoBond®Xtra Midi/Maxi kit from Macherey-Nagel, and verify insert by sequencing. Store plasmid preparations at -20 °C until use.
21. Prepare freeze cultures of relevant recombinant *E. coli* strains by simply adding one-third volume sterile glycerol to log-phase cultures in a test tube and storing at -80 °C in sterile plastic cryotubes. Invert tubes several times before freezing to ensure mixing of culture and glycerol solution.

3.3 Mammalian Cell Culture and Transfection

1. Culture RPE1 cells in T75 flasks in 10 ml growth medium at 37 °C, 5% CO_2, 95% humidity.
2. Split cells when they reach 80–100% confluence. Wash the cells in 37 °C PBS and trypsinize with 2 ml trypsin-EDTA (1%) until the cells detach from the bottom of the bottle.
3. Resuspend in 8 ml 37 °C fresh growth medium and transfer to a new T75 flask or prepare for experiments as follows (*see* **Note 8**).

4. For RNA purification, culture RPE1 cells in growth medium in 10 cm diameter Petri dishes to 80 % confluence, and prepare total RNA according to the NucleoSpin®RNAII protocol from Macherey-Nagel. This includes a DNase step for elimination of contaminating genomic DNA.
5. For plasmid transfection and IFM analysis, seed cells to approximately 30 % confluence on coverslips (*see* **Note 9**) placed in a 6-well culture dish (max four coverslips per well). Use about 1.5 ml cell suspension per well and grow to 60–80 % confluence at 37 °C, 5 % CO_2, 95 % humidity.
6. Prepare transfection mix immediately before transfection: in a sterile microcentrifuge tube, mix 1 μg plasmid with 3 μl FuGENE 6 and 46 μl starvation medium. Incubate for 30 min and then drip onto the cells grown on the coverslips in the 6-well culture dishes (*see* **Note 10**). Incubate cells for 12 h at 37 °C, 5 % CO_2, 95 % humidity (*see* **Note 11**).
7. Remove starvation medium and wash gently in 37 °C PBS. Then add 1.5 ml starvation medium (*see* Subheading 2.2, **item 4**) and incubate cells for another 12 h to induce growth arrest and ciliogenesis (*see* **Note 12**). For SDS-PAGE and western blotting, culture cells to 40–50 % confluence in 6 cm diameter Petri dishes. For transfection with plasmid, *see* **step 6**. For transfection with siRNA, mix 10 μl of mock or DLEC1-specific esiRNA with 10 μl DharmaFECT Duo in 200 μl starvation medium. Incubate for 30 min and drip into well containing 3.8 ml growth medium. Incubate cells for 48 h at 37 °C, 5 % CO_2, 95 % humidity.

3.4 Immunofluorescence Microscopy and Image Analysis

1. Aspirate the starvation medium from cells grown in a 6-well culture dish on glass coverslips (*see* Subheading 3.3, **step 4**) and wash cells gently in 37 °C PBS.
2. Fix cells for 5 min in methanol at −20 °C or in 4 % PFA solution for 15 min directly in the 6-well culture dishes (*see* **Note 13**).
3. For methanol-fixed cells, transfer the coverslips from the 6-well culture dishes using forceps and scalpel onto a piece of Whatman paper, cell side up, and allow to air dry. Then move coverslips to a humidity chamber (*see* **Note 5**) and rehydrate cells by adding a drop of PBS and incubating for 5 min.
4. For PFA-fixed cells, briefly wash cells in PBS and incubate in permeabilization buffer for 12 min. Replace the buffer with PBS and transfer coverslips to a humidity chamber (*see* **Note 5**) using forceps and scalpel.
5. From here on, the same procedure can be used for both methanol- and PFA-fixed cells. Incubation with various solutions is achieved by simply placing a drop of liquid on the top of each coverslip.

6. Immediately following transfer of the coverslips to the humidity chamber, incubate with 100 μl of blocking buffer (*see* Subheading 2.3, **item 7**) for 30 min.
7. Incubate with relevant primary antibodies diluted in blocking buffer (*see* Subheading 2.3, **item 8**) for 1.5 h.
8. Wash 3×5 min with BSA and incubate with appropriate fluorophore-conjugated secondary antibodies diluted in BSA (*see* Subheading 2.3, **step 9**) for 45–60 min.
9. Wash 1×5 min with BSA, stain briefly with DAPI (1:5000 dilution of stock in PBS), and wash 3×5 min in PBS. Please note that the DAPI staining is not depicted in the images shown in Figs. 2 and 3.
10. Mount coverslips by placing them, cells facing down, in a droplet of mounting reagent (*see* Subheading 2.3, **item 11**) on a microscope slide cleaned with 96% ethanol.
11. Remove excess mounting reagent with a paper towel and seal the edges of the coverslips with nail polish.
12. Analyze cells on an epifluorescence microscope with appropriate camera and software. Here, we used a motorized Olympus BX63 upright microscope equipped with a DP72 color, 12.8 megapixel, 4140×3096 resolution camera and differential interference contrast, and Olympus cellSens dimension and Adobe Photoshop CS4 (version 11.0) software for image analysis and processing.
13. For quantification of relative fluorescence intensities, capture images of cells expressing various fusion proteins (Fig. 1) using similar settings.
14. Measure the fluorescence intensity of the each fusion protein at the centrosome/ciliary base as well as in the cytoplasm using the count and measurement function/ measurement and ROI/select measurements Mean (Green) and Max (Green) of the Olympus cellSens software.
15. Calculate the ratio between fluorescence at the centrosome and the cytoplasm and analyze data using Prism Version 6.0 g (GraphPad Software, San Diego). Calculate statistical significance using data from at least three independent experiments by pairwise *t*-test. Error bars denote standard error of the mean. *p*-Values: $^{****}p \leq 0.0001$; $^{***}p \leq 0.001$; $^{**}p \leq 0.01$; $^{*}p \leq 0.05$.

Examples of the results we obtained using the above method, as applied to select ASH domain-containing proteins, are shown in Figs. 2 and 3. Specifically, significant accumulation at the centrosome was detected for the ASH domain-containing TRAPP fusion proteins GFP-TRAPPC8 FL, GFP-TRAPPC8 ASH-Myc,

TRAPPC13 FL-Myc, as well as TRAPPC11 ASH-Myc, whereas the C-terminal Myc-tagged TRAPPC8 FL and TRAPPC8 TPR fusion proteins failed to display significant accumulation at the centrosome (Fig. 2a–d). In addition, GFP-TRAPPC8 FL, but not the other TRAPP fusion proteins, was detected in the nucleus (Fig. 2a). However, western blot analysis of cells expressing the same fusion proteins (see below) showed that all the TRAPPC8 fusion proteins lost their epitope tag when expressed in cells, whereas similar analysis of the TRAPPC13 FL-Myc and TRAPPC11 ASH-Myc fusions yielded bands of expected sizes (Fig. 2e, f and data not shown). The centrosomal localization of the latter two proteins is therefore likely to reflect that of the endogenous proteins. In contrast, because GFP alone also showed prominent centrosomal and nuclear accumulation (Fig. 2d and data not shown), which we attribute to its small size that allows it to diffuse through the nuclear pore complex and the barrier at the ciliary base [15], the observed centrosome and nuclear localization of GFP-TRAPPC8 FL fusion protein (Fig. 2a, d) most likely represents GFP that has been cleaved off (*see* western blot in Fig. 2f). Therefore, we conclude that the generated plasmids encoding epitope-tagged TRAPPC8 fusion proteins are unsuitable for IFM and western blot analysis, at least when expressed under the conditions reported here. We also failed to detect significant centrosome localization of GFP-DLEC1 (data not shown), even though the fusion protein was expressed and stable in cells (Fig. 4a). In contrast, all three NPHP4 fusion proteins analyzed displayed significant enrichment at the centrosome (Fig. 3) and yielded bands of the expected sizes in western blot analysis (data not shown). The latter observation is in agreement with the idea that both the ASH and C2 domains of NPHP4 are able to confer targeting to the centrosome, although it cannot be excluded that amino acid residues outside of these two domains could mediate the centrosome recruitment of the fusion proteins analyzed.

3.5 SDS-PAGE and Western Blotting

1. Wash cells in ice-cold PBS and add boiling lysis buffer to the Petri dish.
2. Scrape cells off using a rubber policeman and transfer the cell lysates to microcentrifuge tubes and keep the samples on ice.
3. Sonicate the cell lysates for 3×10 s. Rest on ice in between to avoid heating of the samples.
4. Centrifuge at $20{,}000 \times g$ for 20 min to precipitate the cell debris. Transfer the supernatant to new microcentrifuge tubes and keep them on ice.
5. Determine the protein concentration in the cell lysates according to the manufacturer's protocol for, e.g., the Bio-Rad *DC* Protein Assay.

6. Dilute the samples in lysis buffer to a protein concentration of 0.5 μg/μl. Mix 28 μl of each of the diluted samples with 2 μl 1 M DTT (0.05 M final) and 10 μl LDS (4×).
7. Place the gel cassette in the electrophoresis chamber and add running buffer to the inner and outer chamber.
8. Incubate samples for 5 min at 95 °C.
9. Load samples (35 μl) and marker (5 μl) on the gel.
10. Run electrophoresis for about 1.5 h at 160 V until the loading dye reaches the bottom of the gel.
11. Break the gel out of the gel cassette using a gel knife and build a sandwich of blotting pads, filter papers, nitrocellulose membrane (all soaked in transfer buffer), and gel in an XCell II™ blot module (Invitrogen™/Thermo Fisher Scientific).
12. Place the blot module in the transfer chamber and add transfer buffer to the inner and the outer chamber. Transfer proteins for 3 h at 25 V.
13. Stain the blot with Ponceau S to check for air bubbles and correct protein transfer.
14. Block the membrane in blocking buffer on a shaking table for 30 min.
15. Incubate the membrane with the primary antibody diluted in blocking buffer in the cold room overnight or for 1–2 h.
16. Wash the membrane in TBS-T, 5 × 5 min on a shaking table.
17. Incubate the membrane with secondary antibody diluted in blocking buffer for 45 min in a humidity chamber.
18. Wash the membrane in TBS-T, 5 × 5 min on a shaking table.
19. Mix equal volumes of each of the Amersham ECL™ Western Blotting Detection Reagents and apply to the membrane for 1 min (see manufacturer's protocol for further details).
20. Detect the chemiluminescence using, e.g., FUSION-FX, Vilber Lourmat.

In summary, the described IFM-based procedure was useful for confirming centrosome localization for five out of eight fusion proteins analyzed, namely, TRAPPC11 ASH-Myc, TRAPPC13 FL-Myc, and all three NPHP4 fusion proteins. The described IFM-based procedure is applicable to other candidate ciliary or centrosomal proteins as well, but should be combined with western blot analysis to ensure integrity of the fusion protein in question in order to avoid misinterpretation of IFM-based data. Also, consideration of fusion protein size is important for data interpretation as proteins smaller than 40 kDa may diffuse passively into cilia [15].

4 Notes

1. Prepare 1000× stock solution of ampicillin (100 mg/ml in a 1:1 solution of ddH_2O and ethanol), kanamycin (50 mg/ml in ddH_2O), and tetracycline (10 mg/ml in ethanol) in sterile plastic tubes. Store at −20 °C protected from light and moisture. Add relevant antibiotics to bacterial growth medium after autoclaving and cooling the medium to 55 °C or below.
2. For routine cloning, we usually use *E. coli* strain DH10α, but for plasmids that are to be digested with *dam* or *dcm* methylation-sensitive restriction enzymes (e.g., *Cla*I), we use the INV110 strain. Competent cells can be purchased from several vendors. Alternatively, they can be prepared as follows: streak *E. coli* cells onto an LB agar plate with or without relevant antibiotics and incubate overnight at 37 °C. Transfer one colony from the plate to 50 ml of liquid LB broth until an OD_{600} of 0.4–0.8 is reached. Centrifuge at 1500 × *g* for 5 min and gently resuspend pelleted cells in 5 ml of ice-cold, sterile 50 mM $CaCl_2$, and let cell suspension rest on ice for 30 min. Centrifuge at 1500 × *g* for 5 min and gently resuspend pelleted cells in 3 ml of ice-cold, sterile 50 mM $CaCl_2$ containing 20% glycerol. Store at −80 °C in aliquots of 100 μl.
3. Available from ATCC. The hTERT-RPE1 cell line is an adherent, diploid retinal epithelial-like cell line that was immortalized by transfection of the RPE-340 cell line with the pGRN145 hTERT-expressing plasmid (available from ATCC). The cells readily undergo ciliogenesis upon reaching confluence and following serum deprivation.
4. Stir the solution overnight for the *N*-propyl gallate to dissolve. The solution can be stored for several days at 4 °C.
5. To prepare the humidity chamber, one can place a moist piece of Whatman paper in an empty Petri dish, and put a piece of Parafilm on the top onto which coverslips (for IFM analysis) or nitrocellulose membranes (for western blot analysis) are placed, cells/proteins facing up.
6. The optimal annealing temperature for each primer set can be calculated using OligoCalc, available at http://www.basic.northwestern.edu/biotools/oligocalc.html. For the elongation step, we recommend using ca. 1 min per 1000 bp.
7. The optimal ratio of vector to insert can be calculated using the ligation calculator on http://www.insilico.uni-duesseldorf.de/Lig_Input.html.
8. Cells should not be passaged more than 20 times.

9. For IFM analysis, prepare coverslips by immersion in concentrated HCl for 1 h under gently agitation followed by ten washes in ddH_2O. Sterilize the acid-washed coverslips by rinsing 10 times in 96% ethanol. Store in 70% ethanol and air dry immediately before use.
10. The 30 min incubation step is especially important for larger plasmids.
11. Transfection time can vary, depending on plasmid and cell type used. For transient transfection of RPE1 cells, we do not recommend transfection times longer than 24 h (including subsequent starvation period) as this will typically lead to overexpression and mislocalization of the exogenous proteins.
12. Although 12 h starvation may not be sufficient to induce ciliogenesis in all cells, we found that longer starvation time may compromise viability of transfected cells. Also note that RPE1 cells grown to more than 80% confluence may enter growth arrest and form cilia despite the presence of serum.
13. The optimal fixation method used may differ depending on the protein of interest and antibodies used.

Acknowledgments

This work was supported by the University of Copenhagen Excellence Programme for Interdisciplinary Research and the Danish Council for Independent Research (1331-00254). S.K.M. was partially supported by a Ph.D. fellowship from the Department of Biology, University of Copenhagen. We thank Sophie Saunier for the gift of human NPHP4 cDNA, Søren L. Johansen for technical assistance, Kenneth B. Schou for assistance with DLEC1 and NPHP4 PCR and cloning, and Louise Lindbæk for assistance with image analysis.

References

1. Arnaiz O, Cohen J, Tassin AM, Koll F (2014) Remodeling Cildb, a popular database for cilia and links for ciliopathies. Cilia 3:9. doi:10.1186/2046-2530-3-9
2. Ponting CP (2006) A novel domain suggests a ciliary function for ASPM, a brain size determining gene. Bioinformatics 22(9):1031–1035. doi:10.1093/bioinformatics/btl022
3. Schou KB, Morthorst SK, Christensen ST, Pedersen H (2014) Identification of conserved, centrosome-targeting ASH domains in TRAPPII complex subunits and TRAPPC8. Cilia 3:6. doi:10.1186/2046-2530-3-6
4. Tarr DE, Scott AL (2005) MSP domain proteins. Trends Parasitol 21(5):224–231. doi:10.1016/j.pt.2005.03.009
5. Bork P, Holm L, Sander C (1994) The immunoglobulin fold. Structural classification, sequence patterns and common core. J Mol Biol 242(4):309–320. doi:10.1006/jmbi.1994.1582
6. Westlake CJ, Baye LM, Nachury MV, Wright KJ, Ervin KE, Phu L, Chalouni C, Beck JS, Kirkpatrick DS, Slusarski DC, Sheffield VC, Scheller RH, Jackson PK (2011) Primary cilia membrane assembly is initiated by Rab11 and transport protein particle II (TRAPPII) complex-dependent

trafficking of Rabin8 to the centrosome. Proc Natl Acad Sci U S A 108(7):2759–2764. doi:10.1073/pnas.1018823108

7. Gillingham AK, Munro S (2000) The PACT domain, a conserved centrosomal targeting motif in the coiled-coil proteins AKAP450 and pericentrin. EMBO Rep 1(6):524–529. doi:10.1093/embo-reports/kvd105
8. Kishi K, van Vugt MA, Okamoto K, Hayashi Y, Yaffe MB (2009) Functional dynamics of Polo-like kinase 1 at the centrosome. Mol Cell Biol 29(11):3134–3150. doi:10.1128/MCB.01663-08
9. Januschke J, Reina J, Llamazares S, Bertran T, Rossi F, Roig J, Gonzalez C (2013) Centrobin controls mother-daughter centriole asymmetry in Drosophila neuroblasts. Nat Cell Biol 15(3):241–248. doi:10.1038/ncb2671
10. Pazour GJ, Agrin N, Leszyk J, Witman GB (2005) Proteomic analysis of a eukaryotic cilium. J Cell Biol 170:103–113
11. Uhlen M, Fagerberg L, Hallstrom BM, Lindskog C, Oksvold P, Mardinoglu A, Sivertsson A, Kampf C, Sjostedt E, Asplund A, Olsson I, Edlund K, Lundberg E, Navani S, Szigyarto CA, Odeberg J, Djureinovic D, Takanen JO, Hober S, Alm T, Edqvist PH, Berling H, Tegel H, Mulder J, Rockberg J, Nilsson P, Schwenk JM, Hamsten M, von Feilitzen K, Forsberg M, Persson L, Johansson F, Zwahlen M, von Heijne G, Nielsen J, Ponten F (2015) Proteomics. Tissue-based map of the human proteome. Science 347(6220):1260419. doi:10.1126/science.1260419
12. DiPetrillo CG, Smith EF (2010) Pcdp1 is a central apparatus protein that binds Ca(2+)-calmodulin and regulates ciliary motility. J Cell Biol 189(3):601–612. doi:10.1083/jcb.200912009
13. Brown JM, Dipetrillo CG, Smith EF, Witman GB (2012) A FAP46 mutant provides new insights into the function and assembly of the C1d complex of the ciliary central apparatus. J Cell Sci 125(Pt 16):3904–3913. doi:10.1242/jcs.107151
14. Lechtreck KF, Witman GB (2007) Chlamydomonas reinhardtii hydin is a central pair protein required for flagellar motility. J Cell Biol 176(4):473–482. doi:10.1083/jcb.200611115
15. Kee HL, Dishinger JF, Blasius TL, Liu CJ, Margolis B, Verhey KJ (2012) A size-exclusion permeability barrier and nucleoporins characterize a ciliary pore complex that regulates transport into cilia. Nat Cell Biol 14(4):431–437. doi:10.1038/ncb2450
16. Zhang D, Aravind L (2012) Novel transglutaminase-like peptidase and C2 domains elucidate the structure, biogenesis and evolution of the ciliary compartment. Cell Cycle 11(20):3861–3875. doi:10.4161/cc.22068
17. Sang L, Miller JJ, Corbit KC, Giles RH, Brauer MJ, Otto EA, Baye LM, Wen X, Scales SJ, Kwong M, Huntzicker EG, Sfakianos MK, Sandoval W, Bazan JF, Kulkarni P, Garcia-Gonzalo FR, Seol AD, O'Toole JF, Held S, Reutter HM, Lane WS, Rafiq MA, Noor A, Ansar M, Devi AR, Sheffield VC, Slusarski DC, Vincent JB, Doherty DA, Hildebrandt F, Reiter JF, Jackson PK (2011) Mapping the NPHP-JBTS-MKS Protein Network reveals ciliopathy disease genes and pathways. Cell 145(4):513–528. doi:10.1016/j.cell.2011.04.019

Chapter 3

Morphological and Functional Characterization of the Ciliary Pocket by Electron and Fluorescence Microscopy

Rania Ghossoub*, Louise Lindbæk*, Anahi Molla-Herman*, Alain Schmitt, Søren Tvorup Christensen*, and Alexandre Benmerah*

Abstract

In many vertebrate cell types, the proximal part of the primary cilium is positioned within an invagination of the plasma membrane known as the ciliary pocket. Recent evidence points to the conclusion that the ciliary pocket comprises a unique site for exocytosis and endocytosis of ciliary proteins, which regulates the spatiotemporal trafficking of receptors into and out of the cilium to control its sensory function. In this chapter, we provide methods based on electron microscopy, 3D reconstruction of fluorescence images as well as live cell imaging suitable for investigating processes associated with endocytosis at the ciliary pocket.

Key words Primary cilia, Ciliary pocket, Clathrin-dependent endocytosis, Early endosomes, Cellular signaling

1 Introduction

Primary cilia are 9+0 microtubule (MT)-based, membrane-enclosed projections that emanate as solitary organelles on the surface of most quiescent cell types in vertebrates [1]. The ciliary axoneme is nucleated from the centrosomal mother centriole (basal body), which docks to the plasma membrane via its distal appendage proteins also known as transition fibers [2]. The membrane surrounding the axoneme is continuous with the plasma membrane, but it has a unique lipid composition. Moreover, the ciliary membrane harbors a unique complement of receptors and ion channels that enable the cilium to function as a sensory organelle that relays signals from the extracellular environment to control developmental processes and tissue homeostasis. The sensory capacity of the cilium is maintained through structural and

*These authors contributed equally to the manuscript.

Peter Satir and Søren Tvorup Christensen (eds.), *Cilia: Methods and Protocols*, Methods in Molecular Biology, vol. 1454, DOI 10.1007/978-1-4939-3789-9_3, © Springer Science+Business Media New York 2016

functional barriers above the basal body. This sets up a transition zone or molecular filter that gates the selective passage of receptors and ion channels into and out of the ciliary compartment. Accordingly, defects in formation of primary cilia or in the trafficking of receptors into and out of the cilium lead to a variety of genetic disorders and diseases known as ciliopathies, such as congenital heart disease, cognitive disorders, craniofacial and skeletal patterning defects, and kidney diseases [3–7].

The region between the plasma membrane and the ciliary membrane is infolded to produce a ciliary pocket in many cell types. This invagination comprises an interphase for the actin cytoskeleton and may function as a major site for exocytic and endocytic events for the sorting and targeting of proteins to and from the ciliary base [8, 9]. Indeed, the infolded plasma membrane, also known as the periciliary membrane, comprises a unique site for clathrin-dependent endocytosis, which critically regulates a series of signaling events associated with ligand-induced receptor internalization and subsequent recycling or degradation in the late endosomes/lysosomes compartment [10]. This raises the possibility that endocytosis at the ciliary pocket regulate the level of many different ciliary signaling systems, such as those operated by G protein-coupled receptors [11], receptor tyrosine kinases [12], TRP ion channels [13], and receptors for extracellular matrix proteins [14]. Indeed, we have previously showed that transforming growth factor beta (TGFβ) signaling and activation of Smad transcription factors is associated with clathrin-dependent endocytosis of TGFβ receptors at the ciliary pocket in fibroblasts and in stem cells undergoing cardiomyogenesis [15].

In this chapter we provide distinctive protocols for investigating the morphology and function of the ciliary pocket in RPE1 and NT2 cells, which are cell lines commonly used in laboratories worldwide. However, our protocols apply to many other cell types such as connective tissue cells, which can be cultured and analyzed by procedures similar to those presented here. Specifically, we present methods based on electron microscopy, 3D reconstruction of fluorescence images as well as live cell imaging suitable for investigating cellular and signaling processes associated with endocytosis at the ciliary pocket.

2 Materials

2.1 Cells and Cell Culture Media

1. RPE1 cells. A human retinal pigment epithelial cell line that stably expresses human telomerase reverse transcriptase (hTERT-RPE1; CLONTECH Laboratories, Inc.; ATCC CRL4000™).
2. NT2 cells. A human pluripotent embryonal carcinoma stem cell line (NTERA-2cl.D1; ATCC CRL1973™).

3. Dulbecco's Modified Eagle's Medium DMEM-F12 1:1 GlutaMAX supplement (Thermo Fisher Scientific, 31331-028) supplemented by 10% fetal bovine serum (FBS, Thermo Fisher Scientific, 10270-106) for basic RPE1 cell culture conditions and supplemented with 0.5% FBS for low serum containing media.
4. OPTI-MEM (Thermofischer Scientific, 31985-047) was used for transfections in RPE1 cells.
5. DMEM (Thermo Fisher Scientific, 41965-039) was used for transferrin uptake experiments in RPE1 cells and NT2 cells.
6. DMEM (ATCC, 30-2002) supplemented by 10% FBS (Sigma-Aldrich, F9665) and 1% pen/strep for basic cell culture conditions in NT2 cells.
7. DMEM supplemented by 10% FBS without pen/strep was used for transfections in NT2 cells (*see* **Note 1**).

2.2 Buffers and Solutions for Fluorescence Microscopy

1. 1× phosphate-buffered saline (PBS) (pH 7.4).
2. 1× phosphate-buffered saline with Ca^{2+} and Mg^{2+} (PBS(Ca/Mg)) (pH 7.4).
3. Transferrin internalization buffer (TIB): DMEM with 1 mg/mL bovine serum albumin (BSA).
4. Fixation solution: 4% paraformaldehyde (PFA).
5. Permeabilization buffer: Triton X-100 (0.1 or 0.2%) and BSA (1 mg/mL) in PBS. Buffer should be passed through a sterile filter for long-term storage to avoid contamination. Store at −20 °C in aliquots and keep at 4 °C upon use.
6. Blocking buffer: PBS with BSA (1 or 2 mg/mL).
7. Quenching solution: PBS–NH_4Cl (50 mM).
8. Mounting media for NT2 cells: 4 mL glycerol (5 g) and 500 μL 10× PBS. Fill up to 5 mL with double distilled water (ddH_2O). Add 0.1 g n-propyl gallate, dissolve overnight at 4 °C while rotating. Store covered in tin foil at 4 °C.
9. Mounting media for RPE1 cells: PBS–glycerol mix (1:1) using the SlowFade Light Antifade Kit containing DAPI from Molecular Probes (Thermo Fisher Scientific, S36938).

2.3 Equipment

1. Incubators with standard culture conditions (37 °C and 5% CO_2).
2. Glass coverslips, 12 mm (NeuVitro, GG1212).
3. Plain glass microscope slides (1 mm, Pearl, 7101).
4. Humidity chambers made by placing damp filter paper in the lid of a petri dish, under a layer of Parafilm®. Coverslips can be placed on top of the Parafilm® layer, and the bottom of the petri dish can be used as the lid of the chamber.

5. Eight wells μ-slides (Ibidi, catalog #80821).
6. Ultramicrotome (Reichert Ultracut S).

2.4 Ligands, Antibodies, Staining Reagents, Plasmids, Conjugates, and Transfection Reagents

1. Ligand: Transforming growth factor beta-1 (TGFβ-1) (R&D Systems, 240-B-010).
2. Primary antibodies (species, reference number, and dilution are indicated):
   Acetylated-α-tubulin (AcTub), mouse monoclonal clone 6-11B-1 (Sigma, T7451) (1:10,000).
   ADP-ribosylation factor-like 13b (ARL13b), rabbit polyclonal (Proteintech 17711-1-AP) (1:600).
   Clathrin, rabbit polyclonal (AbCam Ab21679) (1:500).
   Clathrin assembly lymphoid myeloid leukemia (CALM), goat polyclonal (Santa Cruz, sc-6463) (1:300).
   Transforming growth factor beta receptor I (TGFβ-RI, V22) rabbit polyclonal (Santa Cruz sc398) (1:200).
3. Secondary antibodies and conjugates:
   Alexa 350-conjugated donkey anti-mouse IgG (Invitrogen, A10035) (1:600).
   Alexa 350-conjugated donkey anti-rabbit IgG (Invitrogen, A10039) (1:600).
   Alexa 488-conjugated donkey anti-mouse IgG (Invitrogen, A21202) (1:600).
   Alexa 488-conjugated donkey anti-rabbit IgG (Invitrogen, A21206) (1:600).
   Alexa 488-conjugated donkey anti-goat IgG (Invitrogen, A11055) (1:600).
   Alexa 568-conjugated donkey anti-goat IgG (Invitrogen, A11057) (1:600).
   Alexa 568-conjugated donkey anti-mouse IgG (Invitrogen, A10037) (1:600).
   Alexa 555-conjugated transferrin from Molecular Probes (Thermo Fisher scientific, T-35352).
   Texas Red-conjugated transferrin from Molecular Probes (Thermo Fisher scientific, T-2875).
4. Expression plasmids (available from our laboratories upon request):
   Rab8 fused to GFP (Rab8-GFP) and clathrin light chain fused to DsRed were kindly provided by Arnaud Echard (Institut Pasteur, Paris, France) and Thomas Kirchhausen (Immune Disease Institute, Boston, USA), and have been described previously [16].
   pEGFP-F plasmid (Clontech, 6074-1) expressing farnesylated green fluorescent protein (GFP-F).
   GFP-2×FYVE (kindly provided by Harald Stenmark (Institute for Cancer Research, Oslo, Norway)).

Cherry-2×FYVE (kindly provided by Harald Stenmark (Institute for Cancer Research, Oslo, Norway)).

5. Transfection reagents: Fugene 6 (Promega, E2691).
6. Nuclear staining: 4′,6-diamidino-2-phenylindole dihydrochlorid (DAPI) (Invitrogen, D1306).
7. Nail polish (Electron Microscopy Science, ref 72180).

2.5 Electron Microscopy

1. Fixation: H_2O is double-distilled apyrogenic sterile water rinsing and irrigation (Dominique Dutscher Products : ref 069802A). Phosphate buffer is Sorensen's Sodium-Potassium Phosphate Buffer, prepared as followed:

   Solution A: 35.61 g $Na_2HPO_4{\cdot}2H_2O$—make up to 1000 mL with distilled H_2O, stir until dissolved.
   Solution B: 27.6 g $NaH_2PO_4.H_2O$—make up to 1000 mL with distilled H_2O, stir until dissolved.
   Add 40.5 mL of solution A to 9.5 mL of solution B to give 50 mL 0.2 M phosphate buffer pH 7.4.
   Glutaraldehyde EM grade 25% (Electron Microscopy Science, EMS ref 16210) should be diluted in PBS.
   Osmium tetroxide 4% (EMS ref 19180) stock should be diluted 1:1 in H_2O then diluted 1:1 in phosphate buffer 0.2 M to obtain Osmium 1% in phosphate buffer 0.1 M perform post fixation.
   Phosphate buffer 0.1 M is used for washing coverslips after post fixation.

2. Embedding: Epon is prepared with Mollenhauer's Kit with Epon-812 (EMS ref 13940). Embedding is done using gelatin capsule type 04 (EMS ref 70105).
3. Sectioning: Ultrathin sections are performed on a Reichert Ultracut S (Leica Microsystems) with a Diatome Diamond Knife 2 mm 45°.
4. Specimen grids: Gilder 200 mesh Standard Square Mesh Nickel Grids (EMS ref G200Ni).
5. Miscellaneous: 7% uranyl acetate (EMS ref 22400) in water. Lead citrate (EMS ref 17800).
6. Observation/acquisition: JEOL 1011 transmission electron microscope with a GATAN CCD camera Erlangshen 1000 with GATAN Digital Micrograph Software. Generated images were recorded in DM4 format, then converted in TIFF.

2.6 Preparation of Coverslips for Immunofluorescence Microscopy

For NT2 cell growth, 12 mm coverslips for immunofluorescence microscopy analysis are treated with acid prior to growing cells on them (*see* **Note 2**).

1. Incubate coverslips in 32% HCl for 60 min. Mix frequently by swirling bottle around. This step should be performed in the fume hood.
2. Discard HCl (can be saved and reused) and add water, continue working in the fume hood. Swirl and discard. You can now move away from the hood for washing coverslips 15 times in ddH_2O.
3. Subsequently, coverslips are washed 15 times in 96% ethanol.
4. Store in 70% ethanol.

3 Methods

All experiments are carried out at room temperature unless otherwise stated.

3.1 Induction of Ciliogenesis in RPE1 Cells

1. Set down one coverslip per well in a regular 24-well plate under the hood. Coverslips can be autoclaved to avoid any contamination.
2. To optimize cell culture and transfection steps, fix your coverslips to the bottom of the wells by filling each well with 0.5 mL of PBS and then aspirate it with a 200 μL tip pipet to create a vacuum between the coverslip and the plate.
3. Seed 250,000 cells per well and grow them in basic cell culture conditions for 24 h.
4. Wash the cells twice in PBS. To avoid cell detachment, bend the plate and use a 200 μL tip pipet to aspirate the medium without touching the cells.
5. Add low serum containing media for an additional 24–48 h period to allow primary cilia formation.
6. Monitor the efficiency of primary cilia formation by staining with acetylated α-tubulin or ARL13b antibodies (typically ~80% of cells are ciliated).

3.2 Induction of Ciliogenesis in NT2 Cells

Seed cells as described for RPE1 cells under basic cell culture conditions. Primary cilia will form when cells reach confluence of approximately 80% or above, and can be visualized using specific antibodies against structural proteins and/or ciliary receptors, e.g., ARL13b. NT2 cells form rather long primary cilia, which may at times connect to cilia from adjacent cells, as shown in Fig. 1a.

3.3 Transmission Electron Microscopy (TEM) Protocol for Flat-Embedding

1. Seed RPE1 cells on coverslips and grow them in 24-well plates to induce ciliogenesis as described above.
2. Wash cells twice with PBS (Ca/Mg) to avoid cell detachment.

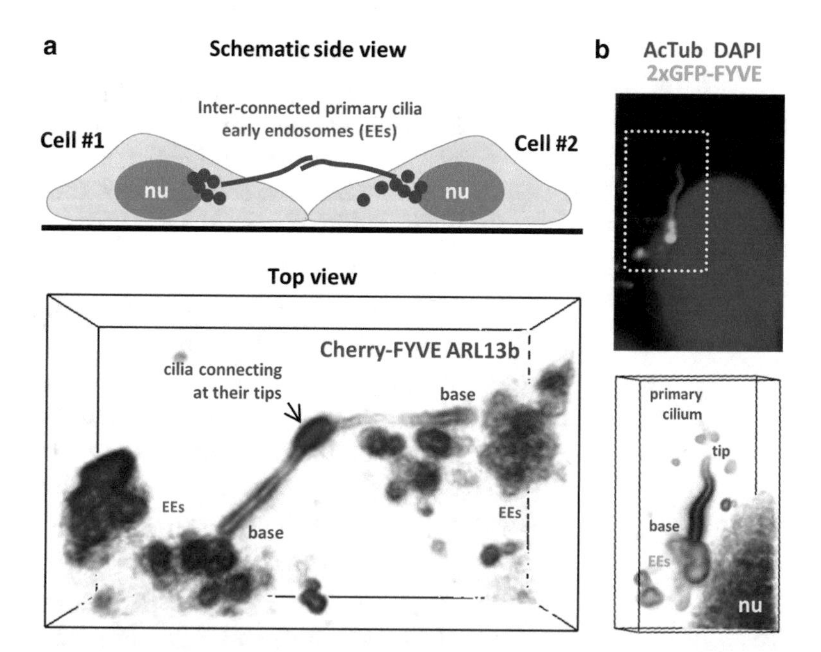

Fig. 1 Fluorescence microscopy analysis of the localization of early endosomes (EEs) around the ciliary pocket region in NT2 cells. (**a**) The localization of endosomes was observed in cells transiently transfected with plasmids expressing either Cherry-FYVE (**a**) or 2×GFP-FYVE (**b**) and primary cilia were stained with either anti-ARL13b or anti-acetylated α-tubulin (AcTub). In some cases, the nucleus was stained with DAPI (**b**). Occasionally, the tip of primary cilia from adjacent cells is connected to each other as shown in (**a**). Endosome localization is shown both in epifluorescence (**b**, *upper panel*), and using 3D reconstruction (*lower panels* in (**a**) and (**b**))

3. Fix the cells on coverslips with 3% glutaraldehyde (200 μL/well) for 1 h.
4. Wash cells twice with PBS (Ca/Mg).
5. Postfix your samples in 1% osmium tetroxide in 0.1 M phosphate buffer.
6. Dehydrate your samples in successive baths of 70, 90, and 100% ethanol (three baths for 100% ethanol).
7. Incubate the cells in 50% ethanol and 50% Epon for 20 min.
8. Incubate the cells in 25% ethanol and 75% Epon for 20 min.
9. Incubate the cells in 100% Epon for 20 min.
10. Fill gelatin capsules with freshly prepared epoxy resin and polymerized at 60 °C for 24 h.
11. Gently remove the coverslip from the wells, pay attention not to break them.
12. Remove the Epon excess on the coverslip by gently holding the coverslip with the forceps and letting it flow into the well, only leaving a thin layer of Epon on the cells.

13. Place the coverslip on a glass slide, with the cells facing up.
14. Cover the coverslip with the Epon-filled capsule. One capsule is usually enough; putting two or more could lead to difficulties later to separate them. It can be also useful to place the capsule where the cells are more concentrated (this can be simply found by placing the slide/coverslip under a binocular).
15. Polymerize at 60 °C for 24 h.
16. Remove the gelatin capsule from the coverslip by heating the glass slide at 70 °C for 5 min.
17. Trim the block using a clean razor blade and cut your samples in 90 nm thin sections with an ultramicrotome. Ultrathin sections are put on specimen grids. Do not try to find region of interest by using 5 μm thin sections; the cell monolayer is too thin.
18. Perform serial cutting in the orientation showed in Fig. 2a. Keep grids in the cutting order for observation.
19. Stain your sections with uranyl acetate (10 min, followed by H_2O 3×5 min) and Reynold's lead citrate (5 s, followed by H_2O 3×5 min).
20. Perform observation beginning with the region corresponding to the adherent surface of the cells, i.e., the first section, then the following, to be sure not to miss the region of interest. Cilia are usually found close to the nucleus (on the "side")

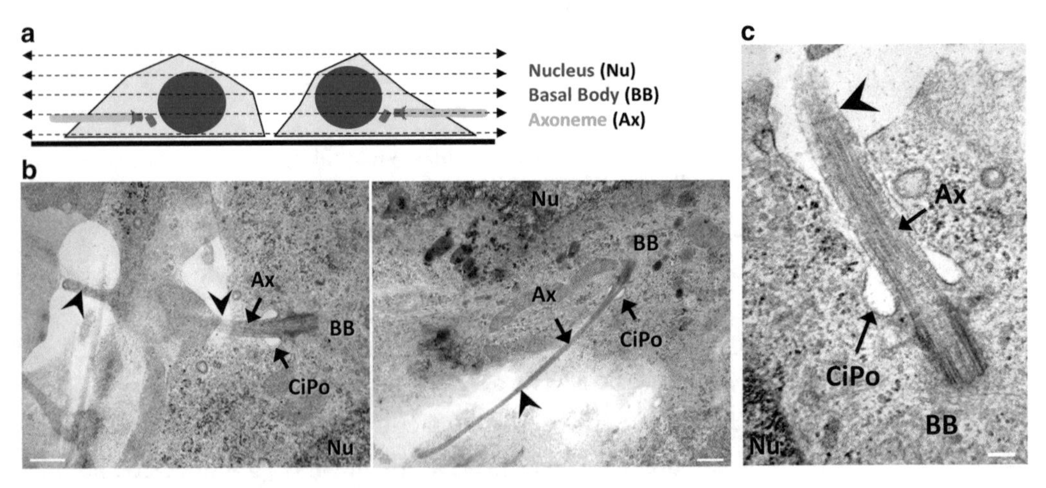

Fig. 2 Transmission electron microscopy analysis of the ciliary pocket in RPE1 cells, which were grown on coverslips to confluence, serum-starved, fixed and treated for microscopy. (**a**) Sections were realized parallel to the adherent surface as depicted in the scheme and explained in the methods. (**b**) Examples of cilia from cells starved for 24 (*left*) or 72 h (*right*) in which both the basal body (BB) and the axoneme (Ax) are clearly visible. All observed cilia in RPE1 cells present a similar organization, with the basal body not at the same plane as the plasma membrane but inside the cytoplasm and a membrane domain around the proximal region of the axoneme. (**c**) A representative TEM image is showed to highlight the shape and organization of the ciliary pocket (CiPo) relative to the axoneme (Ax) and the basal body (BB). *Arrowheads* indicate the extracellular, distal part of the axoneme. Scale bars: 500 nm in (**b**), 100 nm in (**c**). Adapted from figures previously published in ref. 16

as indicated by our analysis of cilia positioning in RPE1 cells [17] and as shown in Fig. 2.

3.4 Transfection of Cells with Plasmid DNA in RPE1 Cells

1. Seed 100,000 cells per well in a 24-well plates containing a coverslip and grow them as indicated above for 24 h.
2. Prepare the transfection mix. For four wells, mix 1 μg of DNA with 100 μL of Opti-MEM and 3 μL of Fugene 6.
3. Wash the cells twice with PBS (Ca/Mg) as described above.
4. Aspirate the PBS and add 500 μL of low serum containing Opti-MEM (0.5% FBS) to each well.
5. Add 25 μL of transfection mix per well. Move gently your plate in perpendicular directions in order to disperse the transfection mix in the wells. Avoid circular movements.
6. Incubate the cells with the transfection mix for 24 h before fixation and/or live imaging analysis.

3.5 Transfection of Cells with Plasmid DNA in NT2 Cells

1. Grow cells to approximately 80% density under basic culture conditions.
2. Prepare the transfection mix as described for RPE1 cells above.
3. Wash cells in PBS and add 100 μL DMEM (10% FBS) without antibiotics to each well. Proceed as above and transfect cells using Fugene 6 and DNA in a 3:1 dilution. Dilute reagents in DMEM without serum or antibiotics.

An example of transient plasmid transfection is shown in Fig. 1, where NT2 cells express Cherry-FYVE (Fig. 1a) and 2×GFP-FYVE (Fig. 1b), that mark early endosomes (EEs) in close proximity to the pocket region of the primary cilium stained with either ARL13b or acetylated α-tubulin. EEs are enriched in phosphatidylinositol-3 phosphate, which anchors FYVE zinc finger domain-containing proteins, such as SMAD anchor for receptor activation (SARA) that conveys SMAD2/3 to activated TGF-β receptors and promotes TGF-β-RI-mediated phosphorylation of SMAD2/3 [18].

3.6 Transferrin Internalization in RPE1 Cells

1. Incubate RPE1 cells grown on coverslips (previously transfected or not, see above) for 20 min at 37 °C in serum free DMEM (*see* **Note 3**).
2. Wash the cells twice in TIB previously warmed to 37 °C.
3. Incubate the cells for 1 min at 37 °C in TIB containing 6 μg/mL Alexa 555-conjugated transferrin (Tf). This allows accumulation of receptor-bound Tf into clathrin-coated pits and vesicles (CCPs/CCVs).
4. Wash your cells rapidly in cold PBS (4 °C) to stop membrane trafficking.
5. Fix cells in 4% PFA (10 min at 4 °C, then 20 min at room temperature).

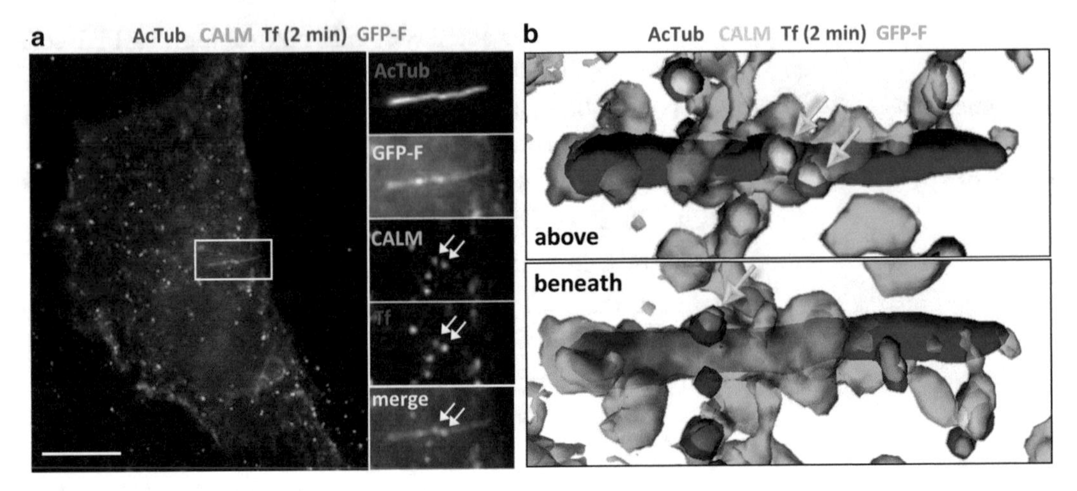

Fig. 3 Fluorescence microscopy analysis of the localization of clathrin-coated pits (CCPs) at the ciliary pocket in RPE1 cells. Cells were transiently transfected with the farnesylated green fluorescent protein (GFP-F) encoding plasmid for 24 h and then incubated for 2 min at 37 °C with Alexa 555-conjugated transferrin (Tf). Cells were immediately fixed and processed for immunofluorescence using antibodies against acetylated-α-tubulin (AcTub) to stain the axoneme, and CALM to stain clathrin-coated pits (CCPs). Epifluorescence (**a**) and 3D reconstruction (**b**) images of a representative cilium are shown. CiPo-associated CCPs present along the axoneme (*arrows*) are stained for both CALM and Tf. GFP-F is prominently targeted to the membrane systems around the cilium and the ciliary pocket. Adapted from figures previously published in ref. 16

6. Process for immunofluorescence as described above.

An example of RPE1 cells transfected with pEGFP-F, treated to follow transferrin internalization at the ciliary pocket, and fixed and stained for acetylated α-tubulin and CALM is shown in Fig. 3a.

3.7 Transferrin Internalization in NT2 Cells

1. Wash NT2 cells grown on coverslips (previously transfected or not, see above) twice in serum free DMEM, pre-heated to 37 °C.
2. Incubate cells for 20 min at 37 °C in serum free DMEM.
3. Wash the cells twice in TIB pre-heated to 37 °C.
4. Incubate the cells for 10 min in TIB preheated to 37 °C containing 15 mg/mL Texas Red-conjugated transferrin (Tf), to allow receptor internalization (*see* **Note 4**).
5. Wash rapidly your cells in TIB preheated to 37 °C. Let cells sit in the incubator for 10 min to allow surface-bound receptor to be internalized.
6. Wash your cells in cold PBS (4 °C) to stop membrane trafficking.
7. Fix the cells in 4% PFA (15 min at 4 °C). Proceed with immunofluorescence as described below.

An example of NT2 cells pulsed for 10 min with Tf is depicted in Fig. 4, where Tf is largely confined to EEs marked by

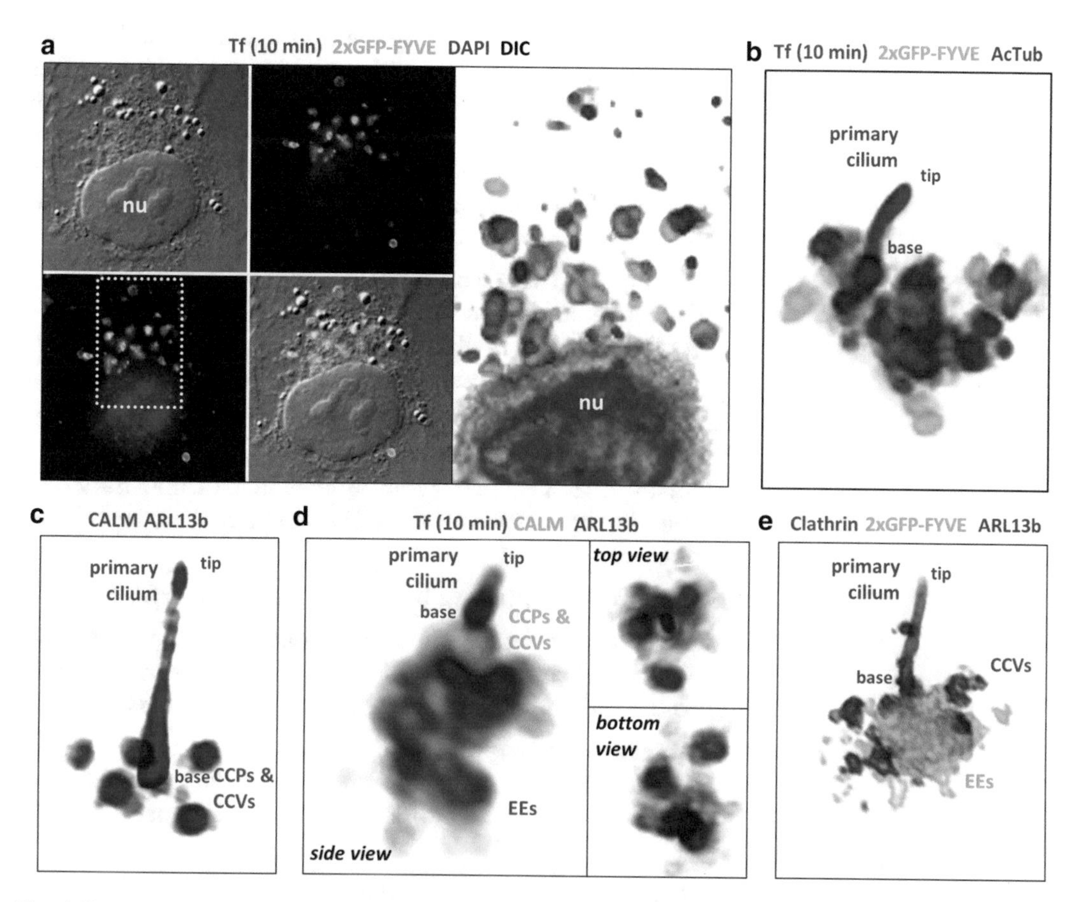

Fig. 4 Fluorescence microscopy analysis on clathrin-coated pits (CCPs), clathrin-coated vesicles (CCVs), and early endosomes (EEs) at the ciliary pocket region in NT2 cells. Cells were transiently transfected with the plasmid expressing 2×GFP-FYVE followed by pulsing with Texas red-conjugated transferrin (Tf) for 10 min. Cells were quickly fixed and prepared for immunofluorescence microscopy. Following Tf incubation, the endocytic pathway can be visualized. (**a**) Epifluorescence and 3D reconstruction of Tf localization to 2×GFP-FYVE positive EEs in an interphase cell. (**b**) 3D reconstruction of Tf accumulation at EEs around the pocket region of the primary cilium. (**c**) 3D reconstructions of co-staining with CALM shows that CCVs are localized around the pocket region of the primary cilium. (**d**) 3D reconstruction shows little or no co-localization between CCVs (CALM) and Tf, which after a 10 min pulse primarily localizes to EEs (**e**). Similarly, CCPs/CCVs show no or little colocalization with EEs marked 2×GFP-FYVE. Primary cilia were stained with either ARL13b or anti-acetylated α-tubulin (AcTub) antibodies

2×GFP-FYVE. Localization of Tf to EEs is shown both in an interphase cell (Fig. 4a) and in a growth-arrested cell (Fig. 4b), where Tf and 2×GFP-FYVE partly co-localizes in multiple endosomes around the pocket region of the primary cilium.

3.8 Preparation of Cells for Immunofluorescence Microscopy in RPE1 Cells

1. Wash the cells twice in PBS as described above.
2. Fix cells in 4% PFA for 20 min at 4 °C (by placing the plate on ice) followed by a 10 min incubation in quenching solution.
3. Wash cells twice with PBS.

4. Incubate the cells with primary antibodies in permeabilization buffer for 30–45 min. Incubation can be made directly in the well; the minimal mix volume is then 200 μL per well to prevent cells from drying out. To save precious antibodies, you can prepare a 40 μL drop containing the antibody/permeabilization mix on Parafilm® attached to the bench, and then gently reverse your coverslip to put the cells in contact with the liquid.
5. Wash the cells twice with blocking buffer; be careful to avoid cell detachment. For cells incubated on a drop, put back the coverslips into the wells properly reversed. In case of doubt about the cells' orientation, cells can be seen on the coverslip by holding it with the forceps and by exposing it to a source of light.
6. Incubate mixed secondary antibodies in blocking buffer for 30 min in the dark to preserve the fluorescence of secondary antibodies.
7. Wash once with blocking buffer and twice with PBS.
8. Lay down your cells on microscope slides in mounting medium, cell surface facing down. Gently press the coverslip onto the slide until it is held in place by surface tension.
9. Add nail polish by putting small drops around the coverslip and then complete the circle once drops dried. Avoid small holes that will favor oil contact with mounting media and the cells.
10. Store at 4 °C, keep in the dark.

3.9 Preparation of Cells for Immunofluorescence Microscopy in NT2 Cells

1. Wash the cells twice with PBS as described above.
2. Fix the cells in 4% PFA for 15 min (*see* **Note 5**).
3. Wash the cells twice in PBS.
4. Incubate your cells in permeabilization buffer for 12 min.
5. Incubate your cells in blocking buffer for 30 min or at 4 °C overnight.
6. Incubate your cells in primary antibodies in blocking buffer for 90 min or 4 °C overnight (*see* **Note 6**).
7. Wash the cells three times, each 5 min, with blocking buffer avoiding cell detachment. Transfected NT2 cells are easily detached upon mechanical stress.
8. Incubate for 45 min in blocking buffer containing mixed secondary antibodies.
9. Wash once (5 min) in blocking buffer and twice (2 × 5 min) in PBS.
10. Lay down your cells on microscope slides in mounting media. Apply light pressure, getting rid of excess mounting medium.

11. Add nail polish around the edge of the coverslip. Allow for 20 min to dry.
12. Store at 4 °C, keep in the dark.

An example of immunofluorescence staining with antibodies directed against CALM and clathrin, which marks clathrin-coated pits (CCPs) and vesicles (CCVs) at the ciliary pocket, is shown in Fig. 4c–e. As expected, CCVs display little co-localization with EEs as evidenced by co-staining with either 10 min Tf pulsing (Fig. 4d) or expression of 2×GFP-FYVE (Fig. 4e). This gives a good overview on the spatiotemporal machinery of endocytosis at the primary cilium, which can be elaborated by live-cell imaging, e.g., Fig. 6.

3.10 Fluorescence Microscopy and 3D Image Reconstruction in RPE1 Cells

1. Acquire Z-stacks epifluorescence images every 200 nm along the *Z* axis using a piezo-electric (PIFOC, E662-LR controller, Physik Instrumente) 100× objective (plan-apo; Axiovert 100 M, Zeiss) with a MicroMAX camera (Princeton Instruments).
2. Three methods of deconvolution of Z-stacks can be recommended including MetaMorph (3D deconvolution option), ImageJ plugin "Deconvolutionlab – EPFL" http://bigwww.epfl.ch/algorithms/deconvolutionlab/ or Huygens https://svi.nl/HomePage.
3. Obtain 3D reconstruction of deconvoluted images with the Imaris software (Bitplane). Images can be extracted from Imaris and then used to obtain the final pictures used in the figures.

An example of RPE1 cells transfected with pEGFP-F, treated to follow transferrin internalization, and fixed and stained for acetylated tubulin and CALM is shown in 3D reconstruction images in Fig. 3b.

3.11 Fluorescence Microscopy and 3D Image Reconstruction in NT2 Cells

Capture NT2 cell images on a fully motorized Olympus BX63 upright microscope with a DP72 color, 12.8-megapixel, 4140 × 3096-resolution camera and differential interference contrast (DIC). The software used include Olympus CellSens dimension, which is able to perform 3D blending projections on captured z stacks, 3D animation videos, and slice views (*see* Figs. 1, 4, and 5).

TGFβ-1 Stimulation and Localization of TGFβ-RI to the Ciliary Pocket

1. Grow NT2 cells to a density of approximately 80%.
2. Wash cells twice in PBS and add serum-depleted DMEM (37 °C) for 30 min.
3. Add TGFβ-1 to the media at a final concentration of 2 ng/mL for 10–30 min.
4. Proceed with immunofluorescence microscopy as described in Subheading 3.9 using antibodies against Transforming growth factor beta receptor I (TGFβ-RI), clathrin to mark CCPs/CCVs, and acetylated α-tubulin to mark the primary cilium.

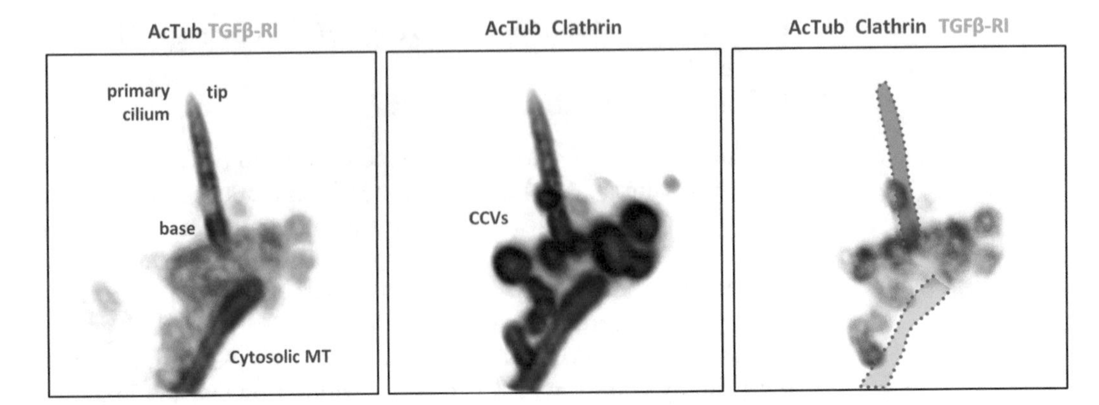

Fig. 5 Fluorescence microscopy and 3D reconstruction analysis of the localization of Transforming growth factor beta receptor I (TGFβ-RI) to clathrin-coated vesicles (CCVs) at the pocket region of the primary cilium (stained with anti-acetylated α-tubulin, AcTub) in NT2 cells. Stimulation with TGFβ-1 results in accumulation of the receptor around the base of the cilium as well as along microtubules (MTs) that project from the ciliary base into the cytosol. The right panel shows a merged image exclusively with co-localization between the receptor and clathrin

Figure 5 shows an example of the co-localization of TGFβ-RI to CCPs and CCVs at the pocket region of the primary cilium and in conjunction with microtubules that spreads out from the ciliary base (*see* **Note 7**). After TGFβ-1 stimulation, the receptor is preferentially internalized in CCVs at the pocket region, and localization along the cytosolic microtubules may indicate trafficking of receptor-positive CCVs along these microtubules for later fusion into EEs, where activation of SMAD2/3 takes place.

3.12 Live Cell Imaging for Dynamic Analysis of Clathrin or Actin at the Ciliary Pocket

1. Seed 60,000 RPE1 cells per well of uncoated eight wells μ-slides.
2. Carry out transfection as described above.
3. Culture cells in the transfection media for at least 24 h before analysis.
4. Acquire images successively every 3 seconds for 3–20 min using an Apo 100× NA 1.43 objective with an inverted epi-illumination microscope (Axiovert 200M, Zeiss) with CCD camera (CoolSNAP HQ; Photometrics) placed within a temperature-controlled enclosure set at 37 °C. Shutters, filters, camera, and acquisition were controlled by MetaMorph. Generate the final movies and derived pictures using ImageJ (http://rsbweb.nih.gov/ij/index.html).

An example of RPE1 cells transfected with Rab8-GFP and Clathrin-DsRed endocytic dynamic activity associated with a primary cilium is shown in Fig. 6.

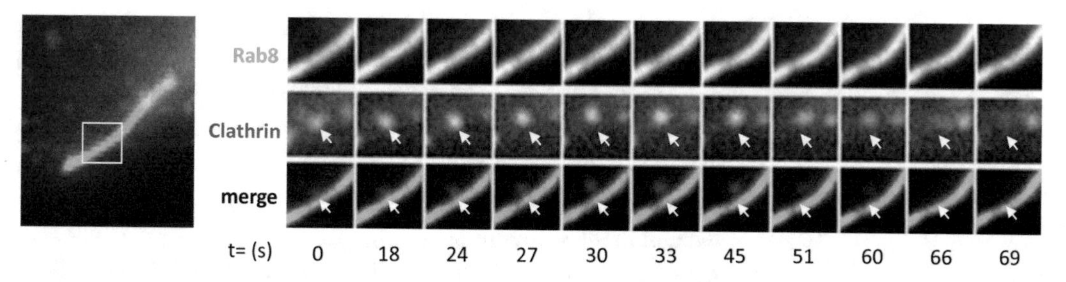

Fig. 6 Live-cell imaging of RPE1 cells were transiently co-transfected with plasmids encoding for Rab8-GFP (*green*) and clathrin-DsRed (*red*) fusions and analyzed by fluorescence microscopy. Live cells were imaged at 37 °C (one picture for each channel every 3 s for 4 min). The first taken picture is shown (*left*). An image stream of a clathrin spot (*red, arrow*), which disappeared during the acquisition, is shown (*right*). An *arrow* indicates the initial position of the followed clathrin spot. Adapted from figures previously published in ref. 16

4 Notes

1. NT2 cells may be difficult to transfect; however, we find increased efficiency when transfecting in media without antibiotics. Cell density does not appear to be very important for these particular constructs; however, we see a clear positive effect when using 1:3 DNA to Fugene amount.
2. Preparation of coverslips using acid wash is an optional step, often used to increase cell attachment to surfaces. We find that this helps our growth rate and cell dispersion when culturing NT2 cells for immunofluorescence experiments. Further, we find an increase in transfection efficiency, possibly due to a decrease in cell loss during immunofluorescence microscopy preparation steps.
3. This step eliminates endogenous receptor-bound transferrin, and therefore increases binding and endocytosis of fluorophore-conjugated transferrin.
4. For NT2 cell experiments, we have used Texas Red-conjugated transferrin. This can be replaced by Alexa 555 conjugated transferrin used in our RPE1 experiments. Concentration may be adjusted, and lower amounts may be used. Further, chasing with non-labeled Tf may be included to optimize the distinct visualization of CCPs, CCVs, and EEs at the ciliary pocket.
5. Variations in protocol for preparation of immunofluorescence microscopy slides between RPE1 and NT2 is a result of different lab techniques, and one may very likely be interchanged with another. For instance, fixation for 15 min followed by 12 min in permeabilization buffer and 30 min in blocking buffer, as we describe for NT2 cells, is also successfully for RPE1 cells.

6. 90 min or shorter is often sufficient for visualization using a fluorescence microscope, however, some antibodies bind more weakly than others, and optimization steps often includes longer incubation times. Therefore, incubating overnight may increase signal strength.
7. When using antibodies directed against microtubules (MTs), we will inevitably obtain staining of other MTs than those constituting the primary cilium. For instance, using AcTub allows us to visualize stable MTs such as the mitotic spindle, midbodies, as well as some cytosolic MTs, the latter is seen in Fig. 5.

Acknowledgments

We would like to thank Pierre Bourdoncle (Cellular imaging Facility, Cochin Institute) for his precious help on deconvolution and 3D reconstruction of fluorescence images on RPE1 cells. This work was supported by the University of Copenhagen Excellence Programme for Interdisciplinary Research and the Danish Council for Independent Research (1331-00254).

References

1. Satir P, Christensen ST (2007) Overview of structure and function of mammalian cilia. Annu Rev Physiol 69:377–400. doi:10.1146/annurev.physiol.69.040705.141236
2. Pedersen LB, Schroder JM, Satir P, Christensen ST (2012) The ciliary cytoskeleton. Compr Physiol 2(1):779–803. doi:10.1002/cphy.c110043
3. Czarnecki PG, Shah JV (2012) The ciliary transition zone: from morphology and molecules to medicine. Trends Cell Biol 22(4):201–210. doi:10.1016/j.tcb.2012.02.001
4. Madhivanan K, Aguilar RC (2014) Ciliopathies: the trafficking connection. Traffic 15(10):1031–1056. doi:10.1111/tra.12195
5. Koefoed K, Veland IR, Pedersen LB, Larsen LA, Christensen ST (2014) Cilia and coordination of signaling networks during heart development. Organogenesis 10(1):108–125. doi:10.4161/org.27483
6. Valente EM, Rosti RO, Gibbs E, Gleeson JG (2014) Primary cilia in neurodevelopmental disorders. Nat Rev Neurol 10(1):27–36. doi:10.1038/nrneurol.2013.247
7. Waters AM, Beales PL (2011) Ciliopathies: an expanding disease spectrum. Pediatr Nephrol 26(7):1039–1056. doi:10.1007/s00467-010-1731-7
8. Ghossoub R, Molla-Herman A, Bastin P, Benmerah A (2011) The ciliary pocket: a once-forgotten membrane domain at the base of cilia. Biol Cell 103(3):131–144. doi:10.1042/BC20100128
9. Benmerah A (2013) The ciliary pocket. Curr Opin Cell Biol 25(1):78–84. doi:10.1016/j.ceb.2012.10.011
10. Mikkelsen EJ, Albert LG, Upadhyaya A (1988) Neuroleptic-withdrawal cachexia. N Engl J Med 318(14):929. doi:10.1056/NEJM198804073181416
11. Schou KB, Pedersen LB, Christensen ST (2015) Ins and outs of GPCR signaling in primary cilia. EMBO Rep 16(9):1099–1113
12. Christensen ST, Clement CA, Satir P, Pedersen LB (2012) Primary cilia and coordination of receptor tyrosine kinase (RTK) signalling. J Pathol 226(2):172–184. doi:10.1002/path.3004
13. Phua SC, Lin YC, Inoue T (2015) An intelligent nano-antenna: Primary cilium harnesses TRP channels to decode polymodal stimuli. Cell Calcium 58(4):415–422. doi:10.1016/j.ceca.2015.03.005
14. Seeger-Nukpezah T, Golemis EA (2012) The extracellular matrix and ciliary signaling. Curr Opin Cell Biol 24(5):652–661. doi:10.1016/j.ceb.2012.06.002

15. Clement CA, Ajbro KD, Koefoed K, Vestergaard ML, Veland IR, Henriques de Jesus MP, Pedersen LB, Benmerah A, Andersen CY, Larsen LA, Christensen ST (2013) TGF-beta signaling is associated with endocytosis at the pocket region of the primary cilium. Cell Rep 3(6):1806–1814. doi:10.1016/j.celrep.2013.05.020
16. Molla-Herman A, Ghossoub R, Blisnick T, Meunier A, Serres C, Silbermann F, Emmerson C, Romeo K, Bourdoncle P, Schmitt A, Saunier S, Spassky N, Bastin P, Benmerah A (2010) The ciliary pocket: an endocytic membrane domain at the base of primary and motile cilia. J Cell Sci 123(Pt 10):1785–1795. doi:10.1242/jcs.059519
17. Vanneste Y, Michel A, Dimaline R, Najdovski T, Deschodt-Lanckman M (1988) Hydrolysis of alpha-human atrial natriuretic peptide in vitro by human kidney membranes and purified endopeptidase-24.11. Evidence for a novel cleavage site. Biochem J 254(2):531–537
18. Tang WB, Ling GH, Sun L, Liu FY (2010) Smad anchor for receptor activation (SARA) in TGF-beta signaling. Front Biosci (Elite Ed) 2:857–860

Chapter 4

Methods to Study Interactions Between Ciliogenesis and Autophagy

Birgit Hegner Satir and Olatz Pampliega

Abstract

Autophagy is a catabolic pathway for the degradation and recycling of intracellular components, contributing to maintain cell homeostasis. Changes in autophagy activity can be monitored by a variety of biochemical and functional assays that should be used in combination. Recently, it has been described that signaling from the primary cilium modulates autophagy. This novel and reciprocal interaction will impact diverse aspects of the cell biology in healthy and pathophysiological conditions. Here, we describe methods to monitor autophagy activity in cilia mutants, as well as the use of autophagy mutants to monitor ciliogenesis.

Key words Primary cilia, Autophagy, mCherry-GFP-LC3, Immunofluorescence, Western blotting, LC3-II

1 Introduction

Autophagy is the intracellular pathway for protein degradation as well as recycling of proteins organelles in the lysosome. The main components of autophagy are conserved through evolution, although diversification of this catabolic pathway has led to the description of three types of autophagy in mammals, namely macroautophagy, chaperone-mediated autophagy (CMA), and microautophagy [1]. Increasing numbers of reports highlight the central role of autophagy in the maintenance of cell homeostasis during health and disease, and recently it has been described that signaling from the primary cilium modulates autophagy, in a reciprocal interaction where autophagy also participates in the modulation of ciliogenesis (Fig. 1) [2, 3].

The rapid growth of the autophagy field has led to a fast evolution in the concepts and methods that govern the study of this pathway. Nevertheless, one of the central steps in macroautophagy (here referred as autophagy) is de novo formation of a new organelle, the autophagosome, which can be identified by the presence of a

Peter Satir and Søren Tvorup Christensen (eds.), *Cilia: Methods and Protocols*, Methods in Molecular Biology, vol. 1454, DOI 10.1007/978-1-4939-3789-9_4,

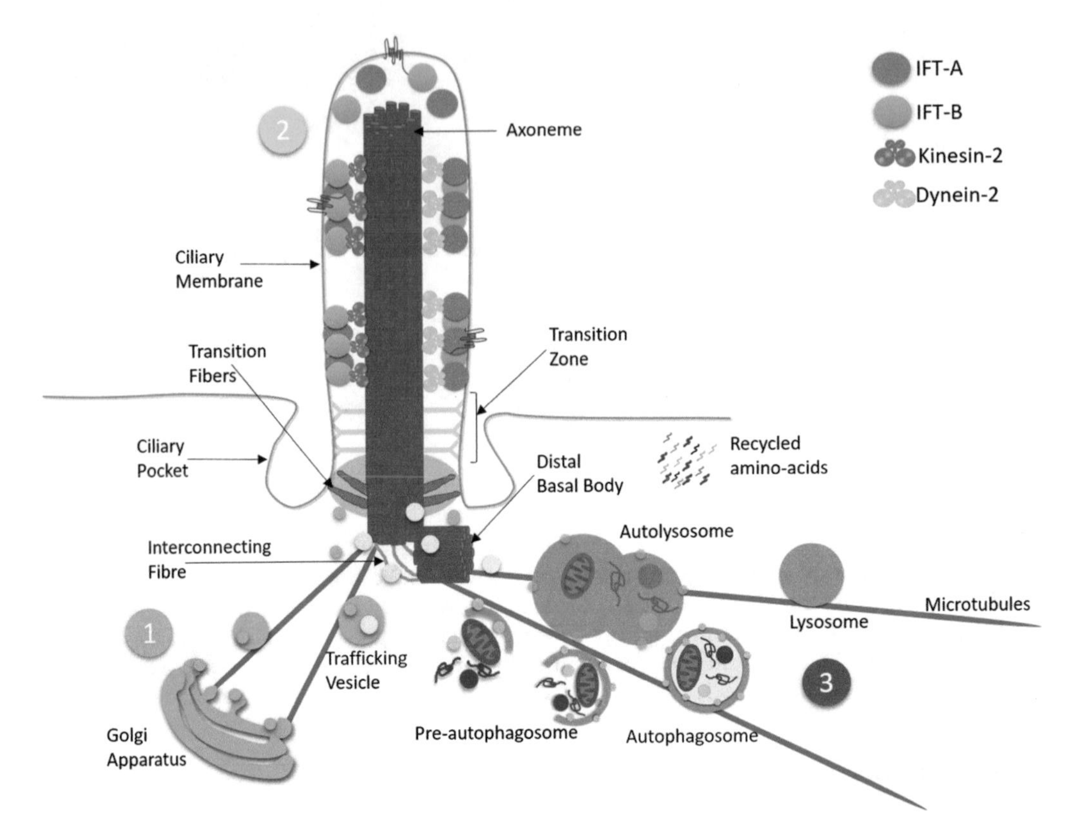

Fig. 1 Ciliary signaling activates autophagy by recruiting ATG proteins to the ciliary base. (*1*) Trafficking from Golgi provides the elements for ciliogenesis and autophagy. (*2*) Ciliogenesis requires an active anterograde and retrograde intraflagellar transport for increasing ciliary length. (*3*) The arrival of autophagy proteins to the ciliary base increases autophagosome formation, leading to an overall activation of the autophagic capacity of the cell

lipidated form of microtubule-associated protein 1A/1B-light chain 3 (LC3) in its membrane, known as LC3-II that acts as a marker of the organelle and of the autophagy process itself. LC3-II is conjugated with phosphatidylethanolamine (PE) and is present on autophagosomal and isolation membranes, while LC3-I is the cytosolic form of the protein [4]. In fact, most of the methods to track autophagy nowadays are based on monitoring changes in LC3-II. It is important to keep in mind that autophagy is a multistep and dynamic process, and therefore, besides end-point quantifications of LC3-II, the methods to study autophagy have to estimate the activity or flux generated. Only an appropriate assessment of autophagy flux will allow for a correct interpretation of the results [5].

Here we describe the most suitable autophagy methods for the study of ciliogenesis and its related signaling functions in mouse embryonic fibroblasts (MEFs) that are defective in ciliary formation or autophagy. We provide guidelines for monitoring autophagy using cilia mutants, as well as methods to study ciliogenesis in a context of a defective autophagy. We use the classical biochemical LC3 flux assay [6], where samples are incubated in the presence and absence of lysosomal inhibitors prior to the quantification of LC3-II levels by immunoblotting. Alternatively, we monitor in real time autophagic activity by the fluorescent tandem fusion protein mCherry-GFP-LC3 [7]. Genetic mutants defective in essential ciliary components such as the intraflagellar transport protein 88, IFT88, which is required for ciliary assembly [8], are used to study the effect of ciliogenesis on autophagy. Similarly, genetic mutants of essential autophagy genes, such as *Atg5* that encodes the autophagy related protein 5 (ATG5), or blockage of early steps of autophagosome formation by chemical inhibitors, such as the phosphatidylinositol 3-kinase (PI3K) inhibitor 3-methyladenine (3-MA), are used to study the role of autophagy on ciliogenesis [2].

2 Materials

2.1 Cell Cultures and Ciliary Formation

1. Wild type (wt) mouse embryonic fibroblasts (MEFs) or other cell types capable of growing primary cilia.
2. Ciliary growth mutants: Tg737 *orpk* kidney epithelial cells (KECs), IFT20−/− and IFT88−/− MEFs.
3. Autophagy mutants: *Atg5*−/− MEFs (*see* **Note 1**).
4. Complete Dulbecco's Modified Eagle Medium (DMEM) and serum free DMEM.
5. Phosphate buffered saline (PBS) buffer.
6. 24-well plastic plates containing 10 mm Ø glass coverslips for immunocytochemistry assays, 12-well plastic plates for immunoblot assays using MEFs, and 6-well plastic plates for immunoblot assays using KECs.
7. Cell scraper.
8. Micropipettes.
9. Plate reader for measuring absorbance in protein quantification.

2.2 Reagents for Biochemical LC3 Flux Assay

1. Lysosomal inhibitors: Use a final concentration of 100 μM leupeptin. Prepare 10 mM leupeptin working aliquots (100×), and store them at −20 °C.
2. Lysosomal inhibitors: Use a final concentration of 20 mM ammonium chloride (NH_4Cl). Prepare immediately prior to use 2 M ammonium chloride in DMEM (100×).
3. Reagents for protein isolation. RIPA buffer: 150 mM NaCl, 1 % NP-40, 0.5 % NaDoc, and 0.1 % SDS in 50 mM Tris, pH 8. Store at 4 °C. Phosphatase inhibitors and protease inhibitors (cOmplete™ Mini EDTA-free; Roche). Store at 4 °C.

4. Reagents for protein quantification by Lowry method. Round-bottom 96-well plate. Solution A: 40 g/l Na_2CO_3, and 4 g/l NaOH, store at room temperature. Solution B: 0.5% $CuSO_4$, and 1% Na Citrate store at room temperature. 1 N Folin, store at room temperature protected from light. 1 mg/ml bovine serum albumin (BSA), store at −20 °C.
5. Buffers used for SDS-PAGE. Resolving buffer: 1.5 M Tris–HCl, pH 8.8, store at 4 °C. Stacking buffer: 0.5 M Tris–HCl, pH = 6.8, store at 4 °C. Thirty percent acrylamide–bis solution (29.2:0.8, acrylamide–bis). 10% sodium dodecyl sulfate (SDS). 10% ammonium persulfate (APS). *N,N,N′,N′* Tetramethylethylenediamine (TEMED). Running buffer: 25 mM Tris, 190 mM glycine, and 0.1% SDS, pH 8.3, store at room temperature.
6. Sample buffer: 4% SDS, 10% 2-mercaptoethanol, 20% glycerol, 0.004% bromophenol blue in 0.125 M Tris–HCl, pH 6.8.
7. Nitrocellulose membranes for gel transfer and western blotting.
8. Transfer buffer: 25 mM Tris, 190 mM glycine, and 15% methanol, pH 8.3.
9. Buffers for western blotting. TBST: 20 mM Tris pH 7.5, 150 mM NaCl, and 0.01% Tween 20. Ponceau solution: 0.01% Ponceau-S in 5% acetic acid. Blocking buffer: 5% milk in TBST, store at 4 °C no longer than 48 h. Dilution of antibodies: 3% BSA in TBST, store at 4 °C.
10. Loading control antibodies: mouse anti-beta actin, and mouse anti-tubulin antibody. Secondary antibodies: goat anti-mouse IgG and goat anti-rabbit IgG (H + L) conjugated with horseradish peroxidase (HRP).
11. Scanning and densitometer apparatus for western blot film. ImageJ software for densitometry analysis.

2.3 Reagents for mCherry-GFP-LC3 Fluorescent Autophagy Reporter and Immunocytochemistry

1. Reagents for transient transfection: Opti-MEM™ media (Gibco) (*see* **Note 2**), store it at 4 °C, and warm up to 37 °C prior to use. Lipofectamine® 2000 (Invitrogen), store it at 4 °C. Purified mCherry-GFP-LC3 DNA, store it at −20 °C.
2. Reagents for cell immunofluorescence: Phosphate buffered saline (1× PBS): 137 mM NaCl, 2.7 mM KCl, 10 mM Na_2HPO_4, 1.8 mM KH_2PO_4, pH = 7.4, store at room temperature. Four percentage paraformaldehyde (4% PFA) in PBS, store working aliquots at −20 °C, and once thawed maintain it at 4 °C. Blocking and permeabilization solution: 0.01% Triton™ X-100 and 5% serum in PBS (use the serum from the animal species where the primary antibody has been produced). Solution for dilution of antibodies: 3% bovine serum albumin (BSA) in PBS, store at 4 °C. DAPI containing mounting media.

3. Primary antibodies for immunocytochemistry: mouse anti-acetylated α-tubulin (Sigma, #T6793), rabbit anti-ATG16L1 (MBL, #PM040), rabbit anti-LC3B (Cell Signaling, #2775).
4. Secondary antibodies for immunocytochemistry: goat anti-mouse IgG H&L Alexa Fluor® 488 (ThermoFisher, #A-11001), goat anti-rabbit IgG H&L Alexa Fluor® 488 (ThermoFisher, #A-11034), goat anti-mouse IgG H&L Alexa Fluor® 594 (ThermoFisher, #A-11005), goat anti-rabbit IgG H&L Alexa Fluor® 594 (ThermoFisher, #A-11032).

2.4 Chemicals for the Inhibition of Autophagy

1. Inhibition of autophagosome formation: 10 mM 3-methyladenine in DMEM. Prepare it immediately prior to use (*see* **Note 1**).

3 Methods

All experiments were carried out at room temperature unless otherwise stated.

3.1 Serum Starvation

1. Use wild-type (WT), ciliary formation or *Atg5* mutant cells. Plate the cells in the suitable plate according to the experiment to be performed (*see* Subheading 2.1).
2. When the cells reach 70–90% confluence, remove the culture media and replace it for serum free DMEM.
3. Repeat **step 2** three times before leaving the cells in serum free DMEM.
4. Incubate for 24 h at 37 °C, 5% CO_2.

3.2 Visualizing Cilia by Immunocytochemistry

1. Use WT, ciliary formation or *Atg5* mutant cells. Plate the cells in 24-well plates containing 10 mm Ø glass coverslips (*see* **Note 3**).
2. Starve half of the wells as per Subheading 3.1. For the other half of the wells, replace the media with fresh complete DMEM, 10% FBS.
3. Incubate for 24 h at 37 °C, 5% CO_2.
4. Remove the culture media and wash once with 1× PBS.
5. Add 0.5 ml of 4% PFA and incubate for 15 min.
6. Wash three times with PBS, 10 min at RT for each of the washes. Fixed samples can be stored at 4 °C (*see* **Note 4**).
7. Incubate for 30 min at RT in blocking-permeabilization solution (*see* Subheading 2.3).
8. Wash three times in PBS.
9. Incubate for 1 h with 1:1000 mouse anti-acetylated α-tubulin.
10. Wash three times in PBS, 5 min each wash.

11. Incubate the cells for 30 min with 1:500 fluorophore-conjugated anti-mouse IgG, i.e., goat anti-mouse IgG H&L Alexa Fluor® 594.
12. Wash three times in PBS, 10 min each wash.
13. For double immunofluorescence labeling, repeat **steps 7** through **10** using a primary antibody raised in a different species, i.e., rabbit anti-ATG16L1 as a primary antibody, and fluorescent goat anti-rabbit IgG H&L Alexa Fluor® 488 secondary antibody. Use a different color fluorophore with no overlapping in the excitation–emission spectra.
14. Mount the stained coverslips onto glass slides.
15. Acquire images of the primary cilia in a confocal microscope or conventional epifluorescence microscope provided with deconvolution software. We recommend that images are acquired at least with the 63× objective in order to have enough resolution. For a complete imaging of the cilia, acquire Z-stack images (*see* **Note 5**).
16. Starting from the Z-stack, generate a maximal projection image using ImageJ/Fiji. This feature will allow for an accurate quantification of the ciliary length and number of ciliated cells in a given sample. From the Z-stack image, go to Image → Stacks → Z Project. Choose "Max Intensity". A new image will be created with the maximal projection. Save the image. Use this image for cilia quantification (*see* **Note 6**) (Fig. 2).

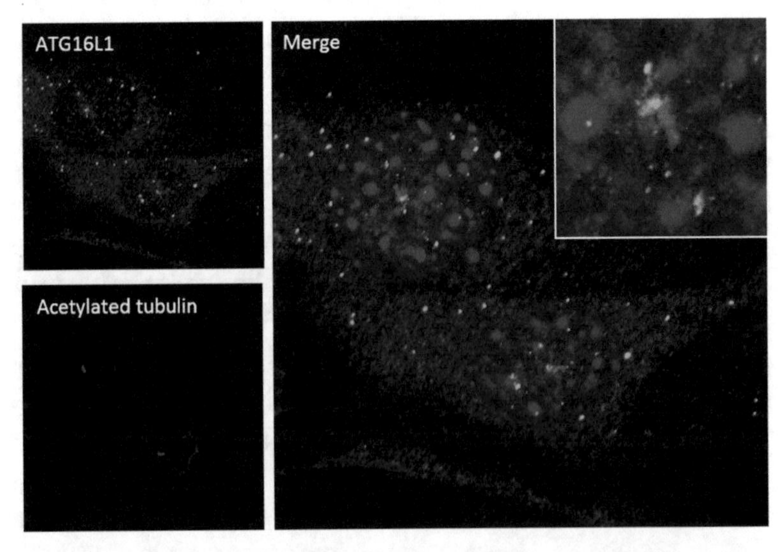

Fig. 2 Autophagy proteins localize at ciliary structures. The early autophagosomal marker ATG16L1 (*green*) has a punctated distribution in fibroblasts and is present at the base of the ciliary axoneme, which is labeled with anti-acetylated tubulin (*red*). *Inset* shows a magnification of the localization of ATG16L1 at the ciliary base

17. Use ImageJ/Fiji to measure ciliary length (*see* **Note 7**). Draw a line over the scale bar incrusted on the image. While the selection is active, go to "Analyze → Set Scale". Write the length of the scale bar in the "known distance" box. Go to Analyze → Measure. The first measurement will be the length of your scale bar.
18. Draw a line over the whole length of the cilia. While the selection is active, go to Analyze → Measure. Copy into an Excel sheet the "Length" values obtained in the "Results" window. Repeat the process for all the primary cilia present in the image. Quantify at least 33 cilia per experimental condition in each replicate of the experiment. A total of 100 cilia per experimental conditions should be quantified before the statistical analysis.
19. Calculate the statistics.

3.3 Correlation of Ciliogenesis and Autophagy Measurements Using Biochemical LC3 Flux Analysis

Autophagic activity can be measured by quantifying the turnover of the autophagic marker LC3-II in the presence and absence of lysosomal inhibitors using western blotting methods. Cilia mutants can affect autophagic measurements quantified as LC3 flux. While LC3-I is cytosolic, LC3-II is conjugated to phosphatidylethanolamine (PE) and is present in isolation membranes and autophagosomal compartments. The amount of LC3-II correlates with the number of autophagosomes, and it is therefore the major marker for autophagy (Fig. 3).

1. Plate cells as per Subheading 2.1 (**items 1–6**).
2. Divide the cultures and serum starve cells as in Subheading 3.1.
3. Add 20 mM NH_4Cl (*see* **Note 8**) and 200 μM leupeptin (final concentrations) to half of the serum-free wells, as well as to half of the serum-containing wells. Incubate the plates at 37 °C for 4 h.

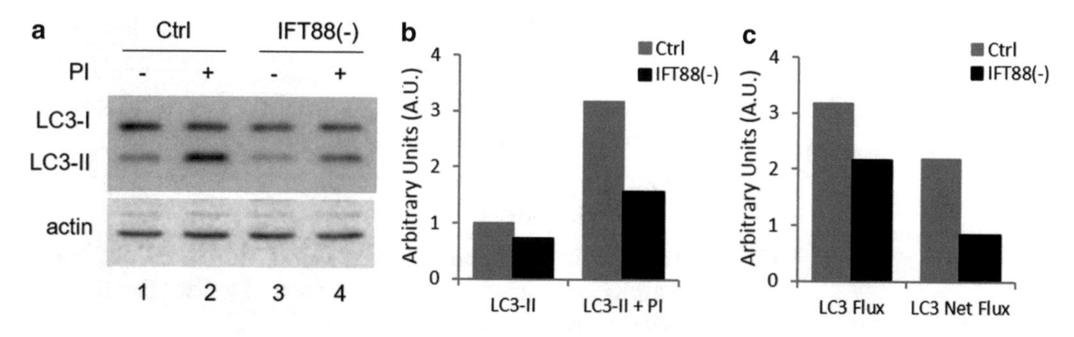

Fig. 3 Autophagy flux measured by western blot. (**a**) Total cell lysates from control and IFT88 knockout cells are incubated in the absence and presence of protease inhibitors (PI), and subjected to protein electrophoresis and western blotting. After, membranes are incubated with antibodies against the autophagosome marker LC3 and actin. (**b**) Graph representation of the densitometric analysis of LC3-II bands. (**c**) Graph representation of the calculated LC3 Flux and LC3 Net Flux

4. Place the plates on ice and wash twice with 1× PBS. Collect the samples by using a cell scraper, and pellet the cells at 300 × *g*, 4 °C for 10 min. Remove the supernatant carefully (*see* **Note 9**).
5. Resuspend the cells by mixing up and down with the pipette in RIPA buffer containing protease and phosphatase inhibitors (*see* **Note 10**) and incubate the samples on ice for at least 30 min.
6. Centrifuge cells at 15,000 × *g*, 4 °C for 15 min and transfer the supernatant to a new tube. The samples can be used immediately; otherwise store them at −20 °C.
7. Quantify protein levels by Lowry method [9]. Quantify each sample in duplicate. Dilute the sample into a final volume of 25 μl per well (i.e., 2 μl of sample and 23 μl of double distilled water, ddH_2O). For the standard curve add 0, 2, 5, 10, 15, and 25 μl of 1 mg/ml BSA into their corresponding total 25 μl. Mix Solution A and B in a 50:1 ratio, using 150 μl of the mixture per sample. Add 15 μl of 1 N folin to each well. Mix well. Incubate for at least 30 min. Read the absorbance at 675 nm using a plate reader. Use the values from the BSA samples to plot a standard curve using linear regression. Use the formula from the standard curve to calculate the protein concentration in the sample.
8. Use 14% polyacrylamide gels. For the resolving gel, mix 2.4 ml of ddH_2O with 2 ml of resolving buffer, 3.5 ml of acrylamide, 80 μl of 10% SDS, 80 μl of 10% APS, and 8 μl of TEMED. After solidification prepare the stacking mixture as follows: 3 ml of ddH_2O, 1.25 ml of stacking buffer, 0.6 ml of acrylamide, 50 μl SDS, 50 μl APS, and 5 μl of TEMED. Let solidify for 30 min to 1 h (solidification time is temperature dependent).
9. Take 10–50 μg of protein per sample (*see* **Note 11**), and dilute it in sample buffer. Boil the samples for 5 min; load them immediately into the polyacrylamide gels (*see* **Note 12**).
10. Electrophorese the samples at constant 90 V until the dye front has reached the bottom of the gel (*see* **Note 13**).
11. Take out the gels and equilibrate them in 15% methanol containing transfer buffer.
12. Mount the transfer sandwiches using nitrocellulose membranes and insert them in the transfer tanks. Run the transference overnight at 4 °C, at constant 300 mA.
13. Take out the nitrocellulose membranes and soak them in Ponceau solution to check for a correct transfer of the proteins from the gel to the membrane. Wash out the Ponceau solution in TBST.

14. Incubate the membranes in blocking buffer for 1 h.
15. Incubate the membranes with rabbit anti-LC3 antibody 1:1000 overnight at 4 °C. Wash them three times in TBST for at least 5 min each.
16. Incubate the membranes with goat anti-rabbit HRP 1:1000 secondary antibody for 1 h.
17. Wash three times in TBST for at least 10 min each and develop the membrane using enhanced chemiluminescence (ECL). Remove the membrane from the TBST and place it on top of a clean surface (i.e., a plastic track). Spread on top of the membrane 1 ml of ECL mixture. Cover the membrane with a transparent, clean, plastic sheet. Insert the membrane into a film developer. Follow the instructions given by the machine. The camera inside the unit detects the chemiluminescence emanating from the membrane, and transforms the signal into a digital image (for interpretation of LC3 bands *see* **Note 14**).
18. Incubate the membranes with mouse anti-β-actin antibody 1:10,000, for 1 h.
19. Wash the membranes three times in TBST for at least 5 min each and incubate with goat anti-mouse HRP 1:1000 secondary antibody, for 1 h.
20. Wash three times in TBST for at least 10 min each.
21. Develop the membrane signal using ECL as instructed in **step 17**.
22. Quantify the LC3-II bands and the loading controls using ImageJ. Use an image where the signal of the bands is not saturated. Open the image in TIF format. Go to "Edit→ Invert". Then "Process→Remove background". Using the *Rectangular* tool, draw a box in the band you want to quantify. While the box is active, go to "Analyze→Measure" (or press Ctrl+M). Repeat this process for all the bands that need to be quantified. Additionally, quantify the intensity of the background signal by drawing a box in an empty place. Keep the Integrated Density data (IntDen) for each band. Copy the Results in an Excel sheet. Substrate the background IntDen value from all the bands IntDen data.
23. Normalize the LC3-II value of each sample to its corresponding loading control.
24. To calculate LC3 flux in each experimental condition, divide the LC3-II value in the presence of lysosomal inhibitors by the value in the absence of inhibitors. An increase in LC3 flux will reflect a higher autophagic activity (*see* **Note 15**) (*see* Fig. 3).
25. Plot LC3 flux against percentage of ciliated cells and average ciliary length (for Subheading 3.2) for each preparation and analyze. Depending on the cell type used, a relationship between autophagy and ciliogenesis may be revealed [2].

3.4 A Complementary Method for Measuring Autophagy: mCherry-GFP-LC3 Fluorescent Autophagy Reporter

The mCherry-GFP-LC3 autophagy reporter is based on the fact that GFP signal is sensitive to the acidic environment of the lysosomal lumen. Therefore, the expression of the dual autophagy reporter in autophagosomes will be visualized as yellow colocalization of GFP^+ and $mCherry^+$ puncta, whereas upon fusion of the autophagosome with the lysosome, GFP^+ signal will be quenched and only red $mCherry^+$ puncta will be visible. This method enables measurements of both autophagy induction and flux simultaneously without the need to use lysosomal inhibitors. Moreover, it has applicability in high throughput screenings (*see* **Note 16**) (Fig. 4). Use autoclaved, sterile tubes and reagents. Maintain the sterility of the procedure by working under a vertical laminar flow hood.

1. One day before transfection, plate the cells onto 10 mm Ø glass coverslips (24-well plates) to be 80–90% confluent at the time of transfection.
2. For each sample dilute 1 μg of reporter plasmid DNA into 50 μl of Opti-MEM. In another tube, mix 2 μl of Lipofectamine 2000 into 50 μl Opti-MEM per sample. Incubate both mixtures separately for 5 min, not exceeding 25 min.
3. Combine the DNA-containing and Lipofectamine-containing mixtures in a single tube and mix gently. Incubate for 20 min. Add 100 μl of the mixture drop wise to each sample (*see* **Note 17**).
4. Incubate transfected cells for 48 h to allow the regulation of the overexpressed protein before performing the experiment.
5. Wash half of the cells three times with serum free DMEM, and leave them with serum free DMEM. Add fresh, serum containing DMEM to the other half of the cells. Incubate for 4–6 h to maximally activate autophagy.
6. Wash out the media with 1× PBS and add 0.5 ml 4% PFA per well and incubate for 15 min to fix the cells.
7. Wash three times with PBS, 10 min for each of the washes. Fixed samples can be stored at 4 °C (*see* **Note 4**).
8. Mount the fixed coverslips into microscope slides using DAPI-containing mounting media and let the slides solidify for at least 6 h. Optimally use the microscope slides the day after mounting.
9. Acquire the images with a confocal microscope, or conventional epifluorescence microscope provided with deconvolution software. We recommend that the images are acquired with a 63× objective in order to have enough resolution for the quantification of the puncta. Avoid cells where the signal of the plasmid is still too high (*see* Fig. 4).
10. For puncta quantification, use the "Green and Red Puncta Colocalization" macro available for ImageJ. Download and install the required plugins and script from the "ImageJ Documentation Wiki" website (imagejdocu.tudor.lu), under the "Plugins" section.

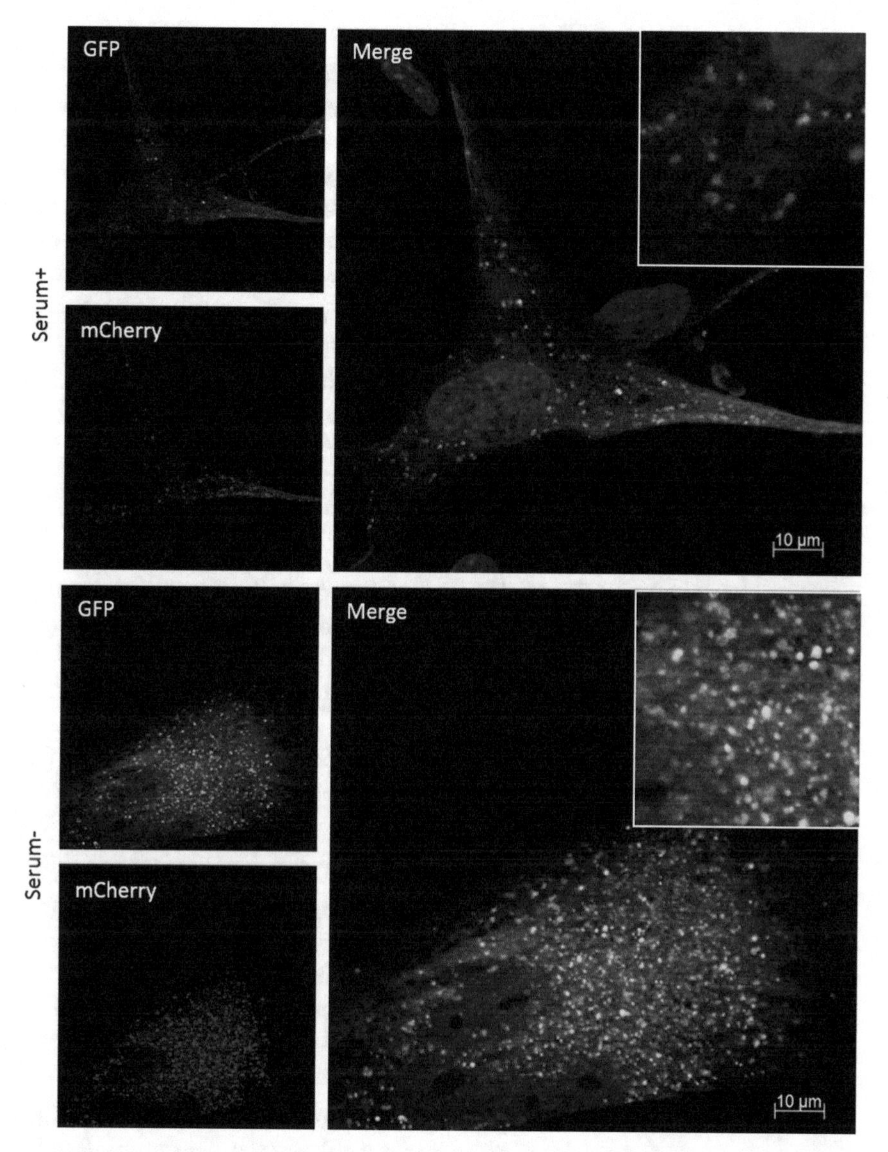

Fig. 4 Autophagy flux measured by mCherry-GFP-LC3 overexpression. Mouse fibroblasts transfected with the autophagy dual reporter mCherry-GFP-LC3 are incubated in the presence and absence of serum for 6 h. Cells were fixed in 4 % PFA, and images taken in an epifluorescence microscope. The increase in *yellow* and *red* puncta upon serum starvation indicates an increase in both autophagosome formation and autophagic function. Cell nuclei are stained with DAPI

11. Open the RGB image for the analysis. Use TIF format. Using the freehand selection tool, draw the perimeter of the cell. While the selection is active, go to "Analyze → Measure". From the Results window, copy the data "area of the cell" in an Excel document.
12. Still with the selection active, go to "Edit → Clear Outside", and then "Edit → Invert". By clicking on the image, the active selection will disappear. Go to "Edit → Invert" to have an image with a single cell.

13. Start the "Green and Red Puncta Colocalization" macro by going to "Plugins → Macros → Run". Open the "Green and Red Puncta Colocalization" macro. The macro will split the image into individual blue, green, and red channels, and process the green and red ones. It will save them under the names "Red.tif" and "Green.tif" and ask the user to open them individually. After a series of processes, the "Results" window will appear. Copy it in an Excel sheet for further analysis.
14. Repeat **steps 10** and **11** for all the cells in each image, and for the rest of the images. Quantify at least 33 cells per experimental condition in each replicate of the experiment. A total of 100 cells per experimental conditions should be quantified before the statistical analysis.
15. From the "Results" window, keep "count" (number of puncta) and "total area" (area occupied by the puncta) values from the "Colocalized points (8-bit)" (yellow puncta), and "Composite (RGB) (red)". Yellow puncta reflect autophagosomes prior to the fusion with the lysosome. "Composite (RGB) (Red)" data is obtained from the red single channel image, and therefore it will reflect the total amount of puncta that are red at some point, including those that colocalize with green puncta. Thus, to obtain the "Only Red" puncta value, substrate "Colocalized points (8-bit)" to "Composite (RGB) (red)" (*see* **Note 18**).
16. Divide each cell area by the total area of the image. Calculate the relative area that the puncta occupy in each cell.
17. Perform the statistics for the "number of puncta" and "cell area occupied by the puncta". Plot the data concerning yellow, red and total puncta in the same graph. Changes in yellow puncta will reflect formation of autophagosomes, while the variation in red puncta will reflect changes in the fusion rate of autophagosomes with lysosomes (*see* **Note 19**) (*see* Table 1).
18. Plot data against percentage of ciliated cells and cilia length to correlate changes in autophagy with ciliogenesis.

Table 1
Possible readouts for the changes observed in the number of yellow and red puncta after mCherry-GFP-LC3 reporter transfection and imaging

Yellow (APH)	Red (APH-Lys)	Readout
↓	↓	Total autophagic capacity is down
↓	↑	Increased flux, faster than the rate of formation of new APH
↔	↔	Basal autophagic activity is maintained
↑	↓	Blockage of autophagosome maturation into autophagolysosomes
↑	↑	Induction of total autophagy

4 Notes

1. As an alternative to the use of genetic mutants, autophagy can be blocked chemically. For example, formation of autophagosomes requires the activity of the PI3KIII, which can be inhibited by 3-MA, wortmannin or LY294002. However, these compounds block both the activity of PI3KIII and PI3KI, and therefore are not autophagic specific. We recommend short treatments with 3-MA, not longer than 6 h, to avoid the effect over the PI3KI.
2. Opti-MEM™ medium is a low serum, commercially available minimal essential medium (MEM) to be used during transfection protocols. It contains insulin, transferrin, hypoxanthine, thymidine, and trace elements, and it uses a sodium bicarbonate buffer system (2.4 g/L), for which it requires a 5–10% CO_2 environment to maintain physiological pH.
3. Cilia can be visualized in differential interference (DIC) or by immunolabeling as in Subheading 3.2.
4. Alternatively cells can be fixed in cold methanol, which removes the cytosolic fluorescent signal and allows a cleaner visualization of membrane bound proteins and cytoskeleton. However, fixation in methanol alone can damage organelles, and therefore it should be carefully used. An approach that combines fixation in formaldehyde followed by methanol fixation preserves the immunoreactivity of intracellular proteins and it is recommended [10].

   Methanol also acts as a permeabilization agent. In this case, cool down an aliquot of 100% methanol to −20 °C. Keep the samples on ice during the procedure. Wash the coverslips once with cold PBS. Incubate the samples in cold methanol at −20 °C for 5 min. Wash the sample three times in PBS.
5. In the acquisition of Z-stack images, the lower and upper limits of the stack should comprise all the focal planes where there is any primary cilia staining.
6. As an alternative to ImageJ, most microscope acquisition software include the option to generate a maximal projection image from the original file. This option is usually straightforward and it is included as a post-processing option.
7. Ciliary length can be measured using the microscope acquisition software. Most acquisition software provide a "length" tool that will automatically convert pixel information in length measures using the metadata contained in the original file format. It has to be noted that immunostaining against acetylated tubulin visualizes the ciliary axoneme, and that the length of the axoneme can vary between cilia. Alternatively, length of the cilia can be measured by immunostaining against ciliary membrane proteins, such as Arlb13b or DNAH5 [11].

8. It has to be taken into consideration that NH_4Cl could affect ciliary length. As a control, in LC3 flux experiments ciliary length should be measured in NH_4Cl-treated samples.
9. Cell pellets can be stored at −80 °C until further analysis. This is especially useful for time-point experiments, when samples are collected at different incubation times. All the samples collected for the same experiment should be processed at the same time to minimize introducing experimental errors.
10. RIPA containing protease and phosphatase inhibitors should be prepared fresh each time. Dilute the stock of protease and phosphatase inhibitors into the RIPA volume needed for each experiment. Preferentially, use a small volume of RIPA for cell resuspension (50–100 μl) in order to accurately determine protein concentration. Using small volumes will allow for later dilution if necessary.
11. The amount of protein used for LC3-II western blotting depends on the cell line used. Fibroblasts contain higher amounts of LC3, and therefore using 10 μg of total protein is enough.
12. LC3 is a lipidated protein; do not freeze samples in sample buffer as this will interfere with the correct resolution of the protein in the denaturating gel. Store the total lysates in RIPA, and prepare the amount of protein to be loaded for the electrophoresis fresh each time.
13. Run the electrophoresis at slow voltage to allow for a better resolution of the LC3 bands. Make sure LC3-I and LC3-II are well resolved in order to accurately quantify the densitometry of the bands.
14. Although the lipidated form of LC3 (LC3-II) has a higher molecular weight than the cytosolic form (LC3-I), LC3-II runs faster (14 kDa) than LC3-I (16 kDa) probably because its hydrophobicity.
15. In addition, the "net LC3 flux" can be calculated, which takes into account the initial amount of autophagosomes in the cell (levels of LC3-II in the absence of lysosomal inhibitors). In this case, substrate LC3-II value in the absence of lysosomal inhibitors to the LC3-II levels in the presence of inhibitors. Steady state levels of LC3-II will correlate with the number of autophagosomes in the cell.
16. The dual autophagy reporter is a powerful tool that allows the study of autophagic activity in a dynamic way. However, it has to be kept in mind that LC3 overexpression could mask defects in the machinery and signaling that are needed for LC3 expression, as it has been reported that overexpression of LC3 by fluorescent reporters results in a higher autophagic activity [12].

17. Media can be changed 4–6 h after transfection. It is recommended for sensitive cell lines with high cytotoxicity to Lipofectamine.
18. Despite the fact that the autophagy reporter should only give yellow and red puncta, there is always a certain amount of "Only green" puncta that could be due to aggregated forms of GFP.
19. Expressing the results as "Number of puncta" will reflect the total autophagic capacity of the cell. However, expressing changes relative to the area is optimal when the experimental conditions change considerably the cell area.

References

1. Orhon I, Dupont N et al (2015) Autophagy and regulation of cilia function and assembly. Cell Death Differ 22(3):389–397
2. Pampliega O, Orhon I et al (2013) Functional interaction between autophagy and ciliogenesis. Nature 502(7470):194–200
3. Tang Z, Lin MG et al (2013) Autophagy promotes primary ciliogenesis by removing OFD1 from centriolar satellites. Nature 502(7470): 254–257
4. Esteban-Martinez L, Boya P (2015) Autophagic flux determination in vivo and ex vivo. Methods (San Diego, CA) 75: 79–86
5. Klionsky DJ, Abdalla FC et al (2012) Guidelines for the use and interpretation of assays for monitoring autophagy. Autophagy 8(4): 445–544
6. Mizushima N, Yoshimori T (2007) How to interpret LC3 immunoblotting. Autophagy 3(6):542–545
7. Pankiv S, Clausen TH et al (2007) p62/SQSTM1 binds directly to Atg8/LC3 to facilitate degradation of ubiquitinated protein aggregates by autophagy. J Biol Chem 282(33):24131–24145
8. Pazour GJ, Dickert BL et al (2000) Chlamydomonas IFT88 and its mouse homologue, polycystic kidney disease gene tg737, are required for assembly of cilia and flagella. J Cell Biol 151(3):709–718
9. Peterson GL (1977) A simplification of the protein assay method of Lowry et al. which is more generally applicable. Anal Biochem 83(2):346–356. doi:10.1016/0003-2697(77)90043-4
10. Schnell U, Dijk F et al (2012) Immunolabeling artifacts and the need for live-cell imaging. Nat Methods 9(2):152–158
11. Omran H, Loges NT (2009) Chapter 7 – Immunofluorescence staining of ciliated respiratory epithelial cells. In: Stephen MK, Gregory JP (eds) Methods in cell biology, vol 91. Academic, New York, NY, pp 123–133
12. Hung SY, Huang WP et al (2015) LC3 overexpression reduces Abeta neurotoxicity through increasing alpha7nAchR expression and autophagic activity in neurons and mice. Neuropharmacology 93:243–251

Chapter 5

Recombinant Reconstitution and Purification of the IFT-B Core Complex from *Chlamydomonas reinhardtii*

Michael Taschner and Esben Lorentzen

Abstract

Eukaryotic cilia and flagella are assembled and maintained by intraflagellar transport (IFT), the bidirectional transport of proteins between the ciliary base and tip. IFT is mediated by the multi-subunit IFT complex, which simultaneously binds cargo proteins and the ciliary motors. So far 22 subunits of the IFT complex have been identified, but insights into the biochemical architecture and especially the three-dimensional structure of this machinery are only starting to emerge because of difficulties in obtaining homogeneous material suitable for structural analysis. Here, we describe a protocol for the purification and reconstitution of a complex containing nine *Chlamydomonas reinhardtii* IFT proteins, commonly known as the IFT-B core complex. In our hands, this protocol routinely yields several milligrams of pure complex suitable for structural analysis by X-ray crystallography and single-particle cryo-electron microscopy.

Key words Intraflagellar transport, IFT-B, Protein purification, IFT-B core complex, X-ray crystallography

1 Introduction

Cilia and flagella are evolutionarily conserved antenna-like eukaryotic organelles with essential functions in motility, sensory reception, and signaling [1]. They can be found on organisms ranging from unicellular green algae (e.g. *Chlamydomonas reinhardtii*) to mammals, and share a conserved architecture with a microtubule-based axoneme originating from a modified centriole (the "basal body") and an overlying ciliary membrane. Another important conserved feature of cilia across the eukaryotic kingdom is the fact that they are assembled by intraflagellar transport (IFT), the bidirectional movement of proteinaceous particles between the flagellar base and tip [2]. This process was first discovered in *Chlamydomonas* flagella by differential interference contrast (DIC) microscopy more than two decades ago [3], and soon thereafter the main components of this transport system were shown to be an anterograde motor (heterotrimeric kinesin 2) [4, 5], a retrograde

Peter Satir and Søren Tvorup Christensen (eds.), *Cilia: Methods and Protocols*, Methods in Molecular Biology, vol. 1454, DOI 10.1007/978-1-4939-3789-9_5,

motor (cytoplasmic dynein-2) [6–9], and the multi-subunit IFT complex that is believed to act as an adapter between the IFT motors and the more than 600 ciliary cargo proteins [10].

Native IFT complexes were initially purified from *Chlamydomonas* flagella and described as either a 13-subunit [11] or 15-subunit [12] complex, but the discoveries of additional subunits have since then brought the total number of IFT proteins to at least 22 (reviewed in [13]). It was shown to dissociate at low salt concentration into two distinct sub-complexes, IFT-A and IFT-B [12], which are now believed to have 6 and 16 subunits, respectively. The IFT-B complex was later shown to contain a salt-stable 9-subunit IFT-B "core" complex and several weakly associated "peripheral" subunits [14]. Further attempts to biochemically elucidate the architecture of these complexes have revealed a number of direct protein–protein interactions and smaller sub-complexes [15–20], although the overall structure of the IFT complex is still unknown. We have previously reported the recombinant reconstitution of a stable 9-subunit *Chlamydomonas* IFT-B core complex [21, 22] and have presented a detailed interaction map of this assembly. The availability of purified recombinant material also paved the way for the determination of high-resolution crystal structures of various smaller sub-complexes [22–24]. Despite this progress, our structural understanding of most IFT complex subunits and of the IFT complex is still very limited. Important unanswered questions are (1) What is the overall structure of the fully assembled IFT-A and IFT-B complexes? (2) How does the IFT complex interact with the anterograde and retrograde motors? (3) What is the molecular basis for how the IFT complex can recognize hundreds of ciliary cargo proteins? (4) How is IFT regulated at the ciliary base and tip? With respect to the regulatory aspect it is important to note that two small GTPases (IFT27 and IFT22) exist in the IFT-B core complex and are prime candidates for playing a regulatory role in the IFT process.

To provide answers to the questions listed above it is desirable to have purified IFT complexes at hand for structural studies (crystallography, single-particle EM, small-angle X-ray scattering), biochemical investigation of motor and cargo binding, biochemical characterization of small IFT GTPases in the context of their native environment (i.e. the assembled IFT complex), and similar experiments. While the recombinant reconstitution of the entire IFT complex will require significantly more work, the 9-subunit IFT-B core complex can form a starting point for in-depth analysis. Whereas our previously reported purifications involved expression of some proteins in the insect cell system using baculoviruses [21, 22] and yielded relatively little final protein complex, we recently optimized the procedure, which now can be done exclusively with proteins and complexes expressed in *E. coli* and results in several milligrams of homogeneous IFT-B core complex.

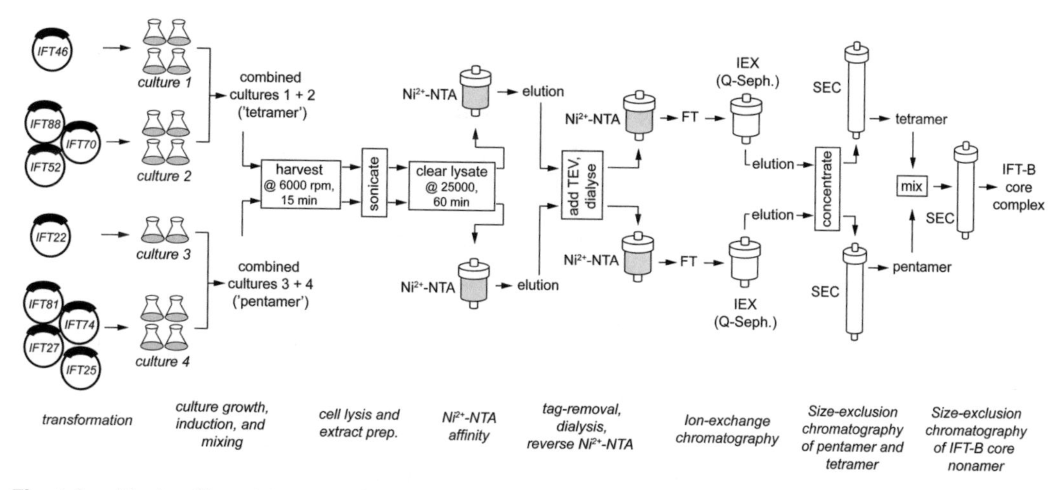

Fig. 1 Graphical outline of the procedure (see text for details)

Here, we provide a detailed protocol for the purification of the 9-subunit *Chlamydomonas* IFT-B core complex from individually purified tetrameric (IFT88/70/52/46) and pentameric (IFT81/74/27/25/22) sub-complexes, each formed by co-lysis of two separate *E. coli* cultures. Three of the nine proteins (IFT74, IFT52, IFT25) are expressed as N- or C-terminally truncated forms to increase expression and to prevent protein degradation. The protocol has been optimized for the purification from several liters of bacterial culture (*see* Fig. 1 for an outline of the procedure) and in our hands typically gives at least 5 mg of pure protein. Most likely the procedure can be scaled down if less complex is needed for the task at hand, but we have not tested such small-scale purifications ourselves. The protocol is written for the simultaneous purification of both sub-complexes followed by reconstitution of the 9-subunit core, but the two sub-complexes can be purified sequentially if various pieces of equipment such as incubators and chromatography systems are limiting. We also highlight potential points in the protocol at which the material can be frozen for future use.

2 Materials

2.1 Bacterial Strains, Plasmids, Antibiotics, and Media

1. TB medium (for 10 l of bacterial expression): mix 120 g Bacto tryptone, 240 g of Yeast extract, and 40 ml of 100% glycerol in a total volume of 9 l of ddH_2O, make 10 × 900 ml aliquots and autoclave (121 °C, 20 min).
2. 10× Phosphate stock (for 10 l of bacterial expression): dissolve 23.14 g of KH_2PO_4 and 164.3 g of $K_2HPO_4{\cdot}3H_2O$ in 1 l of ddH_2O and autoclave (121 °C, 20 min).

3. Antibiotic stocks (all 1000×):
   Ampicillin (Amp): 100 μg/ml in H_2O.
   Kanamycin (Kan): 50 μg/ml in H_2O.
   Streptomycin (Strep): 50 μg/ml in H_2O.
   Chloramphenicol (Cm): 34 μg/ml in 70% ethanol.
4. L-Agar plates (Amp, Strep/Amp/Kan, Strep/Amp/Kan/Cm).
5. Expression plasmids (available from our lab upon request).
   "A-46": pEC-A-CrIFT46.
   "A-22": pEC-A-CrIFT22.
   "S-70": pEC-S-CrIFT70.
   "K-52": pEC-K-His-TEV-IFT52 (1–430).
   "A-88": pEC-A-His-TEV-IFT88.
   "A-81": pEC-A-His-TEV-IFT81.
   "K-74": pEC-K-His-TEV-IFT74 (128-C).
   "S-25": pEC-S-His-TEV-IFT25 (1–136).
   "Cm-27": pEC-Cm-IFT27.
6. Competent *E. coli* cells (we use BL32 (DE3), otherwise different strains suitable for expressing transgenes from T7 promoters which are sensitive to Amp, Kan, Strep, and Cm).

2.2 Equipment

1. 37 °C incubator for bacterial plates.
2. Shaker incubator with enough space for eight 2000 ml Erlenmeyer flasks and temperature control between 18 and 37 °C.
3. Sterile 2000 ml Erlenmeyer flasks.
4. Centrifuge.
5. Rotor suitable for centrifugation of large volumes at low speed (e.g. Beckman JLA 8.1000).
6. Rotor suitable for centrifugation of small volumes at high speed (e.g. Beckman JA-25.50).
7. Appropriate centrifugation containers for both rotors.
8. Sonication device with an appropriate tip for volumes of above 100 ml (e.g. Bandelin Sonopuls HD 3200 with VS 70T sonotrode).
9. 5 μm filters (bottle-top or syringe-driven).
10. Ni^{2+}-NTA columns (e.g. Roche Complete His-Tag Purification Column, 5 ml).
11. Peristaltic pumps.

12. Liquid chromatography system with gradient mixer (e.g. Äkta Prime or Äkta Purifier).
13. SDS-PAGE equipment.
14. Dialysis membrane (we commonly use Spectra/Por 4 Dialysis Membrane with a Molecular Weight Cut-Off (MWCO) of 12–14 kDa from Spectrum Laboratories Inc.).
15. Dialysis clamps.
16. Ion-exchange column (Cation-exchanger; we use a HiTrap Q HP column from GE Healthcare, 5 ml).
17. Centrifugal filters for protein concentration (we use Amicon Ultracel Filters with a cut-off of 100 kDa).
18. SEC column (we use a HiLoad Superdex200 column from GE Healthcare, 125 ml).
19. Injection loop (0.5–2 ml volume).
20. Device for determining protein concentration (we use a Nanodrop ND-1000 spectrophotometer).

2.3 Buffers and Chemicals

1. 1 M Isopropyl β-D-1-thiogalactopyranoside (IPTG), sterile-filtered.
2. β-Mercaptoethanol.
3. 1 mg/ml DNase.
4. Protease inhibitor tablets (e.g. Roche Complete tablets, EDTA-free).
5. 5 M NaCl.
6. 5 M Imidiazole pH 7.5.
7. 1 M Dithiothreitol (DTT).
8. Lysis Buffer: 50 mM Phosphate buffer pH 7.5, 150 mM NaCl, 1 mM $MgCl_2$, 10% glycerol.
9. High-salt wash Buffer: 50 mM Phosphate buffer pH 7.5, 1000 mM NaCl, 1 mM $MgCl_2$, 10% glycerol.
10. TEV protease (ideally His-tagged, so that it can be removed by the reverse Ni^{2+}-NTA step).
11. Dialysis Buffer: 20 mM Tris–HCl pH 7.5, 50 mM NaCl, 10% glycerol, 5 mM β-mercaptoethanol (freshly added).
12. Q-Buffer A: 20 mM Tris–HCl pH 7.5, 50 mM NaCl, 1 mM $MgCl_2$ 10% glycerol, 1 mM DTT (freshly added).
13. Q-Buffer B: 20 mM Tris–HCl pH 7.5, 1000 mM NaCl, 1 mM $MgCl_2$, 10% glycerol, 1 mM DTT (freshly added).
14. SEC Buffer: 10 mM HEPES pH 7.5, 150 mM NaCl, 1 mM $MgCl_2$, 5% glycerol, 1 mM DTT (freshly added).

3 Methods

3.1 Growth of Bacterial Cultures and Induction of Protein Expression

1. Transform the following single plasmids or plasmid combinations into *E. coli* cells suitable for protein expression (ideally BL21 (DE3)) using either electroporation or chemical transformation (*see* **Note 1**):
   Culture 1: A-46 alone.
   Culture 2: S-70/K-52/A-88.
   Culture 3: A-22 alone.
   Culture 4: A-81/K-74/Cm-27/S-25.
2. Plate all transformations on LB plates containing the appropriate antibiotics.
3. Incubate overnight at 37 °C.
4. The following morning, prepare the medium for growing the large-scale cultures (for each liter of medium mix 100 ml of 10× Phosphate stock with 900 ml of TB medium, then add 1 ml of 1000× antibiotic stock).
   For culture 1: 2 l of medium containing Amp.
   For culture 2: 2 l of medium containing Strep/Kan/Amp.
   For culture 3: 1 l of medium containing Amp.
   For culture 4: 2 l of medium containing Strep/Kan/Amp/Cm.
5. For each culture aliquot 500 ml of the prepared medium into separate 2000 ml Erlenmeyer flasks (*see* **Note 2**).
6. Using a sterile pipette tip, scrape similar amounts of bacterial cells off the individual LB plates containing transformed colonies, resuspend thoroughly in 5 ml TB medium and inoculate the appropriate flasks with 1 ml of the resulting suspension.
7. Incubate the flasks at 37 °C shaking at 180 rpm.
8. Monitor the OD_{600} in regular intervals.
9. Once the OD_{600} reaches a value of 1, reduce the temperature of the incubator to 18 °C and let the cultures cool down for about 1.5 h (*see* **Note 3**).
10. When the temperature reaches 18 °C add 200 μl of 1 M IPTG to each flask (giving a final concentration of 0.4 mM) to induce protein expression.
11. Continue the incubation at 18 °C overnight.

3.2 Cell Lysis and Extract Preparation

1. Combine culture 1 with culture 2 ("tetramer" culture) in one larger beaker and culture 3 with culture 4 ("pentamer" culture) in another large beaker, giving you a total of 4 and 3 l, respectively.
2. Harvest the cells by centrifugation (we use a Beckman JLA 8.1000 rotor with 1 l centrifugation bottles; 8000×*g* for 15 min).

3. Resuspend all the pellets belonging to the same culture ("tetramer" or "pentamer") in 100 ml of Lysis Buffer and transfer to a beaker of appropriate size.
4. At this point the resuspended pellet can be frozen at –80 °C if desired.
5. Add a magnetic stir bar and supplement the cell suspension with 5 mM β-mercaptoethanol, 10 μg/ml DNase, and one protease inhibitor tablet while stirring.
6. Place each beaker on ice and sonicate to open the cells (we use a Bandelin Sonopuls HD 3200 with VS 70T sonotrode to sonicate at 40% output in 1 s pulses (1 s on/1 s off) giving a total energy of 25 kJ).
7. After sonication aliquot the "tetramer" and "pentamer" lysates in appropriate tubes for high-speed centrifugation and balance them properly.
8. Clear the lysate for 1 h at 75000×*g* (we use a Beckman JA-25.50 rotor).
9. Carefully decant the supernatants for the two cultures into clean beakers, making sure not to take along any material from the pellet (*see* **Note 4**).
10. Filter the resulting supernatant through a 5 μm filter to remove any residual particulate material.

3.3 Ni^{2+}-NTA Affinity Purification

1. Equilibrate two 5 ml Ni^{2+}-NTA column with at least 5 column volumes (cV) Lysis Buffer using a peristaltic pump.
2. Load the filtered lysates on the column at a flow rate of 1 ml/min (*see* **Note 5**).
3. Wash the columns with 5 cV (i.e. 25 ml) of Lysis Buffer supplemented with 5 mM β-mercaptoethanol.
4. Wash the columns with 10 cV (i.e. 50 ml) of high-salt wash Buffer supplemented with 5 mM β-mercaptoethanol.
5. Wash the columns with 5 cV (i.e. 25 ml) of Lysis Buffer supplemented with 5 mM β-mercaptoethanol.
6. On a liquid chromatography system (we use an Äkta Prime) equilibrate first pump B with Lysis Buffer supplemented with 500 mM imidazole and 5 mM β-mercaptoethanol, then equilibrate pump A with Lysis Buffer supplemented with 5 mM β-mercaptoethanol.
7. Attach one of the protein-loaded 5 ml Ni^{2+}-NTA columns and elute the bound protein with a 50 ml gradient from A to B, collect 5 ml fractions.
8. Repeat **step 7** for the second column (if both purifications are performed in parallel).
9. Check the identity and integrity of the eluted protein in the fractions by SDS-PAGE (*see* **Note 6**) (*see* Fig. 2a, e).

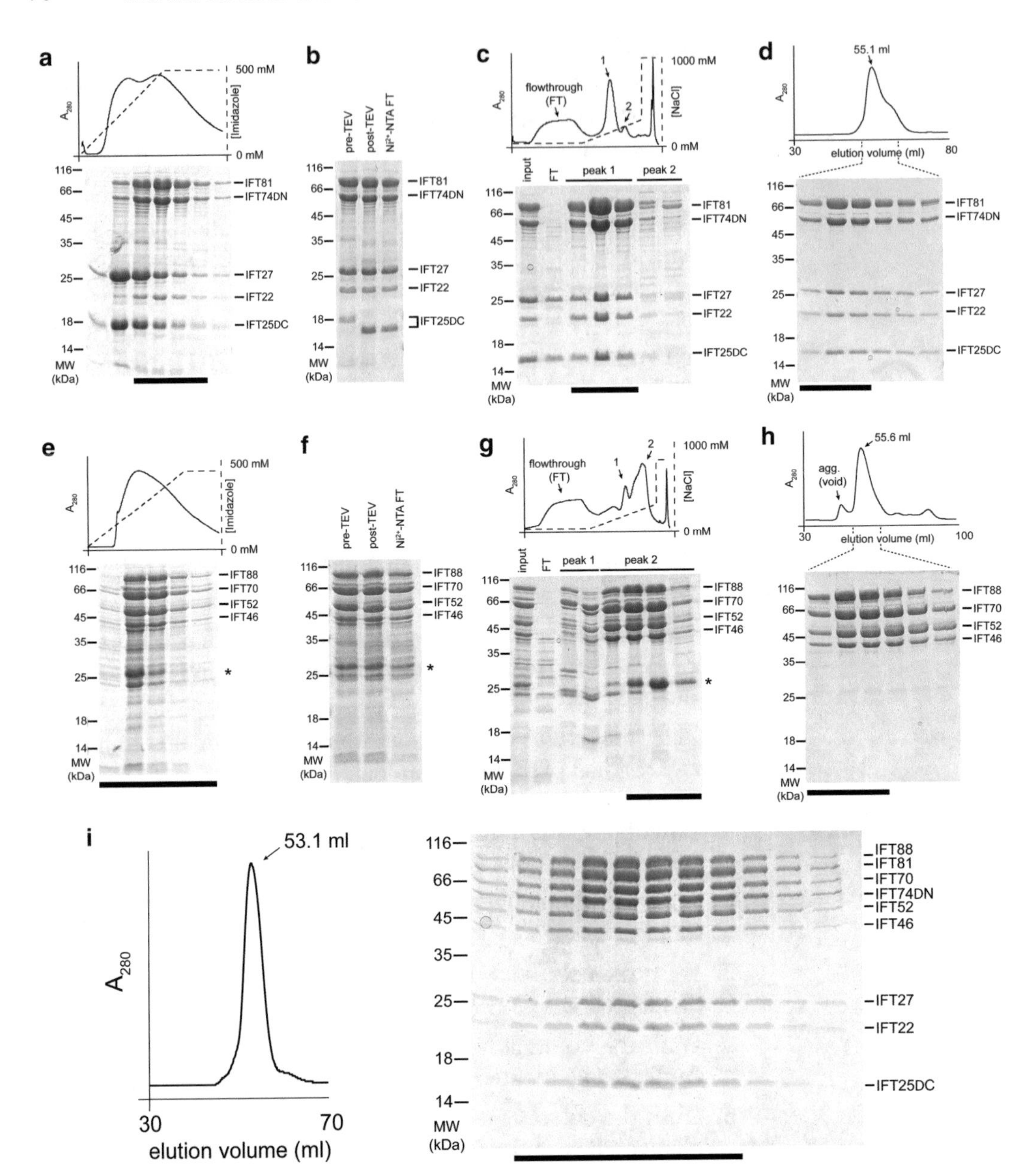

Fig. 2 Coomassie-stained SDS gels and chromatograms of important steps in the protocol (see text for details). *Black bars* below the gels indicate which fractions were pooled for subsequent steps. (**a**–**d**) Steps in the purification of the IFT81/74/27/25/22 pentamer. (**a**) Chromatogram and SDS-PAGE of elution fractions from the Ni^{2+}-NTA purification step. Two complexes are obtained here, an IFT27/25 complex eluting early in the gradient, and the IFT81/74/27/25/22 pentamer eluting later. (**b**) TEV-cleavage of His-tags and reverse Ni^{2+}-NTA. The removal of the tag can be clearly seen for the small IFT25 protein. (**c**) Chromatogram and SDS-PAGE of elution fractions from the Ion-exchange-chromatography step. Whereas the excess IFT27/25 complex does not bind to the column and is found in the flowthrough (FT) fractions, the IFT81/74/27/25/22 pentamer binds and elutes in the first peak (peak 1), whereas contaminants elute later (peak 2). (**d**) SEC profile and SDS-PAGE showing selected fractions across the peak. (**e**–**h**) Steps in the purification of the IFT88/70/52/46 tetramer (a main contaminating band is marked with an *asterisk*). (**e**) Chromatogram and SDS-PAGE of elution fractions from the Ni^{2+}-NTA purification step. (**f**) TEV-cleavage of His-tags and reverse Ni^{2+}-NTA. The removal of the tags from

10. Pool the fractions containing the desired pentameric and tetrameric complexes separately.
11. Take out 20 μl of each pooled eluate and mix with 20 μl 2× SDS loading dye (or an appropriate amount of an SDS-loading dye with higher concentration) as a "Ni^{2+}-elu-pre-TEV" sample.

3.4 Tag-Cleavage, Dialysis, and Reverse Ni^{2+}-NTA (See Note 7)

1. Add TEV protease (0.5 mg of TEV protease per sample) to remove the Hexahistidine tags from IFT88, IFT52, IFT81, IFT74, and IFT25 (*see* **Note 8**).
2. Prepare 2 l of Dialysis Buffer in a 2 l measuring cylinder and add a magnetic stir bar.
3. Cut two pieces (or only one if the two purifications are not performed in parallel) of dialysis membrane to a size appropriate for the volumes of your combined "pentamer" and "tetramer" elutions, respectively.
4. Wash the dialysis membranes briefly in ddH_2O.
5. Seal one end of each membrane with a clamp and put both membranes into the cylinder containing the Dialysis Buffer.
6. Fill the membrane(s) with the (two) combined elution(s) and close the open ends with additional clamps.
7. Stir the buffer at low speed (around 100 rpm) and let the dialysis run overnight at room temperature, giving the protease enough time to cleave all the tags.
8. The following morning carefully transfer the content of each dialysis tube into a fresh 50 ml Falcon tubes.
9. Spin the tubes at high speed ($>8000\times g$) for 10 min to pellet any precipitate which might have formed.
10. Take out 20 μl of each sample and mix with 20 μl 2× SDS loading dye (or an appropriate amount of an SDS-loading dye with higher concentration) as a "Ni^{2+}-elu-post-TEV" sample.
11. On a liquid chromatography system equilibrate first pump B with Dialysis Buffer supplemented with 500 mM imidazole and 5 mM β-mercaptoethanol, then equilibrate pump A with Dialysis Buffer supplemented with 5 mM β-mercaptoethanol.
12. Attach a 5 ml Ni^{2+}-NTA column to the system and wash it with 5 cV (25 ml) of Dialysis Buffer (pump A).

Fig. 2 (continued) IFT88 is not visible due to the high molecular weight of the protein, but in the case of IFT52 a small shift can be seen. (**g**) Chromatogram and SDS-PAGE of elution fractions from the Ion-exchange-chromatography step. Contaminants either do not bind to the column (FT) or elute early in the gradient (peak 1), whereas the IFT88/70/52/46 complex and some excess IFT70/52/46 elute later (peak 2). (**h**) SEC profile and SDS-PAGE showing selected fractions across the peak. (**i**) SEC profile and SDS-PAGE showing fractions across the peak from the reconstitution of the recombinant IFT-B core nonamer

13. Load one of the two samples ("tetramer" or "pentamer") on the column and collect the flowthrough (FT) in 5 ml fractions (the cleaved protein should no longer bind and should instead flow through the column).
14. When all the samples have been loaded wash the column again with Dialysis Buffer (pump A) until the UV signal goes back down.
15. Switch to Dialysis Buffer supplemented with 500 mM imidazole (pump B) and elute the bound protein to clean the column.
16. Pool the FT fractions containing protein, take out 20 μl and mix with SDS-loading dye as "rev-Ni^{2+}-FT" sample.
17. If both samples are purified in parallel, repeat **steps 12–16** for the second sample.
18. Check the three samples ("Ni^{2+}-elu-pre-TEV", "Ni^{2+}-elu-post-TEV", and "rev-Ni^{2+}-FT") for each complex by SDS-PAGE (*see* Fig. 2b, f).

3.5 Ion-Exchange Chromatography

1. On a liquid chromatography system, equilibrate first pump B with Q-Buffer B then equilibrate pump A with Q-Buffer A.
2. Attach the Ion-exchange column.
3. Wash the column with 5 cV (25 ml) of Q-Buffer A.
4. Load the FT fractions of the reverse Ni^{2+}-NTA step on the column (*see* **Note 9**), and start collecting fractions of 5 ml.
5. When everything is loaded wash the column with Q-Buffer A (pump A) until the UV signal goes down to the baseline.
6. Start a gradient of 100 ml volume to a final concentration of Q-Buffer B of 50% (corresponding to a NaCl concentration of ~500 mM).
7. When the proteins are eluted (*see* **Note 10**), the gradient can be stopped and the column can be washed with 100% Q-Buffer B to remove bound nucleic acids and other impurities.
8. If both purifications are performed in parallel, repeat **steps 3–7** for the second sample.
9. Check the proteins in the FT (*see* **Note 11**) as well as in the peaks by SDS-PAGE and pool the fractions containing the complexes of interest (*see* Fig. 2c, g).

3.6 Size-Exclusion Chromatography (SEC)

1. Concentrate the elution fractions from the Ion-exchange step to a final volume between 0.5 and 2 ml, using a centrifugal filter.
2. If necessary the concentrated protein sample(s) can be snap-frozen in liquid nitrogen and stored at −80 °C.
3. Equilibrate a size-exclusion column with 1 cV of SEC Buffer.
4. Inject the sample using a loop of an appropriate size and start collecting fractions of 1 ml.

5. Repeat **step 3** for the second sample (if both purifications are performed in parallel).
6. Check the proteins in the peak fractions (*see* **Note 12**) by SDS-PAGE (*see* Fig. 2d, h).
7. Pool the desired fractions and concentrate again (using a fresh centrifugal filter) to a volume between 0.5 and 1 ml.
8. Determine the exact protein concentration of both "pentamer" and "tetramer" samples and determine the molar concentration of both complexes (*see* **Note 13**).

3.7 Reconstitution of the Nonameric IFT-B Core Complex from Purified Pentameric and Tetrameric Sub-complexes

1. Mix equimolar amounts of both pentameric and tetrameric sub-complexes and incubate at 4 °C for at least 30 min, making sure that the final volume does not exceed 2 ml (*see* **Note 14**).
2. Equilibrate a size-exclusion column with 1 cV of SEC Buffer (*see* **Note 15**).
3. Inject the sample using a loop of an appropriate size and start collecting fractions of 1 ml.
4. Check the proteins in the peak fractions by SDS-PAGE (*see* Fig. 2i).
5. Concentrate the peak fractions to a desired final concentration for downstream applications.
6. Aliquot the sample according to future needs, snap-freeze in liquid nitrogen, and store at −80 °C (*see* **Note 16**).

4 Notes

1. Whereas transformation of single plasmids is easy, co-transformation of multiple plasmids (especially of 4 plasmids as in the case of IFT81/74/27/25) can be very inefficient and depends on the availability of highly competent cells. We recommend preparing glycerol stocks of successfully transformed bacteria to make future purifications easier.
2. Do NOT use baffled flasks, especially for the culture-expressing IFT88. Even though the increased aeration in such flasks helps to increase the cell yield, we find that it severely affects IFT88 expression levels.
3. The cultures will most likely not grow at the same rate due to the presence of different numbers of antibiotics. If some flasks need longer to reach the desired OD_{600}, it will be necessary to move them to an extra shaker at 37 °C until they are ready to be cooled down.
4. The top layer of the pellet can be quite soft. Try to avoid contaminating your extract with this pellet. It is better to sacrifice some extract than to compromise the purity of the protein.

5. Faster flow rates can be used if time is limiting, but a slower flow rate usually results in higher protein yields.
6. For the pentamer purification the early fractions will mainly contain excess IFT27/25 complex, whereas the IFT81/74/27/25/22 complex elutes later. Taking along this excess IFT27/25 is not a problem, as it will be separated from the pentamer at a later stage in the purification.
7. Tag-cleavage for one or both complexes can be omitted if the His-tags are desired on the final protein complex. In this case just proceed with dialysis without TEV addition. The reverse Ni^{2+}-NTA step also has to be skipped if the His-tags are retained. Even if the tags are cleaved the reverse Ni^{2+}-NTA step can be skipped if the washes were performed properly and the purity of the protein complexes is high enough (*see* Fig. 2 for example).
8. TEV protease is not a limiting factor in our purifications. If the cost of this enzyme is an issue, lower amounts can be used but the sample should be checked for complete cleavage (this can be easiest monitored for the small IFT25 protein) before continuing. Addition of more enzyme and increased incubation time might be necessary to obtain complete tag-removal.
9. If the reverse Ni^{2+}-NTA step was not carried out (e.g. because His-tags were retained), proceed here with the dialyzed sample.
10. Two peaks are expected for the "pentamer" sample (at ~12% B and at ~18% B), whereas several peaks with increasing height should be obtained for the "tetramer" sample (the last—and most important—one eluting at ~28% B).
11. For the "pentamer" purification, the flowthrough of the ion-exchange step will contain excess IFT27/25 complex, the amount of which will depend on how much of the early elution fractions of the Ni^{2+}-NTA step were included (*see* Fig. 2a). If purification of this dimeric complex is desirable, the fractions can be pooled and concentrated for further use.
12. For both purifications ("pentamer" and "tetramer"), the proteins should elute in peaks around 55 ml (if a HiLoad Superdex200 column is used) and will almost certainly have "shoulders" toward larger elution volumes. In case of the "pentamer", this is most likely due to different conformational states, and the extent of this "shoulder" varies between purifications. The reason for these different states is currently under investigation. For the "tetramer" purification, the "shoulder" contains some excess IFT70/52/46 complex. In both cases, we recommend to only pool the fractions of the main peak (indicated with a black bar underneath the SDS-gel picture in Fig. 2d, h).

13. The exact determination of the protein concentration and molarity is critical for obtaining a stoichiometric nonameric IFT-B core complex in the final step. When using a Nanodrop ND-1000 Spectrophotometer (or a similar device which measures the absorbance at 280 nm) be aware that the obtained values for the "pentamer" and the "tetramer" sample have to be divided by factors of 0.6 and 0.87, respectively, due to the different extinction coefficients of these complexes.
14. The final volume of the mixed sample will be determined by the final volume of concentration in Subheading 3.6, **step 6**.
15. If a different buffer to our SEC buffer is desirable for downstream applications of the purified nonameric IFT-B core complex, it can be tested here. In our hands, a pH range of 6.0–8.0 and a NaCl concentration of 50–2000 mM do not affect stability of the resulting complex. Glycerol can also be removed.
16. To maintain the reconstituted complex in a good condition, we highly recommend preparing aliquots for single use. Even though in our hands the complex is stable and does not show a significant tendency to aggregate when frozen and thawed once, repeated cycles of freezing and thawing might compromise its integrity.

References

1. Ishikawa H, Marshall WF (2011) Ciliogenesis: building the cell's antenna. Nat Rev Mol Cell Biol 12:222–234
2. Rosenbaum JL, Witman GB (2002) Intraflagellar transport. Nat Rev Mol Cell Biol 3:813–825
3. Kozminski KG, Johnson KA, Forscher P, Rosenbaum JL (1993) A motility in the eukaryotic flagellum unrelated to flagellar beating. Proc Natl Acad Sci U S A 90:5519–5523
4. Cole DG, Cande WZ, Baskin RJ, Skoufias DA, Hogan CJ, Scholey JM (1992) Isolation of a sea urchin egg kinesin-related protein using peptide antibodies. J Cell Sci 101(Pt 2): 291–301
5. Walther Z, Vashishtha M, Hall JL (1994) The Chlamydomonas FLA10 gene encodes a novel kinesin-homologous protein. J Cell Biol 126:175–188
6. Pazour GJ, Wilkerson CG, Witman GB (1998) A dynein light chain is essential for the retrograde particle movement of intraflagellar transport (IFT). J Cell Biol 141:979–992
7. Porter ME, Bower R, Knott JA, Byrd P, Dentler W (1999) Cytoplasmic dynein heavy chain 1b is required for flagellar assembly in Chlamydomonas. Mol Biol Cell 10:693–712
8. Perrone CA, Tritschler D, Taulman P, Bower R, Yoder BK, Porter ME (2003) A novel dynein light intermediate chain colocalizes with the retrograde motor for intraflagellar transport at sites of axoneme assembly in chlamydomonas and Mammalian cells. Mol Biol Cell 14:2041–2056
9. Hou Y, Pazour GJ, Witman GB (2004) A dynein light intermediate chain, D1bLIC, is required for retrograde intraflagellar transport. Mol Biol Cell 15:4382–4394
10. Pazour GJ, Agrin N, Leszyk J, Witman GB (2005) Proteomic analysis of a eukaryotic cilium. J Cell Biol 170:103–113
11. Piperno G, Mead K (1997) Transport of a novel complex in the cytoplasmic matrix of Chlamydomonas flagella. Proc Natl Acad Sci U S A 94:4457–4462
12. Cole DG, Diener DR, Himelblau AL, Beech PL, Fuster JC, Rosenbaum JL (1998) Chlamydomonas kinesin-II-dependent intraflagellar transport (IFT): IFT particles contain proteins required for ciliary assembly in Caenorhabditis elegans sensory neurons. J Cell Biol 141: 993–1008
13. Taschner M, Bhogaraju S, Lorentzen E (2012) Architecture and function of IFT complex

proteins in ciliogenesis. Differentiation 83: S12–S22

14. Lucker BF, Behal RH, Qin H, Siron LC, Taggart WD, Rosenbaum JL, Cole DG (2005) Characterization of the intraflagellar transport complex B core: direct interaction of the IFT81 and IFT74/72 subunits. J Biol Chem 280:27688–27696
15. Mukhopadhyay S, Wen X, Chih B, Nelson CD, Lane WS, Scales SJ, Jackson PK (2010) TULP3 bridges the IFT-A complex and membrane phosphoinositides to promote trafficking of G protein-coupled receptors into primary cilia. Genes Dev 24:2180–2193
16. Baker SA, Freeman K, Luby-Phelps K, Pazour GJ, Besharse JC (2003) IFT20 links kinesin II with a mammalian intraflagellar transport complex that is conserved in motile flagella and sensory cilia. J Biol Chem 278:34211–34218
17. Wang Z, Fan Z-C, Williamson SM, Qin H (2009) Intraflagellar transport (IFT) protein IFT25 is a phosphoprotein component of IFT complex B and physically interacts with IFT27 in Chlamydomonas. PLoS One 4, e5384
18. Fan Z-C, Behal RH, Geimer S, Wang Z, Williamson SM, Zhang H, Cole DG, Qin H (2010) Chlamydomonas IFT70/CrDYF-1 is a core component of IFT particle complex B and is required for flagellar assembly. Mol Biol Cell 21:2696–2706
19. Follit JA, Xu F, Keady BT, Pazour GJ (2009) Characterization of mouse IFT complex B. Cell Motil Cytoskeleton 66:457–468
20. Omori Y, Zhao C, Saras A, Mukhopadhyay S, Kim W, Furukawa T, Sengupta P, Veraksa A, Malicki J (2008) Elipsa is an early determinant of ciliogenesis that links the IFT particle to membrane-associated small GTPase Rab8. Nat Cell Biol 10:437–444
21. Taschner M, Bhogaraju S, Vetter M, Morawetz M, Lorentzen E (2011) Biochemical mapping of interactions within the intraflagellar transport (IFT) B core complex: IFT52 binds directly to four other IFT-B subunits. J Biol Chem 286:26344–26352
22. Taschner M, Kotsis F, Braeuer P, Kuehn EW, Lorentzen E (2014) Crystal structures of IFT70/52 and IFT52/46 provide insight into intraflagellar transport B core complex assembly. J Cell Biol 207:269–282
23. Bhogaraju S, Taschner M, Morawetz M, Basquin C, Lorentzen E (2011) Crystal structure of the intraflagellar transport complex 25/27. EMBO J 30:1907–1918
24. Bhogaraju S, Cajanek L, Fort C, Blisnick T, Weber K, Taschner M, Mizuno N, Lamla S, Bastin P, Nigg EA, Lorentzen E (2013) Molecular basis of tubulin transport within the cilium by IFT74 and IFT81. Science 341: 1009–1012

Chapter 6

Methods for Studying Movement of Molecules Within Cilia

Karl F. Lechtreck

Abstract

The assembly of cilia and eukaryotic flagella (interchangeable terms) requires the import of numerous proteins from the cell body into the growing organelle. Proteins move into and inside cilia by diffusion and by motor-based intraflagellar transport (IFT). Many aspects of ciliary protein transport such as the distribution of unloading sites and the frequency of transport can be analyzed using direct in vivo imaging of fluorescently tagged proteins. Here, we will describe how to use total internal reflection fluorescence microcopy (TIRFM) to analyze protein transport in the flagella of the unicellular alga *Chlamydomonas reinhardtii*, a widely used model for cilia and cilia-related disease.

Key words Cilia, Flagella, Green fluorescent protein, Intraflagellar transport (IFT), Total internal reflection fluorescence microscopy (TIRFM)

1 Introduction

Fluorescence-based microscopy has revolutionized the analysis of proteins in living cells and organisms. Genetically encoded fluorescence proteins (FPs) such as green fluorescence protein (GFP) can be fused to virtually any protein permitting one to track its localization and changes thereof in vivo. However, often noise in the form of autofluorescence or out-of-focus signal from other regions of the cell will reduce the image quality and prevent the imaging of single proteins. These hurdles can be overcome by techniques that will excite fluorophores only within a limited focal plane or by imaging thin portions of the cells such as cell projections. Total internal reflection fluorescence (TIRF) imaging of cilia, thread-like cell extensions of several μm length, combines both of these features: Depending on the illumination angle, TIRF has a reach of 30–300 nm from the surface of the cover glass and thus is a useful tool to image objects close to the cover glass. Cilia and flagella have a diameter of approximately 200 nm, and, when cilia are appressed to cover glass, fluorophores in the entire ciliary volume can be excited and imaged by TIRFM with a high signal-to-noise ratio.

Peter Satir and Søren Tvorup Christensen (eds.), *Cilia: Methods and Protocols*, Methods in Molecular Biology, vol. 1454, DOI 10.1007/978-1-4939-3789-9_6, © Springer Science+Business Media New York 2016

When optimally positioned, the cell bodies will remain out of range of the TIRF excitation depth avoiding autofluorescence, out-of-plane signal, phototoxicity, and photobleaching. TIRF imaging has been applied to study protein transport in cilia of various systems including *Tetrahymena thermophila* and mammalian cells [1, 2]. The flagella of *Chlamydomonas reinhardtii* are excellently suited to study protein transport by TIRFM because the cells naturally adhere via their two 12-µm long flagella to smooth surfaces such as cover glass [3]. This which will immobilize the cells and the flagella will be positioned within the TIRF excitation range while the cell bodies remain out of reach. Further, numerous *C. reinhardtii* mutants for IFT and other ciliary proteins are available; this collection includes conditional mutants in which IFT can be switched-off by a temperature shift [4]. Mutants can be rescued with FP-tagged versions of the affected genes effectively replacing the endogenous protein with a FP-tagged version [5]. Moreover, flagellar assembly can be easily experimentally induced in *C. reinhardtii* allowing for an analysis of flagellar growth, genetically distinct cells can be fused by mating to study flagellar repair and turnover, and flagella can be isolated in bulk to determine their biochemical composition and changes thereof as they are caused by defects in ciliary protein transport in an unbiased manner. Here, we will provide step-by-step instructions on how to image protein transport in the flagella of *C. reinhardtii* using TIRF microscopy.

2 Materials

2.1 Preparing Strains Expressing Fluorescent Protein-Tagged Ciliary Proteins

1. PCR machine, DNA polymerase with proof reading activity, restriction enzymes, agarose electrophoresis, cell electroporator.
2. Templates for fluorescent proteins (Table 1 and http://chlamycollection.org/).
3. M-medium (http://chlamycollection.org/m-or-minimal-media/); TAP-40mM sucrose (sterile filtered; http://www.chlamy.org/TAP.html).
4. Culture plates with selective media, 96- and 24-well plates, toothpicks.
5. Microscope, Western blotting equipment, rabbit polyclonal antibody to GFP (A-11122 Life Technologies), antibodies specific for the target protein.

2.2 Sample Preparation

1. 20 × 60 mm No. 1.5 and 22 × 22 mm cover glass.
2. Syringe with petroleum jelly or silicone vacuum grease.
3. Immobilization buffer (10 mM Hepes, 5 mM EGTA, pH 7.2).

Table 1
Fluorescence proteins successfully expressed in the cytoplasm of *C. reinhardtii*

Color	Name	Max. excitation/ emission (nm)	Brightness (% of EGFP)	Maturation (min)	Stability	Monomer (m)/ dimer (*d*)	Codon usage	Reference
Green	CrEGFP	488/507	100	25	174	m	Cr	[6]
Green	sfGFP	485/510	161	~6	157	m	Hs	[7]
Green	mNeonGreen	506/517	276	<10	158	m	Cr	[7]
Red	mCherry	587/610	47	15-40	96	m	Cr, Hs	[8, 9]
Orange-red	tdTomato	554/581	283	~60	?	d	Cr	[8]
Yellow	Venus	515/528	156	40	15	m	Cr	[8]
Blue-green	mCerulean	433/475	79	?	36	m	Cr	[8]
Blue	mTagBFP	402/457	98	50	34	m	Cr	[8]

Cr, *C. reinhardtii*, Hs, *Homo sapiens*. Stability: time in s to photobleach from 1000 to 500 photons per s per molecule under arc-lamp illumination. Maturation: Time in minutes for fluorescence to obtain half-maximal value after exposure to oxygen (*see* **Note 1**). *Sources*: http://www.my-whiteboard.com/properties-of-bright-and-photostable-monomeric-fluorescent-proteins/ ; http://nic.ucsf.edu/FPvisualization/; http://zeiss-campus.magnet.fsu.edu/articles/probes/anthozoafps.html; http://nic.ucsf.edu/dokuwiki/doku.php?id=fluorescent_proteins; http://www.weizmann.ac.il/plants/Milo/images/flurophoreMaturation140619Clean.pdf

2.3 In Vivo Microscopy

1. TIRF microscope with a 60× or 100× NA 1.49 TIRF objective (NA should be at least 1.45).
2. Lasers of the desired wavelengths, e.g., 488 nm for GFP, sfGFP, and mNeonGreen and 565 nm for mCherry, tdTomato, etc. Corresponding excitation and emission filters.
3. Single-photon EMCCD camera, e.g., Andor iXon 897 or similar.
4. Imaging software, e.g., μManager (https://micro-manager.org/wiki/Download_Micro-Manager_Latest_Release; this site will soon move).

2.4 Analyzing Protein Transport During Flagella Regeneration and Repair

1. Flagella amputation by pH-shock and regeneration: Stirrer, pH meter, 500 mM acetic acid, 250 mM KOH, fresh M-medium, centrifuge, shaker.
2. Long-zero cells (cells with only one flagellum): 1-ml syringe, 26G½ needle.
3. Mating: Strains of opposite mating type (+ and −).
4. M-N medium (http://chlamycollection.org/m-or-minimal-media/), 20% M-N medium supplemented with 10 mM Hepes pH 7.2, 150 mM dibutyryl(db)-cAMP (Sigma) in water, centrifuge, shaker.

2.5 Analysis of Imaging Data

1. ImageJ (http://rsbweb.nih.gov/ij/download.html) or Fiji (http://fiji.sc/Fiji). For ImageJ, you will need to download the Multiple Kymograph plugin (http://www.embl.de/eamnet/html/body_kymograph.html); the LOCI Bioformats plugin (http://loci.wisc.edu/software/bio-formats) will allow you to open files from various imaging softwares in ImageJ.
2. Microsoft Office Excel.

3 Methods

3.1 Construction and Transformation of Expression Vectors for FP Tagging of Flagellar Proteins

C. reinhardtii allows stable nuclear transformation of endogenous and heterogeneous genes [10]. After PCR amplification using either total genomic DNA, total cDNA, or plasmid DNA as a template, the gene of interest (GOI) is fused to GFP or other FPs many of which have been codon-adapted and positively tested in *C. reinhardtii* (Table 1, *see* **Note 1**). Complete synthesis of genes is an affordable alternative, in particular for small genes. In addition to N- or C-terminal fusions, FPs have been also inserted into genes and successfully expressed (Fig. 1). In general, the presence of the GOI and the selectable marker gene on the same plasmid will increase the number of positive transformants. For best results, the construct should be expressed in a mutant null for the corresponding endogenous gene. The FP-tagged GOI should rescue

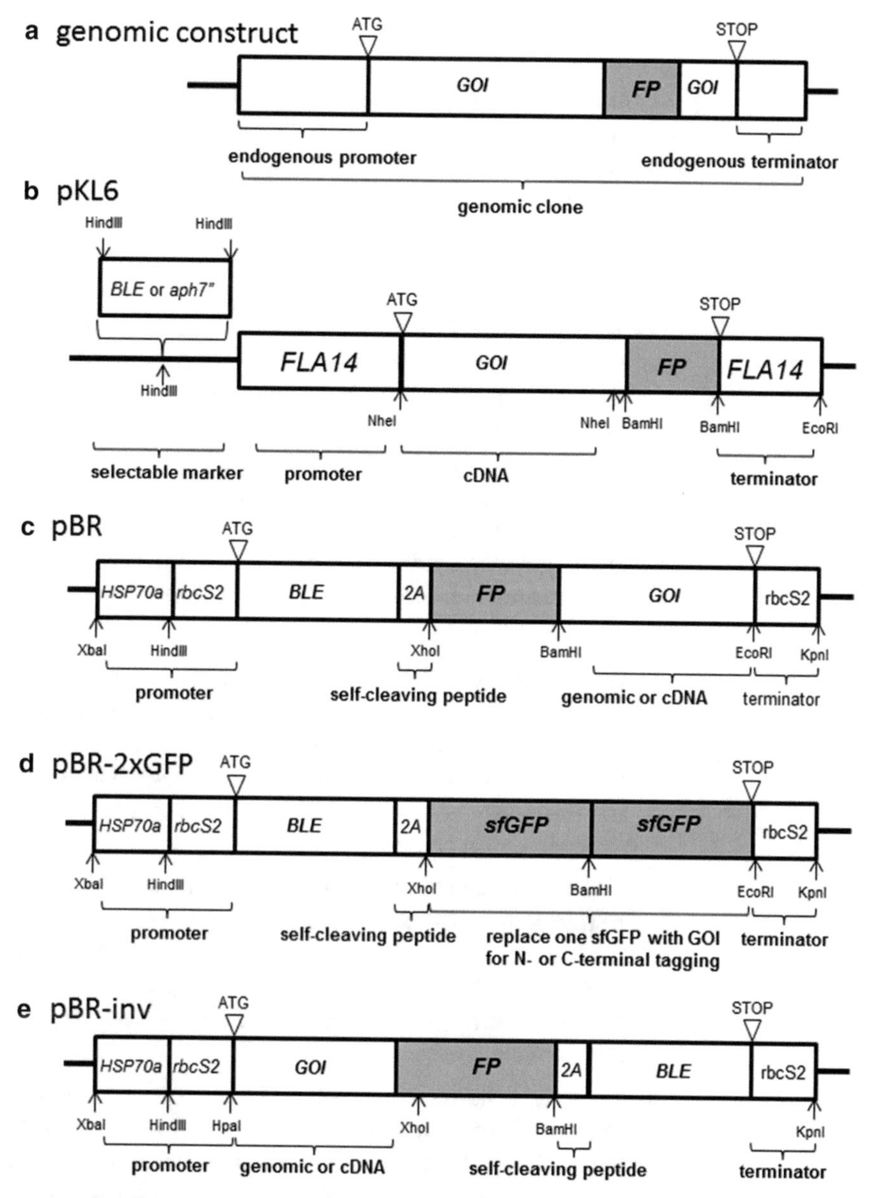

Fig. 1 *Chlamydomonas* expression vectors for FP tagging. Schematic presentation of various vector designs used to express ciliary proteins fused to fluorescent proteins (FP) in *C. reinhardtii*. (**a**) FP gene inserted in the protein coding region of the gene of interest (GOI); this technique was used, e.g., to express DRC4-GFP [11]. (**b**) Expression vector for cDNAs based on the intron-free *FLA14* gene. The GOI can be inserted into a unique NheI site downstream of the ATG of the *FLA14* gene and upstream of the gene encoding the FP. The vector has been used to express IFT20-mCherry and BBS4-GFP and rescue the corresponding mutants [9]. This vector will add the sequence MASLA to the N-terminus of the expressed protein. (**c**) Bicistronic pBR expression vector using the self-cleaving 2A sequence to express two proteins from one mRNA. The vector will add the residue P to the N-terminus of the protein following the 2A sequence. Also, cleavage fails for a significant portion of the encoded protein resulting in a fusion protein enlarged by an additional ~25 kDa. The vector has been successfully used to express alpha-tubulin fused to various FPs [7, 8]. (**d**) pBR-2xsfGFP vector for convenient expression of proteins fused either N- or C-terminally to sfGFP. The vector has been successfully used to express mNeon, mCherry, and sfGFP-tagged radial spoke proteins in *Chlamydomonas*. (**e**) Inverted pBR vector allowing for the expression of proteins with their original N-terminus and C-terminal fusion to FP (*see* **Note 16**). The vector has been successfully used to express IFT20 and phospholipase D with a C-terminal mNeonGreen tag

the mutant phenotype verifying the functionality of the fusion protein; it is then reasonable to assume that the fusion protein mirrors the distribution and behavior of the endogenous protein. Expression in wild-type strains, however, can be also achieved in particular using the recently introduced polycistronic vectors, in which transgene expression is linked to antibiotic resistance using a viral 2A sequence encoding a self-cleaving peptide [8]. The presence of both the endogenous wild-type and FP-tagged transgenic protein warrants special caution when interpreting data.

Various strategies have been successfully used to express IFT and other ciliary proteins as FP-fusion proteins in *C. reinhardtii*; some of the vectors are depicted in Fig. 1.

If possible, linearize the expression vector or cut out the cassette containing the selectable marker and GOI. Purify the DNA by agarose gel electrophoresis and gel elution. Verify fragment size, DNA concentration, and purity by electrophoresis of an aliquot of the eluate (*see* **Note 2**).

1. Wash cells in TAP-40mM sucrose and concentrate cells to ~1×10^8 cells/ml. Pipette 80 μl of cells into pre-chilled electroporation cuvettes and add ~100–800 ng of DNA. Alternatively, premix cells and DNA before transferring the mix into the cuvettes. Always prepare a non-DNA control for each strain to be transformed.
2. Transform cells by electroporation. We use a BTX Electro Cell Manipulator 600 with the following settings: Capacitance & Resistance: 500 V, Resistance Timing: R1 (12Ω), Capacitance Timing: 1000 μF, Set Charging Voltage: 200 V. Adjust volume and settings for other cuvette sizes and electroporators, respectively (*see* **Note 3**).
3. Transfer cells to 10 ml of TAP + sucrose and incubate overnight in light with agitation. Concentrate cells to ~500 μl, mix with ~2 ml of liquid TAP-0.5 % agarose at ~40 °C, and pour mix onto ambient-temperature TAP plates with selective medium. Follow a protocol describing the plating of phages with top agarose. Incubate plates in constant light until colonies are clearly visible (*see* **Note 4**).
4. Using sterile toothpicks, transfer colonies into ~150 μl of M-medium in a 96-well plate. If applicable, check for the rescue of flagellar growth, motility, or phototaxis phenotypes by light microscopy; this should be done ~2 h after picking the colonies and then repeatedly over the next couple of days to ensure consistency of the phenotype (*see* **Note 5**).
5. Screen transformants for expression of the GOI-FP fusion using TIRF microcopy of living cells and/or Western blotting and/or immunofluorescence with anti-GFP or a protein-specific antibody (*see* **Note 6**).

3.2 Sample Preparation

C. reinhardtii swims by means of two motile 9+2 flagella. The observation of protein transport and IFT in cilia requires immobilization of the cells. This can be achieved by gluing the cells to the cover glass using, e.g., poly-L-lysine, by using immotile *paralyzed flagella* (*pf*) mutants strains, or by agarose embedding of the cells. Best results, however, are obtained by allowing the cells to naturally adhere to the untreated cover glass via their two flagella. Gliding motility [12] can be suppressed by removing calcium from the environment using EGTA (*see* **Note 7**).

1. Wash cells into fresh M-medium (*see* **Note 8**).
2. Apply a ring of vacuum grease or petroleum jelly with a diameter of ~15 mm onto a clean 24×60 mm cover glass; this serves as a spacer to avoid compression of cells and to seal the specimen (Fig. 2).
3. Place ~5–11 μl of cell suspension onto the cover glass and allow cells to adhere for 1–5 min depending on cell density (*see* **Note 9**).
4. Apply an equal volume of immobilization buffer (~5–11 μl) onto a 22×22 mm cover glass; use a similar pattern of drops as generated in **step 2**. Lower the 22×22 mm coverslip with the drops hanging down onto the large coverslip to prepare the observation chamber.
5. Close the chamber immediately before imaging of the cells. We use slides for up to 30 min.

3.3 In Vivo Imaging

Imaging proteins inside flagella has to overcome two problems: (1) signals are typically relatively weak because many ciliary proteins are transported in small clusters or as single proteins, and (2) the *C. reinhardtii* cell body is very autofluorescent due to its large chlorophyll-containing plastid. The latter can be surmounted by selective excitation of just the flagella [13]; the former will need a light-sensitive camera (*see* **Note 10**). TIRF microscopes are optimal for imaging proteins in flagella appressed to cover glass. TIRF illumination typically excites fluorophores in a ~30–300 nm region which is well below the optical section thickness of most confocal fluorescence microscopes; signal intensity decreases exponentially with distance from the cover glass effectively eliminating fluorescence from the cell body. We use a home-build through-the-objective TIRF system, which has been previously described in detail [3]; commercially available TIRF systems will work just as well.

1. Place the chamber onto the microscope with the large 24×60 mm coverslip facing the objective; center the spot of adhered cells into the middle of the objective.

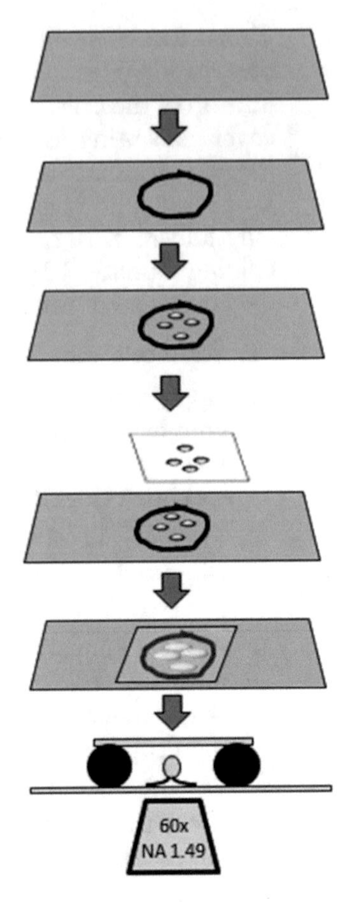

Fig. 2 Assembly of observation chambers for *in vivo* imaging of protein transport in *C. reinhardtii* flagella. See text for details

2. Using the microscope's standard light source focus onto the plane in which the flagella are attached to the cover glass. Ensure that cells attached to the larger bottom cover glass are in focus.
3. Turn-off standard light and switch to TIRF illumination. FP-tagged proteins in the flagella should be visible (*see* **Note 11**).
4. Frame rates of 5–10 fps are sufficient to document protein transport by IFT; higher frame rates (20 fps) are recommended to analyze the diffusional behavior of proteins inside cilia.

3.4 Imaging of Protein Transport During Flagellar Growth and Repair

The transport frequency of tubulin and other axonemal proteins is upregulated during flagellar growth making it easier to observe transport events [7, 14]. Flagella can be easily removed from cells by a pH shock and cells will largely synchronously regenerate flagella within ~80 min. Flagella can be also amputated by mechanical shear. Mating of a cell expressing a FP-tagged ciliary protein to a mutant strain defective in the corresponding endogenous gene

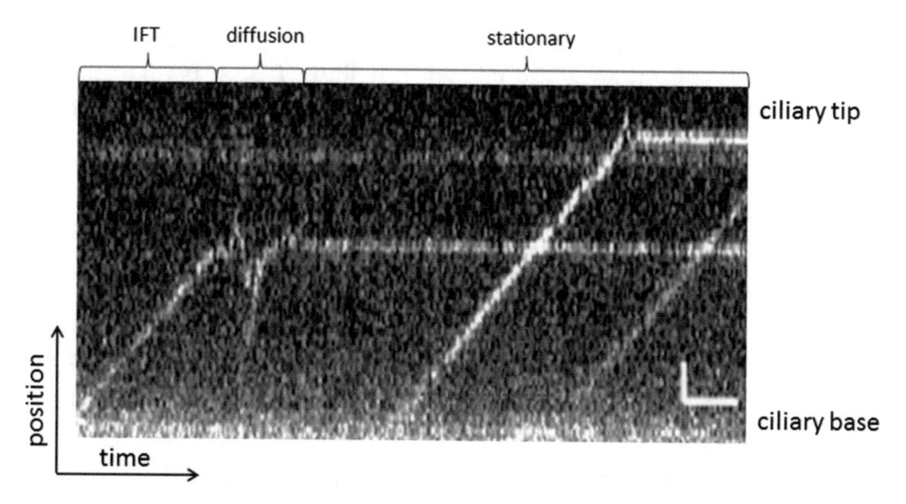

Fig. 3 Determining the velocity of IFT transport based on kymograms. The kymogram depicts varies phases of transport of the axonemal protein DRC4-GFP in an initially DRC4-deficient flagellum of a *pf2* × *pf2* DRC4-GFP zygote. Transport by IFT, unloading, diffusion, and docking (incorporation into the axoneme) can be observed. See text on how to calculate the velocity of particles from the angle θ of the IFT-like trajectories. Reprinted in modified form from Wren et al. [14]

allows for the analysis of protein transport during flagellar repair (*see* **Note 12**). Because the mutant-derived flagella of such zygotic fusion cells initially lack FP-tagged proteins, protein transport in flagella can be imaged with remarkable clarity (Fig. 3).

3.4.1 Flagellar Amputation by pH Shock

1. Place about 60 ml of cell culture into a beaker of appropriate size (~100 ml total volume), add a stir bar, and stir rapidly. Monitor pH using an electrode (*see* **Note 13**).
2. Using a transfer pipet, rapidly add ~1.5 ml of acetic acid (0.5 M) until pH reaches ~4.3. Monitor pH while adding the acetic acid. Do not overshoot 4.0. Wait for 45 s.
3. Add KOH (0.25 M) until pH reaches 6.5–7. Do not overshoot pH 7, the pH will initially increase slowly, but will rise rapidly once the pH is above 6.
4. Immediately place cells in fresh M-medium (by centrifugation at 900 × *g* for 3 min at RT) and store on ice.
5. Verify deflagellation by comparing cells before and after pH shock by light microscopy. For most strains more than 98 % of the cells should have shed their flagella.
6. Allow an aliquot of cells to regenerate flagella in bright light at room temperature with agitation. After ~20 min, check cells for flagellar regeneration and image cells (*see* **Note 14**).

3.4.2 Generation of Long-Zero Cells

1. Forcefully pass the cells for 3–8 times through the needle. Check for spinning cells using a microscope at low magnification. Spinning cells are likely to have lost one of their two flagella.

2. Start imaging after ~10 min. Long-zero cells will rapidly regrow the missing flagellum. Use a standard light source to locate long-short cells which account for ~1 % of the total cells.

3.4.3 Generation of Ciliary Chimeras by Mating

1. Spread cells on M-plate and incubate for 10 days in the culture room with a ~14:10 h light:dark cycle at ~22 °C.
2. After 10 days, place the plates for 1 or 2 days in dim light in the culture room.
3. In the evening before the experiment, transfer cells from the plates into 10-ml of M-N medium in a small Erlenmeyer.
4. Incubate cells overnight in bright light with constant shaking at RT or somewhat warmer.
5. In the morning, mix small aliquots of the cells to check for mating reaction by light microcopy. Well mating cells will form clumps.
6. If the mating is insufficient, place cells in ~5 ml 20 % M-N plus 10 mM Hepes and continue to shake in bright light. Mating ability should peak after ~4 h. Many mutant cells fail to form cilia of proper length when agitated on a shaker. Place such strains in bright light without agitation.
7. Once cell clumping is observed, mix an appropriate volume of the two mating strains. Start imaging after ~5 min and search for quadriflagellate zygotes (*see* **Note 15**). Zygote formation can be triggered by adding db-cAMP (15 mM final concentration).

3.5 Analysis of Transport Velocity

The movement of proteins inside cilia can be described using various parameters such as mode of movement (diffusion or IFT), velocity, frequency, processivity (how long a cargo remains attached to IFT), etc. Unless one uses a trained particle tracking software, kymograms are the best way to determine these features. Kymograms are time–distance plots displaying the protein dynamics within cilia of an entire movie in a single image (Fig. 3).

1. Open movie file in ImageJ or Fiji.
2. Select the straight or segmented line tool by clicking on the icon.
3. Place a line running from the cell body to the tip of the cilium to be analyzed. Adjust line thickness by double clicking the line icon to adjust line width so that the line covers the entire width of the cilium.
4. Select the “Multiple Kymograph” tool under “Analysis” (Fiji) or “Plugins” (ImageJ). Choose “Linewidth” equals 1 and hit Okay.
5. Rotate the kymogram 90° counterclockwise. IFT-like trajectories should be oriented as shown in Fig. 3. Proteins moving anterogradely from the ciliary base to the tip should result in

trajectories running from the bottom left to the top right, and vice versa for retrograde transport.

6. Switch to the Angle tool by clicking on the icon, use the tool to trace a trajectory on the kymogram from the bottom right to top left and measure the angle θ by pressing "CRTL M".
7. Copy the data to Excel. Use the following equation to determine the velocity: Velocity (μm/s) = TAN(π/180 × θ) × fps/(pixels/μm); fps = frames per second.

4 Notes

1. Among the various green FPs, mNeonGreen [15] is currently the best choice: it folds quickly and is considerably brighter (2.6×) than EGFP (Table 1). The maturation time of a FP can be important, e.g., when protein transport during ciliary regeneration is studied: Freshly translated proteins might be transported into cilia before the FP is able to exhibit fluorescence. The red FP mCherry, while significantly dimmer and less stable than GFP, is a good compromise between maturation time and brightness. Also, note that the folding/maturation of many FPs is accelerated at higher temperatures.
2. Using linearized vectors or purified fragments will increase the transformation rate (colonies/μg of DNA). Instead of gel-purified DNA, alcohol-precipitated DNA works as well. The presence of enzymes will affect the transformation rate of electroporation, but not that of the glass-bead method [10].
3. In contrast to glass bead transformation [10], which mostly results in high co-transformation rates (~90%) even when the selectable marker gene and GOI are on separate plasmids, transformation by electroporation performs better when both genes are on the same plasmid. For co-transformation, a ratio of at least 1:5 between the selectable marker plasmid and the GOI plasmid should be used. Because smaller plasmids enter cells easier, the ratio should be increased when the GOI plasmid exceeds the selectable marker plasmid in size. Too much DNA will decrease transformation rates for electroporation and will increase the share of transformants with more than one GOI plasmid integrated into the genome; this can lead, e.g., to rapid silencing of the construct. Also, be aware that DNA inserts randomly into the *Chlamydomonas* genome and thus can cause new mutations. Compare several positive transformants for consistency of phenotype.
4. Transformants should appear after 6–8 days for aph7″/hygromycin and after 8–11 days for BLE/zeocin. Generally, transformants should be absent from the no-DNA control plates

but some flagella-less mutants display an increased resistance against antibiotics.

5. Rescue rates vary greatly from >60 to <1%. pBR vectors allow for expression of tagged genes in wild-type cells; expression of the GOI is typically observed in >50% of the transformants [8]. Due to the linkage of antibiotic resistance to GOI expression, transformants expressing the GOI often grow faster after transfer to the 96-well plates.
6. The FP-fusion protein should be ~27 kDa larger than the target protein alone; expression from pBR vectors can lead to incomplete cleavage of the product resulting in a shBLE-FP-target protein fusion that is ~52 kDa larger than the unmodified protein. Use a strain with verified expression of a FP-fusion protein as a control for Western blotting and especially to adjust the TIRF microscope.
7. It remains to be determined to which extent the removal of external calcium affects protein transport in cilia. We compared the velocity of IFT and the transport frequency of the axonemal protein DRC4 in samples prepared with and without the EGTA-containing immobilization buffer without noting a difference. Transport of the transmembrane protein FMG-1, however, appears to require external calcium [12].
8. This is optional for cells in mid-log growth phase but required for dense cultures because the cells from such cultures tend to not adhere properly to cover glass presumably due to the accumulation of secreted material in the culture medium.
9. For low-density cultures (~10^6 cells/ml) apply one drop into the middle of the ring; for dense suspensions apply several small drops amounting to ~5–11 μl; this will make it easier to find a region in which the cells are nicely spaced (Fig. 2). High cell densities negatively affect cell immobilization.
10. FP-tagged IFT proteins are easy to visualize by various microscopic techniques including standard epifluorescence. Note that cargoes are typically much less abundant in an IFT train than the IFT proteins: a train is composed of dozens of IFT particles, only a few of which might carry the cargo of interest.
11. Laser strength should be as low as possible to avoid photobleaching. Make sure to adjust TIRF illumination to minimize stray light reaching the cell body: the autofluorescence of the cell body should be hardly visible while tagged proteins in the flagella are bright. This might require adjustment of the TIRF angle and centering of the TIRF spot. Often, TIRF illumination is only realized for a portion of the field of view.
12. Many motility mutants of *C. reinhardtii* can be rescued by dikaryon complementation. After fusion of the mutant and the wild-type cell, wild-type protein present in the now shared

cytoplasm will enter the mutant-derived flagella. Repair typically occurs within 20–60 min. However, several mutant phenotypes cannot be rescued in dikaryons or only after the old mutant flagella have been amputated. These include *pf18* and *sup-pf3* [11].

13. Older cultures often have lowered the pH of the medium to ~6 and are best washed into fresh M-medium prior to deflagellation.
14. TIRF imaging requires cells to adhere which works better once flagella exceed the length of 2–3 μm. Cells can be kept on ice for hours without impairing their ability to synchronously regenerate flagella; aliquots can be removed at various time points and incubated to room temperature to initiate flagella formation. Cell wall-deficient strains do not properly regenerate flagella; flagellar regeneration is also slower in many mutants as well as in some strains rescued with FP-tagged IFT proteins.
15. Mating efficiency can vary greatly between strains and experiments (0–85% zygotes); some cell fusion can occur within 1 min after mixing of the gametes but many strains require more time to form zygotes.
16. An unaltered N-terminus can be critical for protein function. Myristoylation, for example, occurs co-translationally on the second glycine and requires removal of the initial methionine by a special peptidase. This reaction is unlikely to work when residues are added to the N-terminus. Myristoylated proteins are enriched in cilia [16].

Acknowledgement

This work was supported by a grant from the National Institutes of Health (GM110413).

References

1. Jiang YY, Lechtreck K, Gaertig J (2015) Total internal reflection fluorescence microscopy of intraflagellar transport in Tetrahymena thermophila. Methods Cell Biol 127:445–456. doi:10.1016/bs.mcb.2015.01.001
2. Ishikawa H, Marshall WF (2015) Efficient live fluorescence imaging of intraflagellar transport in mammalian primary cilia. Methods Cell Biol 127:189–201. doi:10.1016/bs.mcb.2015.01.002
3. Lechtreck KF (2013) In vivo imaging of IFT in Chlamydomonas flagella. Methods Enzymol 524:265–284. doi:10.1016/B978-0-12-397945-2.00015-9
4. Kozminski KG, Beech PL, Rosenbaum JL (1995) The Chlamydomonas kinesin-like protein FLA10 is involved in motility associated with the flagellar membrane. J Cell Biol 131:1517–1527
5. Mueller J, Perrone CA, Bower R, Cole DG, Porter ME (2005) The FLA3 KAP subunit is required for localization of kinesin-2 to the site of flagellar assembly and processive anterograde intraflagellar transport. Mol Biol Cell 16:1341–1354. doi:10.1091/mbc.E04-10-0931
6. Fuhrmann M, Oertel W, Hegemann P (1999) A synthetic gene coding for the green fluorescent protein (GFP) is a versatile reporter in

Chlamydomonas reinhardtii. Plant J 19:353–361. doi:10.1046/j.1365-313X.1999.00526.x

7. Craft JM, Harris JA, Hyman S, Kner P, Lechtreck KF (2015) Tubulin transport by IFT is upregulated during ciliary growth by a cilium-autonomous mechanism. J Cell Biol 208(2):223–237. doi:10.1083/jcb.201409036
8. Rasala BA, Barrera DJ, Ng J, Plucinak TM, Rosenberg JN, Weeks DP, Oyler GA, Peterson TC, Haerizadeh F, Mayfield SP (2013) Expanding the spectral palette of fluorescent proteins for the green microalga Chlamydomonas reinhardtii. Plant J 74: 545–556. doi:10.1111/tpj.12165
9. Lechtreck KF, Johnson EC, Sakai T, Cochran D, Ballif BA, Rush J, Pazour GJ, Ikebe M, Witman GB (2009) The Chlamydomonas reinhardtii BBSome is an IFT cargo required for export of specific signaling proteins from flagella. J Cell Biol 187:1117–1132. doi:10.1083/jcb.200909183
10. Kindle KL (1990) High-frequency nuclear transformation of Chlamydomonas reinhardtii. Proc Natl Acad Sci U S A 87:1228–1232
11. Bower R, Tritschler D, Vanderwaal K, Perrone CA, Mueller J, Fox L, Sale WS, Porter ME (2013) The N-DRC forms a conserved biochemical complex that maintains outer doublet alignment and limits microtubule sliding in motile axonemes. Mol Biol Cell 24:1134–1152. doi:10.1091/mbc.E12-11-0801
12. Collingridge P, Brownlee C, Wheeler GL (2013) Compartmentalized calcium signaling in cilia regulates intraflagellar transport. Curr Biol 23:2311–2318. doi:10.1016/j.cub.2013.09.059
13. Huang K, Diener DR, Mitchell A, Pazour GJ, Witman GB, Rosenbaum JL (2007) Function and dynamics of PKD2 in Chlamydomonas reinhardtii flagella. J Cell Biol 179:501–514. doi:10.1083/jcb.200704069
14. Wren KN, Craft JM, Tritschler D, Schauer A, Patel DK, Smith EF, Porter ME, Kner P, Lechtreck KF (2013) A differential cargo-loading model of ciliary length regulation by IFT. Curr Biol 23:2463–2471. doi:10.1016/j.cub.2013.10.044
15. Shaner NC, Lambert GG, Chammas A, Ni Y, Cranfill PJ, Baird MA, Sell BR, Allen JR, Day RN, Israelsson M, Davidson MW, Wang J (2013) A bright monomeric green fluorescent protein derived from Branchiostoma lanceolatum. Nat Methods 10:407–409. doi:10.1038/nmeth.2413
16. Pazour GJ, Agrin N, Leszyk J, Witman GB (2005) Proteomic analysis of a eukaryotic cilium. J Cell Biol 170:103–113. doi:10.1083/jcb.200504008

Chapter 7

A FRAP-Based Method for Monitoring Molecular Transport in Ciliary Photoreceptor Cells In Vivo

Kirsten A. Wunderlich and Uwe Wolfrum

Abstract

The outer segment of rod and cone photoreceptor cells represents a highly modified primary sensory cilium. It renews on a daily basis throughout lifetime and effective vectorial transport to the cilium is essential for the maintenance of the photoreceptor cell function. Defects in molecules of transport modules lead to severe retinal ciliopathies. We have recently established a fluorescence recovery after photobleaching (FRAP)-based method to monitor molecular trafficking in living rodent photoreceptor cells. We irreversibly bleach the fluorescence of tagged molecules (e.g. eGFP-Rhodopsin) in photoreceptor cells of native vibratome sections through the retina by high laser intensity. In the laser scanning microscope, the recovery of the fluorescent signal is monitored over time and the kinetics of movements of molecules can be quantitatively ascertained.

Key words Ciliary transport, Primary sensory cilia, Ciliated photoreceptor cells, Rhodopsin, Intracellular transport

1 Introduction

Vertebrate photoreceptor cells are highly active polarized sensory neurons (Fig. 1). At one pole, specialized ribbon synapses are present and transmit the electrical signal generated in the photoreceptor cell to second-order neurons of the neuronal retina. The photoreceptive outer segment is located at the opposite pole and represents a highly modified primary sensory cilium, containing hundreds of flattened membrane disks bearing all components of the visual transduction cascade (Fig. 1). The "ciliary" outer segment is linked via its transition zone, also known as the connecting cilium, to the inner segment. The inner segment contains all organelles for cell metabolism and biosynthesis of molecules essential for photoreceptor cell function. To prevent molecular aging, every day one-tenth of the phototransductive membranes of the outer segment are renewed throughout lifetime of the photoreceptor cell [1]. Newly synthesized membrane components are added to the base of the outer

Peter Satir and Søren Tvorup Christensen (eds.), *Cilia: Methods and Protocols*, Methods in Molecular Biology, vol. 1454, DOI 10.1007/978-1-4939-3789-9_7, © Springer Science+Business Media New York 2016

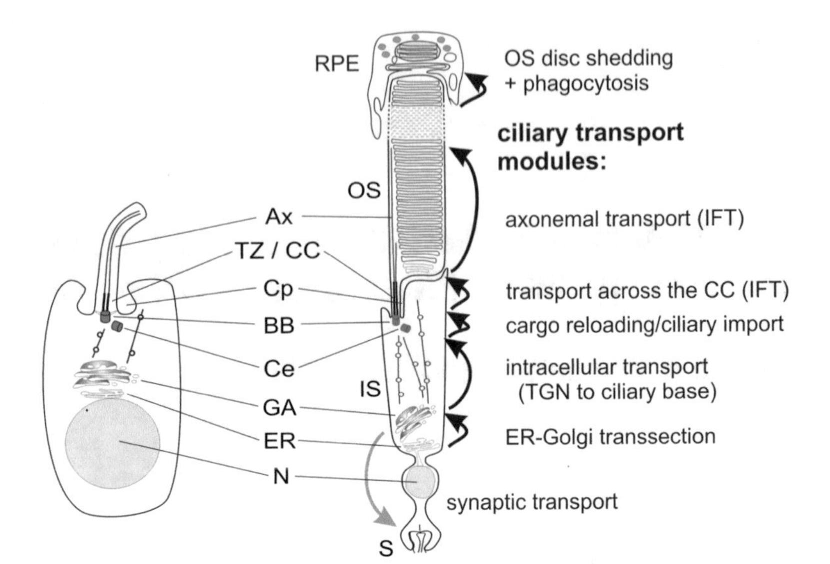

Fig. 1 Comparison of a prototypic ciliated cell with a rod photoreceptor cell and its ciliary transport modules. The photosensitive outer segment (OS) of vertebrate photoreceptor cells represents a highly modified primary sensory cilium, containing thousands of phototransductive membranous disks. The connecting cilium (CC), the equivalent of the transition zone (TZ), links the OS with an inner segment (IS), where all metabolic active organelles of the cell are housed. A basal body complex, the basal body (BB) (mother centriole) in proximity to the adjacent centriole (Ce) (daughter centriole), anchors both, the prototypic cilium as well as the photoreceptor cilium in the cytoplasmic compartment of the cell. OSs are shed from their tip on a daily basis and phagocytosed by cells of the retinal pigment epithelium (RPE). Transport modules are indicated; *arrows* highlight trafficking routes of ciliary cargo. *Ax* axoneme, *Cp* ciliary pocket, *GA* Golgi apparatus, *ER* endoplasmic reticulum, *N* nucleus, *S* synapse, *IFT* intraflagellar transport, *TGN trans*-Golgi network

segment by the expansion of the plasma membrane [2] or by incorporation of vesicles into nascent disk membranes [3]. At their distal tips, packets of one-tenth of the outer segment disks are phagocytosed by the cells of the retinal pigment epithelium (RPE) every day (relation of ~30 rods per RPE cell) [4]. For the maintenance of the continual renewal process high numbers of molecules (e.g. every day 5×10^6–7×10^6 opsin/rhodopsin molecules are transported to the outer segment [5, 6]) have to be delivered into the "ciliary" compartment of the outer segment, which can only be ensured by an effective polarized transport. The polarized composition and the high rates in biosynthesis and targeted transport make photoreceptors an excellent model system for the study of modules of the ciliary transport not only for photoreceptor cilia but also for cilia in general. Although molecular genetics and biochemical studies have provided novel insights into the understanding of the molecular transport of rhodopsin and other ciliary cargoes (e.g. reviewed in [5, 7]), reliable in vivo data are still missing in photoreceptor cells as well as in other systems of vertebrate ciliated cells.

Thus far, only few studies have approached to monitor transport processes in photoreceptor cells in vivo, for example in *Xenopus* photoreceptor cells [8] and in mouse photoreceptors [9]. This chapter describes a method, which we have recently established to elucidate transport modules in living rodent photoreceptor cells. The method is based on the fluorescence recovery after photobleaching (FRAP) and allows the quantitative assessment of the translocation of fluorophore-tagged proteins in native tissue sections in vivo. For this, the fluorescence is irreversibly bleached in the region of interest of the tissue, by high laser intensity, and the recovery of the fluorescent signal is monitored over time and thereby the kinetics of movements of molecules can be ascertained [10].

In the present chapter, we illustrate the FRAP-based method for monitoring ciliary transport in vertebrate photoreceptor cells using the transport of rhodopsin as a showcase. The illustrated method will allow the evaluation of the transport of any other fluorescently tagged ciliary molecule in photoreceptor cells.

2 Materials

1. Ringer solution (≈300 mOsm, pH 7.4; 140 mM NaCl, 2 mM $CaCl_2$, 1 mM $MgCl_2$, 10 mM Hepes in ddH_2O). Can be stored at 4 °C for several weeks. Add fresh before use: glucose 1 g/l.
2. Two percent low-melt agarose in Ringer solution (without glucose; can be stored at room temperature and re-boiled before use).
3. Dissection instruments: round forceps, two pairs of straight forceps, 27 G cannula and eye scissors; silicon-embedding molds (soft plastic chambers), imaging chamber (e.g. Ibidi µ-slide 4-well, Ibidi, GmbH, Martinsried, Germany), a fine, smooth paint brush, superglue, vibratome (Leica VT 1000 S, Leica, Wetzlar, Germany), stereo dissection microscope, coverslips (20 mm × 20 mm, cut in half with a diamond cutter).
4. Inverted confocal laser scanning microscope (i.e. Zeiss LSM710 equipped with Zen-software 2011 including the FRAP-module).
5. Animals: Heterozygous *hRho-eGFP knock-in mice* [11] are kept on a 12 h light–dark cycle, with food and water *ad libitum* (*see* **Note 1**). The fluorophore-tagged rhodopsin is targeted to the outer segments of these mice similar to rhodopsin in wild-type mice. Homozygous mice exhibit retinal degeneration and should not be used [11].

6. All experiments conform to the statement by the Association for Research in Vision and Ophthalmology (ARVO) and by the local ethics committees concerning the care and use of animal research.

3 Methods

Since photoreceptors are likely to survive longer in the dark, avoid exposure to bright light during preparation whenever possible.

1. Start by dissolving 2% low melt agarose in Ringer solution at 100 °C (*see* **Note 2**). Let it cool down, while dissecting the retina.
2. Euthanize the mouse according to the local animal care rules and enucleate the eyes (*see* **Note 3**).
3. Place the eye in a Petri dish containing Ringer solution. Remove the cornea and take out the lens and vitreous. Isolate the retina from the eye cup and dissect into (four) smaller pieces (*see* **Note 4**).
4. Embed the retina vertically in 2% agarose (<40 °C, to avoid thermal damage of biomolecules) in silicon-embedding molds (*see* **Notes 5** and **6**).
5. Once solid, remove the agarose block carefully from the plastic mold. Trim the block and fix to the specimen holder of the vibratome with superglue. Orient so that each section is cut vertically (*see* **Note 7**).
6. Fill the vibratome chamber with ice-cold Ringer solution. Save 50- to 100-μm thick longitudinal sections with a fine, smooth paint brush and transfer it to an imaging chamber (e.g. a 4-well μ-slide) filled with Ringer solution containing 1 g/l glucose. Two-three retina slices will fit into one well. Remove some of the Ringer solution by suction and place half a glass coverslip onto the slices. Place a weight on top, to hold it in place (e.g. small metal nuts). Refill the imaging chamber with Ringer solution containing glucose (*see* **Note 8**).
7. Place the imaging chamber in the object holder of an inverted confocal laser scanning microscope. Image the section (488 nm excitation and 500–550 nm emission) with as little laser power as possible to avoid bleaching or injury to the tissue. Scan as quickly as possible (setting modes: low pixel values, bidirectional scanning). Identify areas that are as vertical as possible. Choose a region of interest, e.g. at the apical inner segment below the connecting cilium (Fig. 2a), and image a z-stack of 5–6 μm above and below your region of interest. As a negative control, use PFA-fixed tissue. Alternatively, bleach a

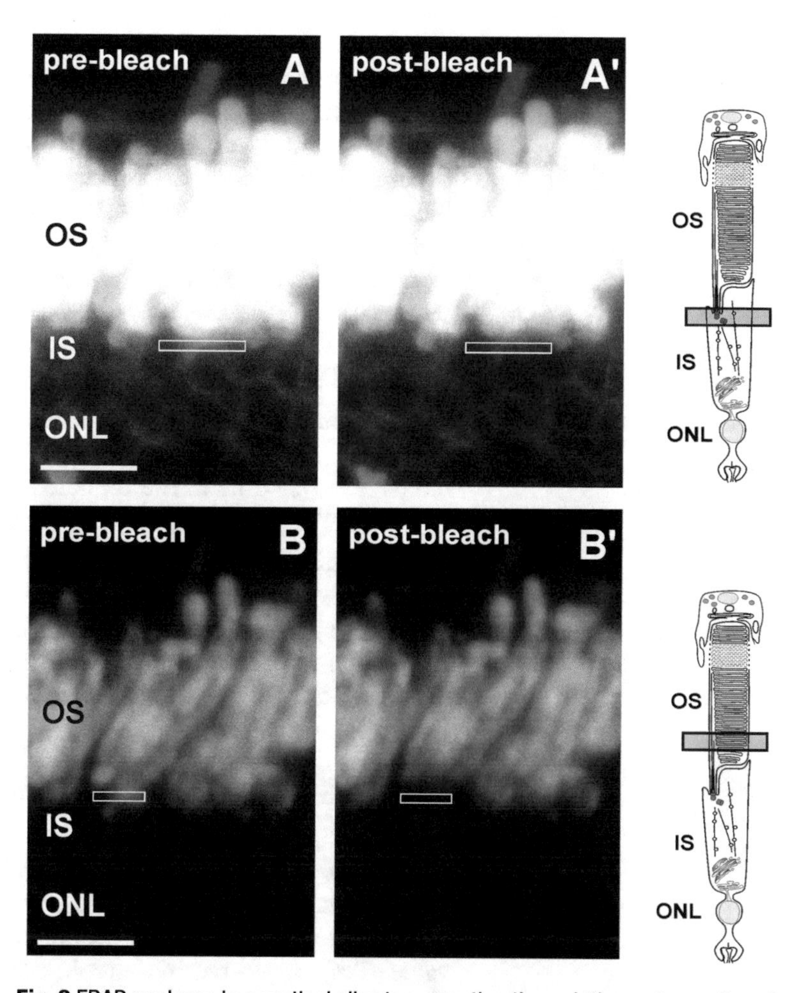

Fig. 2 FRAP analyses in a vertical vibratome section through the mature retina of a heterozygous *hRho-eGFP knock-in* mouse. *Left panels* (**a**, **b**) show images before bleaching (pre-bleached). *Middle panels* (**a′**, **b′**) show images immediately taken after bleaching (post-bleach). *Right panel*, schematic illustration of bleached regions (*gray squares*) in a rod photoreceptor cell. *White squares* in fluorescent images indicate regions of interest (ROI), in which 20 % of fluorescence was bleached by a high-power laser beam. (**a**) ROI within the apical inner segment (IS), bleaching of the apical IS (**a**, **a′**). (**b**) ROI within the basal OS; bleaching of the OS. Note that laser intensities were reduced for measurements in the OS to avoid overexposures, which occur when the intensities in the ISs are in a measurable range. *RPE* retinal pigment epithelium, *IS* inner segment, *S* synapse, *ONL* outer nuclear layer containing photoreceptor nuclei; scale bar, 10 μm

region in the outer segment (Fig. 2b). There is very little fluorescence recovery in the outer segment as the recovery rates here are extremely slow, because translocation of rhodopsin is likely dependent on outer segment disk displacement, which takes place over a longer time period in the range of days (*see* Subheading 1, [12]).

8. Set bleaching parameters in a way that bleaching is clearly visible, for example stop bleaching when the measured intensity drops down to 80%. We use two bleach cycles with 300 iterations each with laser intensity of 100% (488 nm) with a pixel dwell time of 3.15 μs (*see* **Note 9**). Monitor 3–5 frames pre-bleach and several frames (1–2 min) post-bleach as quickly as possible so as not to miss any fast recoveries. After these first fast scans, measurements every 60 s are sufficient. At room temperature, a recovery plateau will be reached after 10–15 min. If the microscope is equipped with a temperature chamber, faster recovery at higher temperatures is likely.
9. Merge z-stack images at each time point summing up the intensities of the slices, align the frames, and measure the intensities over time in the bleached region, a control region, and if possible also a background region (*see* **Note 10**).
10. For evaluation of the data, subtract the background intensities from the according intensities in the control region of interests (ROIs) and bleach ROIs for each time point. Calculate how much the whole frame has bleached by the measurement itself. For this, divide the intensity of the control ROIs (after background substraction) at the first time point by the intensity of each following time point. Multiply this bleach factor with the according intensity of the bleach ROI at each specific time point.
11. The best way to express bleaching and recovery is by normalized values. Typically, those values are expressed in percent of pre-bleach values (pre-bleach values = 100%, first post-bleach value = 0%). If plotted in a graph, the plateau (P1) resembles the mobile fraction; that is to which intensity the fluorescence recovers maximally by translocation of unbleached molecules from the surrounding into the bleach region. This plateau does not necessarily reach the original (pre-bleach) fluorescence intensity levels. The difference to the pre-bleached value resembles an immobile protein fraction in the bleached region that cannot be replaced (Fig. 3a). A second important parameter to interpret FRAP data is the speed of recovery. The time point at which the curve reaches 50% of the intensity level of the plateau is called $t/2$. The smaller this index is, the faster was the recovery.
12. An exponential equation can be used for curve fitting to calculate $t/2$ and the mobile fraction: $y = \mathrm{P1} - \mathrm{P2} \times \exp(-(x - \mathrm{P3})/\mathrm{P4})$ (Fig. 3b). P1 is the y-value at which the curve approaches infinity (plateau; mobile fraction). P2 is a normalization constant. P3 is the shift of the curve in the x direction; the time difference between $x = 0$ and the first point of recovery. P4 determines how fast the curve is growing ($t/2 = \ln(2) \times \mathrm{P4}$).

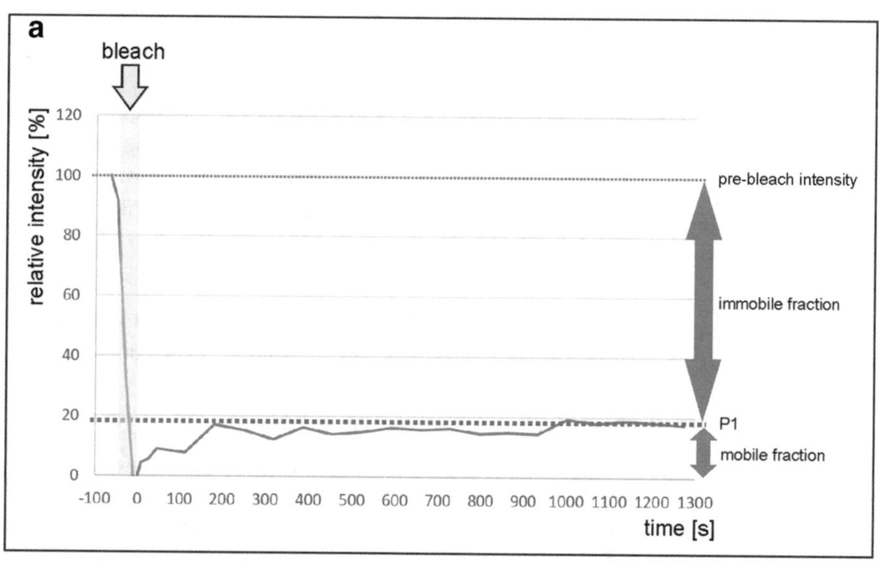

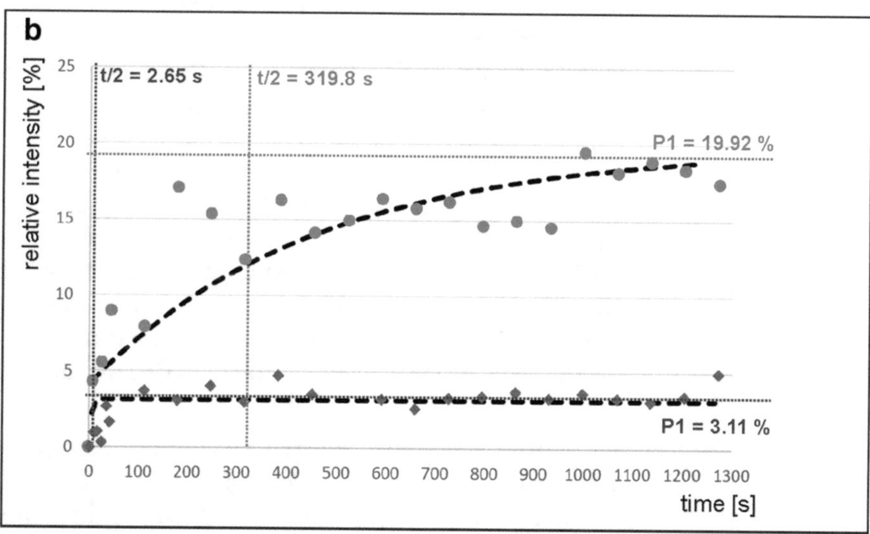

Fig. 3 FRAP analyses of rhodopsin trafficking in murine ciliary photoreceptor cells. (**a**) Representative plot of normalized fluorescence recovery versus time (in seconds) in the apical inner segments of a mature, heterozygous *hRho-eGFP knock-in* mouse (*see* Fig. 2a, a′). Fluorescence intensity values are depicted as percent of pre-bleach intensities (*apical, narrow dotted line*), while the first post-bleach value was set to 0. For illustration purposes, measurement points are connected by the blue line. *Yellow arrow* indicates the time points of bleaching. Fluorescence recovery reaches a plateau (P1; *wide dotted line*) indicating the mobile fraction of eGFP-rhodopsin. The difference from P1 to 100% resembles the immobile fraction. (**b**) Representative plot comparing normalized fluorescence recovery versus time (in seconds) in the apical inner segments (*blue dots*, Fig. 2a) with measurements of a ROI in the basal outer segments (*red diamonds*, Fig. 2b) of a mature adult, heterozygous *hRho-eGFP knock-in* mouse. Post-bleach fluorescence recovery values are depicted as percent of the pre-bleach intensities. The *black dashed lines* indicate the fitted curve after a single exponential fit of each recording. The *horizontal blue dotted line* represents P1 of the inner segment, and *horizontal red dotted line* represents P1 of the outer segment. Time points at which half of the plateau intensities are reached (*t*/2) are indicated by *vertical dotted lines* of according colors

4 Notes

1. Mice can easily be phenotyped by applying some atropine eye drops dilating the pupils. When shining into the eye with a UV torch, rhodopsin-GFP-positive eyes will clearly glow green in a dark surrounding. Wear protective eyewear! Note: The retinas of homozygous animals will degenerate [11]. Use heterozygous animals for the experiments.
2. It is quickest to boil up agarose in a microwave with gentle shaking in between until it is clear. But, be aware of boiling delay!
3. Spread the eyelids with your thumb and index finger to expose the eye, while with the other hand carefully placing curved forceps behind the eye and gripping the optic nerve tightly. Remove the eye by pulling at the optic nerve, while taking care not to squeeze the eye ball itself. Alternatively, cut the nerve with fine scissors.
4. Hold the eye with fine forceps, e.g. at the muscles, and make a small incision at the limbus with the cannula. Cut around the limbus with fine eye scissors and remove cornea and lens. To remove the vitreous and remaining hyaloid vessels, carefully "fish" blindly in the eye cup close to the retina with fine forceps. If any vitreous, vessels or sclera are left, the retina will get sticky to tools and clump together. Remove the sclera by grapping it with both pairs of forceps and pealing it like a banana. Alternatively, try to turn the eyecup inside out. Avoid touching the retina itself as much as possible. Cut off the optic nerve to set the retina free completely.
5. To transfer the retina, either carefully use forceps or a fine paint brush. Alternatively, use a glass Pasteur pipette from the wrong end: break away the long capillary and put the sucker cup on that end, so the wide opening can be used to suck up the retina. To avoid that the retina sticks to the glass, suck up some liquid before. After letting the retina sink to the meniscus, the retina can easily be released into a new container.
6. Sometimes, the embedded retina falls out of the agarose while cutting. Excess Ringer solution around the retina piece might be a reason. To avoid this, first rinse the retina in a separate dish with 2% agarose (you may keep some warm at 37 °C in a glass dish on a heating plate) before transferring it into a silicon mold with fresh 2% agarose. Position the retina and hold it in place either with the tip of forceps or better with a single horse tail hair attached to an exchange tool holder (e.g. microdissecting needle holder), while the agarose is solidifying to avoid that the retina sinks to the bottom of the well. Cooling

down is facilitated, when the silicon molds are placed on a cold wall tile on ice.

7. Slightly roll the agarose block on a piece of tissue before placing it on the specimen holder, as any leftover moisture will delay hardening of the super glue.
8. It will take some trials to find the right settings, so that the retina is well cut and does not fall out of the agarose. We use our vibratome at the speed of 0.25 mm/s and a frequency of approximately 50 Hz.
9. Most laser scanning microscopes have the option to zoom in while bleaching. That way, the bleaching time will be reduced, however, most often at the costs of accuracy. For the described purpose, it is absolutely sufficient. But for very small structures you may need more precise bleaching.
10. The open source software Fiji (ImageJ; http://fiji.sc/Fiji) is well suited for all necessary image processing: Open the image file ("view stack with Hyperstack"; check "split channels" if you monitor more than one channel). Produce maximal projections of the z-stacks at each time point (image → stacks → Z Project, check "sum slices" and "all time frames"). To compensate for small shifts and drifts of your sample during the time of measurement, you have to align your time series (plugins → registration → linear stack alignment with SIFT; default settings). Choose regions of interest (ROIs) that are smaller than your bleach region by drawing a rectangle (leftmost symbol in the tool bar). It is best to measure several small ROIs and calculate the average. That way, you will notice any outstander that may occur due to strong shifts. Add these ROI to the ROI manager (analyze → Tools → ROI manager → Add). By pressing the key "t" as a shortcut, the ROI will be added to the ROI manager, which automatically opens in a separate window. Add also ROIs from neighboring photoreceptors that have a similar orientation and fluorescence intensity but have not been bleached. Ideally, also some ROIs in the background with no fluorescence should be measured. You may move the very same ROI by right click in the center and dragging with the mouse to the new location to keep the same size, adding each desired location by the shortcut "t". Checking "show all" visualizes all set ROIs in the image. Measure the intensity values in those ROIs in all time frames (ROI manager: → More → Multi Measure; check "measure all XY slices" and "one row per slice"). This will open a separate window with your measurement results. You can choose, which results shall be included, e.g. area and mean grey values: → results → set measurements. Under → results → options, you can choose, e.g. file formats for saving, etc.

Acknowledgments

The present study was supported by the German Research Council (DFG) FOR2149/WO548-8, EC FP7/2009/241955 (SYSCILIA), the FAUN Foundation, Nurnberg, and the JGU Research Support (Stage 1). We thank Dr. J.H. Wilson, Houston, TX, for kindly providing the hRho-eGFP knock-in mouse line and Dr. Helen May-Simera for critical comments on the manuscript.

References

1. Young RW (1967) The renewal of photoreceptor cell outer segments. JCell Biol 33: 61–72
2. Steinberg RH, Fisher SK, Anderson DH (1980) Disc morphogenesis in vertebrate photoreceptors. JCompNeurol 190:501–518
3. Usukura J, Obata S (1995) Morphogenesis of photoreceptor outer segments in retinal development. Prog Retin Eye Res 15:113–125
4. Kevany BM, Palczewski K (2010) Phagocytosis of retinal rod and cone photoreceptors. Physiology 25(1):8–15. doi:10.1152/physiol.00038.2009
5. Pearring JN, Salinas RY, Baker SA, Arshavsky VY (2013) Protein sorting, targeting and trafficking in photoreceptor cells. Prog Retin Eye Res 36:24–51. doi:10.1016/j.preteyeres.2013.03.002
6. Wolfrum U, Schmitt A (2000) Rhodopsin transport in the membrane of the connecting cilium of mammalian photoreceptor cells. Cell Motil Cytoskeleton 46(2):95–107. doi:10.1002/1097-0169(200006)46:2<95::AID-CM2>3.0.CO;2-Q
7. Wang J, Deretic D (2014) Molecular complexes that direct rhodopsin transport to primary cilia. Prog Retin Eye Res 38:1–19. doi:10.1016/j.preteyeres.2013.08.004
8. Tian G, Ropelewski P, Nemet I, Lee R, Lodowski KH, Imanishi Y (2014) An unconventional secretory pathway mediates the cilia targeting of peripherin/rds. J Neurosci 34(3):992–1006. doi:10.1523/JNEUROSCI.3437-13.2014
9. Trivedi D, Colin E, Louie CM, Williams DS (2012) Live-cell imaging evidence for the ciliary transport of rod photoreceptor opsin by heterotrimeric kinesin-2. J Neurosci 32(31): 10587–10593. doi:10.1523/JNEUROSCI.0015-12.2012
10. Reits EA, Neefjes JJ (2001) From fixed to FRAP: measuring protein mobility and activity in living cells. Nat Cell Biol 3(6):E145–147. doi:10.1038/35078615
11. Chan F, Bradley A, Wensel TG, Wilson JH (2004) Knock-in human rhodopsin-GFP fusions as mouse models for human disease and targets for gene therapy. Proc Natl Acad Sci U S A 101:9109–9114
12. Sung CH, Chuang JZ (2010) The cell biology of vision. J Cell Biol 190(6):953–963. doi:10.1083/jcb.201006020

Chapter 8

Kymographic Analysis of Transport in an Individual Neuronal Sensory Cilium in *Caenorhabditis elegans*

Robert O'Hagan and Maureen M. Barr

Abstract

Intraflagellar Transport (IFT) is driven by molecular motors that travel upon microtubule-based ciliary axonemes. In the single-celled alga *Chlamydomonas reinhardtii,* movement of a single anterograde IFT motor, heterotrimeric kinesin-II, is required to generate two identical motile flagella. The function of this canonical anterograde IFT motor is conserved among all eukaryotes, yet multicellular organisms can generate cilia of diverse structures and functions, ranging from simple threadlike non-motile primary cilia to the elaborate cilia that make up rod and cone photoreceptors in the retina. An emerging theme is that additional molecular motors modulate the canonical IFT machinery to give rise to differing ciliary morphologies. Therefore, a complete understanding of the trafficking of ciliary receptors, as well as the biogenesis, maintenance, specialization, and function of cilia, requires the characterization of motor molecules.

Here, we describe in detail our method for measuring the motility of proteins in cilia or dendrites of *C. elegans* male-specific CEM ciliated sensory neurons using time-lapse microscopy and kymography of green fluorescent protein (GFP)-tagged motors, receptors, and cargos. We describe, as a specific example, OSM-3::GFP puncta moving in cilia, but also include (Fig. 1) with settings that have worked well for us measuring movement of heterotrimeric kinesin-II, IFT particles, and the polycystin TRP channel PKD-2.

Key words *Caenorhabditis elegans*, Kymograph, Cell biology, In vivo, Sensory non-motile cilia, PKD-2, Polycystins, OSM-3, Kinesin-2, KLP-6, Kinesin-3, Male

1 Introduction

In this chapter, we describe how to measure the speed of fluorescently marked motors, IFT complex components, and receptors, moving in an individual dendrite or sensory cilium of a *C. elegans* male-specific neuron. The method requires a compound microscope equipped with a GFP filter set and a sensitive camera, but does not require confocal microscopy, and is therefore a technique that can be used by most biology laboratories.

After the movement of heterotrimeric kinesin-II was described in *Chlamydomonas* [1–4], elaborations of the basic IFT machinery by accessory motors were discovered. First, the monomeric kinesin-2 OSM-3 was found to cooperate with the canonical IFT

Peter Satir and Søren Tvorup Christensen (eds.), *Cilia: Methods and Protocols*, Methods in Molecular Biology, vol. 1454,
DOI 10.1007/978-1-4939-3789-9_8, © Springer Science+Business Media New York 2016

kinesin to build amphid and phasmid cilia in *C. elegans* [5–7]. The mammalian homolog of OSM-3, called KIF17, has been reported to be involved in ciliary targeting of olfactory channels [8] and photoreceptor development and maintenance [9], although this conclusion is controversial [10]. Our work further supports this model of the origin of ciliary diversity; we have found an interplay between the canonical kinesin-II, the kinesin-2 OSM-3/KIF17, and a kinesin-3, KLP-6/KIF28 [11], in male-specific neuronal sensory cilia in *C. elegans* [12]. Kymographic analysis of ciliary motor motility was invaluable in testing the modification of IFT by accessory motors.

We will describe the acquisition and analysis of streaming video using Metamorph software, but free ImageJ software (http://imagej.nih.gov/ij/) is an effective substitute. We describe the technique in the study of male-specific neuronal cilia and dendrites, but the method is easily adapted to study transport in axons or in other cell types, using other molecular markers or using other cell-type-specific promoters. We will explain how we grow transgenic *C. elegans* nematodes, mount them for imaging, immobilize them, acquire time-lapse images at high magnification on a compound upright microscope, and how to convert the images into kymographs that are used to measure the in vivo velocity of proteins.

This technique has been used fruitfully and extensively to track movement of tagged proteins in the amphid channel cilia [4, 6, 7], which comprise a bundle of ten cilia originating from eight different neurons [13]. In such studies, a GFP-tagged molecule is expressed in all amphid channel neurons. Because the bundled cilia are spatially very close together, all ten cilia are analyzed *en masse* using kymography [4, 6, 7]. Any differences that might exist between cilia are therefore undetectable. Differences in IFT have been found in individual cilia types [12, 14]. The kymographic analysis of molecular movement in individual CEM neuronal cilia, as described here, confers the advantages of studying molecules in a single cilium [12].

This method relies on tracing a line over the cilium of interest in a stack of sequential images and then plotting all the pixel values along that line for each frame of the video, generating a "kymograph" in which the horizontal axis is the distance along the line and the vertical axis is time. If fluorescently marked particles are stationary, a vertical stripe is seen. However, moving particles will appear as diagonal lines in the kymograph. The characteristics of these lines can be used to determine the in vivo velocity, direction, frequency, and run length of motor complexes. This in vivo method will not necessarily elucidate single-molecule characteristics of motors, such as processivity or binding affinity, which may be best studied in vitro. However, this method will report accurately the activity of the tagged molecules inside cells in a living organism.

Once you reliably detect the characteristic movement of a tagged molecule of interest, you can then assess the function of associated proteins by using genetic crosses to mate the transgene or extrachromosomal array encoding your tagged molecule into various mutant backgrounds and looking for changes in velocity, frequency, distance, or direction. It is important to use the same transgene or array when making comparisons between different genetic backgrounds, developmental stages, or conditions. For example, we used this technique to show that posttranslational glutamylation of microtubules regulates activity of OSM-3 kinesin-2 motors in CEM cilia [15]. Combination of the imaging of motor activity with the powerful genetics toolkit of *C. elegans* can be used to facilitate study of the function of ciliary transport regulators such as IFT components, transition zone components, microtubule-associated proteins, and many others.

2 Materials

2.1 Materials and Equipment Needed to Culture C. elegans

1. Incubator.
2. Bunsen burner.
3. 2-l Erlenmeyer flask.
4. Aluminum foil.
5. 55 °C water bath.
6. Autoclavable ingredients for NGM agar plates: NaCl, agar, peptone, H_2O.
7. Sterile ingredients to add to autoclaved NGM agar: sterile 1 M $CaCl_2$, cholesterol in ethanol (5 mg/ml concentration; not autoclaved), 1 M $MgSO_4$, and filter-sterilized 1 M KPO_4 buffer pH 6.0 (108.3 g KH_2PO_4, 35.6 g K_2HPO_4, H_2O to 1 l).
8. 60 mm sterile petri dishes.
9. OP50 *E. coli* strain: obtained from *Caenorhabditis* Genetics Center (cgc.cbs.umn.edu). This strain is the standard food for *C. elegans.*
10. LB broth for growing bacteria: tryptone, yeast extract, NaCl, 1 N NaOH.
11. 250 ml autoclavable bottles.
12. Transgenic *C. elegans* strain expressing ciliary marker of interest: In this protocol, we will use the strain PT2098 of genotype *pha-1(e2123ts) III; him-5(e1490) V; myEx685 [klp-6p::osm-3::gfp+pBX]* [12] as an example (*see* **Note 1**).

2.2 Mounting Worms for Microscopy

1. Heating block at 65 °C (with adaptor for 16 × 100 mm glass tubes).
2. Agar.
3. Deionized water.
4. Borosilicate glass 16 × 100 mm culture tubes.

5. Worm pick (Fig. 1) Flatten the end of a 3 cm long ~30 G platinum wire by tapping lightly with a hammer, and trim to spatula shape under a stereodissecting microscope using a razor blade. Break about 1 cm off the tapered end of a glass

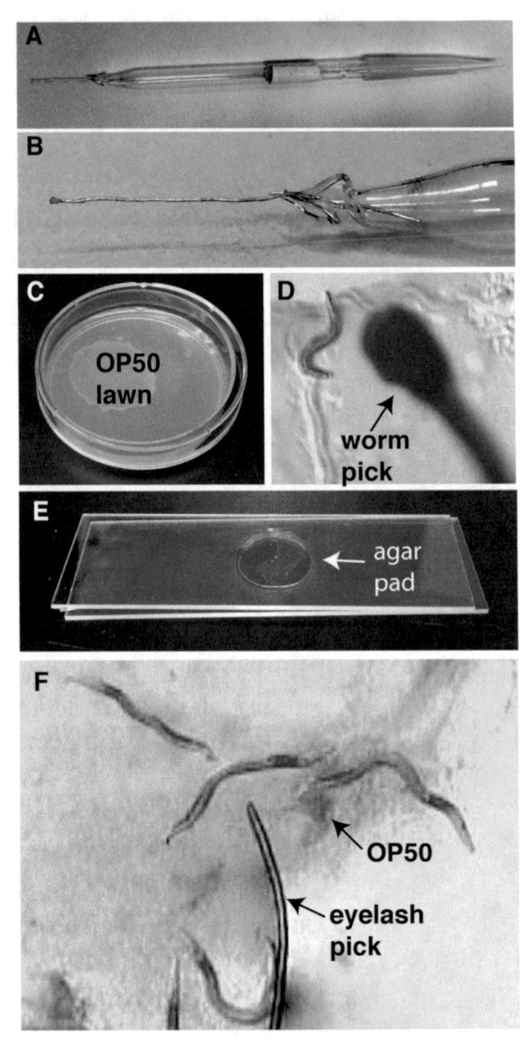

Fig. 1 Tools for nematode handling and growth. (**a**) The worm pick is made from a glass Pasteur pipette with a platinum wire fastened in the tapered end. (**b**) Detail of platinum wire in worm pick. The tip is shaped like a spatula. (**c**) NGM (nematode growth media) plate seeded with the *E. coli* strain OP50 to grow a bacterial lawn that serves as food for growing nematodes. Plate shown is 60 mm diameter. (**d**) View of a NGM plate with worms through the stereodissecting microscope. The visible L4 larval hermaphrodite has left tracks in the OP50 lawn. The spatula-shaped tip of the worm pick is visible. (**e**) An agar pad made by sandwiching a drop of agar between two slides. (**f**) Mounting male worms for imaging, viewed through stereodissecting microscope. Worms have been picked by dabbing with OP50 and transferred into the levamisole + M9 solution on the agar pad. The eyelash pick (eyelash or eyebrow hair glued to a toothpick) is used to sweep away the dark OP50 bacteria to improve image quality and to position the animals

Pasteur pipette. Use a tweezer to hold the platinum wire above the flame of a Bunsen burner while softening the end of the Pasteur pipette in the same flame. When wire glows red and glass is soft enough, insert wire into pipette and allow glass to harden around it. The spatula end of the wire is used to pick up worms either to place worms on OP50-seeded NGM plates or to pick worms off of plates to mount for imaging.

6. Glass slides and coverslips (we use thickness "1").
7. Eyelash pick use Elmer's Glue or similar to fasten an eyelash or eyebrow hair to the end of a toothpick.
8. Stereodissecting microscope (Zeiss Stemi SV 6 or similar).
9. M9 Buffer 3 g KH_2PO_4, 6 g Na_2HPO_4, 5 g NaCl, 1 ml 1 M $MgSO_4$, H_2O to 1 l. Sterilize by autoclaving.
10. Levamisole (sigmaaldrich.com), 10 mM in M9 buffer. We keep a 10× solution (100 mM levamisole in M9) frozen at −20 °C, thaw, and dilute to working concentration of 10 mM in M9 the day of the experiment.

2.3 Worm Imaging

1. Compound microscope equipped with 100× oil immersion objective, camera, and acquisition software: Streaming video is acquired using either a Zeiss Axioplan 2 microscope and Photometrics Cascade 512B EMCCD camera or a Zeiss Imager D1M microscope with a Retiga-SRV Fast 1394 Q-Imaging digital camera with 100× 1.4NA oil Zeiss Plan-APOCHROMAT objective. We use Metamorph software (Version 7.6.1.0, MDS Analytical technologies) to control image acquisition.

2.4 Analysis

1. Metamorph software: Includes kymograph module, which we use to generate and analyze kymographs made from streaming video. Velocities and other data are stored by data logging in Excel (Microsoft Office) software.
2. Prism 5 software: We perform all statistical analyses with Prism 5 (www.graphpad.com). Velocity measurements are typically subjected to one-way ANOVA analysis with a Bonferroni post-hoc test to compare all genotypes in pairwise fashion.

3 Methods

3.1 Preparing NGM (Nematode Growth Media) Plates

1. Mix 3 g NaCl, 17 g agar, and 2.5 g peptone in a 2 l Erlenmeyer flask.
2. Add 975 ml H_2O. Cover mouth of flask with aluminum foil.
3. Autoclave on liquid cycle for 50 min.
4. Cool flask in 55 °C water bath for 15 min.
5. Then add 1 ml sterile 1 M $CaCl_2$, 1 ml 5 mg/ml cholesterol in ethanol (not autoclaved), 1 ml 1 M $MgSO_4$, and filter-sterilized

25 ml 1 M KPO_4 buffer pH 6.0 (108.3 g KH_2PO_4, 35.6 g K_2HPO_4, H_2O to 1 l). Swirl to mix well.

6. Using sterile procedures, dispense the NGM agar solution into Sterile Disposable Petri Plates. Plates should be about 2/3 full of agar. Typically, we dispense ~9 ml into 60 mm plates. Optionally, use a peristaltic pump to dispense.
7. Leave plates at room temperature for 2–3 days before use to allow for detection of contaminants and to allow excess moisture to evaporate. Plates stored in an air-tight container at room temperature will be usable for several weeks.

3.2 OP50 Bacterial Culture

1. Prepare LB broth by combining 10 g tryptone, 5 g yeast extract, 10 g NaCl, 1 ml 1 N NaOH, and H_2O to 1 l. Stir and distribute into ten separate 250 ml bottles. Autoclave and store at room temperature.
2. Pick a single colony from a streak of OP50 bacterial culture as supplied to inoculate 100 ml bottle of sterile LB broth. Grow overnight at 37 °C, and store at 4 °C.

3.3 Nematode Culture

1. Seed NGM plates with ~100 μl of OP50 culture in LB. Dry overnight to form a bacterial lawn upon which the nematodes will feed.
2. Use the worm pick to transfer worms to a fresh NGM plate containing an OP50 lawn (Fig. 1), and incubate at 20 °C for 2–3 days (*see* **Note 2**). Picking worms is most easily done by placing the spatula-shaped tip of the worm pick onto the edge of the OP50 bacterial lawn. A blob of bacteria will adhere to the bottom of the pick and serve as "stickum" with which to pick up *C. elegans*. To pick up an animal, gently dab the stickum onto it and quickly withdraw. A single or several worms will stick to the OP50 stickum. Then gently touch the tip to a fresh NGM plate and wait for animals to crawl off the pick onto the agar. The platinum tip of the pick is flamed in the Bunsen burner to glow red, thus sterilizing it, allowed to cool for a few seconds, and dipped in the OP50 bacterial lawn before each new attempt at picking.

3.4 Preparing Slides and Mounting Worms for Imaging

The day before you plan to image motility of GFP-tagged proteins, use the worm pick to transfer only L4 (fourth larval stage) males (Fig. 2) to a new OP50-seeded NGM plate (*see* **Note 3**).

When animals are in their first day of adulthood (18–24 h after picking), image animals as follows:

1. Make a 4% agar solution by gently melting 0.125 g of agar in 3 ml of water in a glass culture tube over the Bunsen burner. Keep the culture tube moving to avoid both rapid boiling (which can propel hot agar out of the tube or crack the tube)

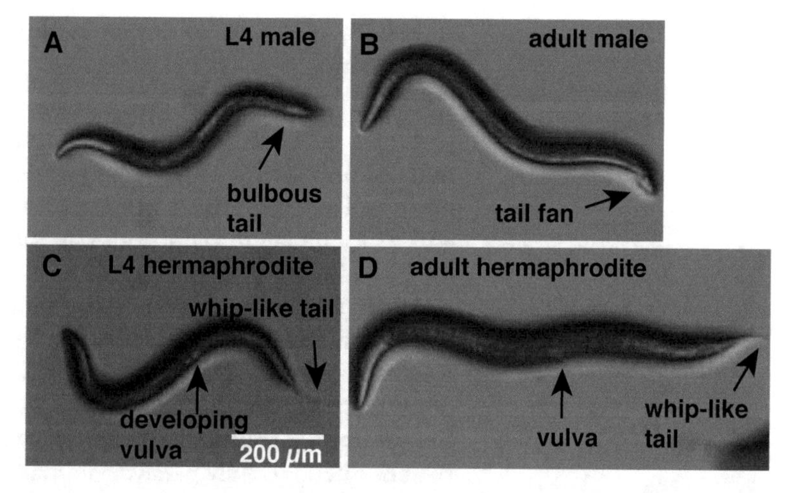

Fig. 2 Morphological features of male and hermaphrodite nematodes viewed through the stereodissecting microscope. (**a**) The day before imaging, pick L4 males. The L4 male is identified by the bulbous tail and lack of a developing vulva. (**b**) L4 males should uniformly be young adult males on the day of imaging. Confirm that all animals picked the previous day are young adult males, characterized by a fully developed tail fan and lack of a vulva. (**c**) L4 hermaphrodites can be distinguished from males by a clear "spot" that marks the developing vulva and a whip-like tail. (**d**) If you have accidentally picked any hermaphrodites the day before imaging, you will find adult hermaphrodites among your males. Adult hermaphrodites are characterized by a visible vulva and a whip-like tail. They are larger than adult males, and eggs may be visible inside the animal (not shown). If any adult hermaphrodites are visible on plates with your adult males, the males should not be imaged, as exposure to hermaphrodites might influence male neuronal activity and change the characteristics of ciliary motility, leading to erroneous data. All animals are positioned with anterior pointing to *left*

and burning of agar. Place tube with melted agar solution in heating block at 75 °C.

2. Lay out 12 glass slides in two rows on the bench.
3. Slightly enlarge the opening of a 200 μl plastic pipette tip by trimming the end with a pair of scissors or blade (to reduce clogging as the agar solidifies), and draw up ~125 ml of hot agar solution using a P200 pipetman.
4. Moving quickly, place a drop of hot agar on a glass slide, cover with a second glass slide, and apply gentle finger pressure to create a thin agar pad. Repeat for the rest of slides. Avoid trapping air bubbles in agar. With a little practice, you can make five or six agar pad slides before the pipette becomes clogged.
5. Slide the top glass slide off the "sandwich" carefully to avoid tearing the agar pad. This can be done soon after the agar solidifies, or top slides can be left in place for up to an hour to prevent agar pads from drying out before use. Place the

slide with agar side facing up on the stereodissecting microscope stage.

6. Add 6 μl of M9 plus levamisole (10 mM) solution to the agar pad on a slide. Levamisole is an acetylcholine receptor agonist that immobilizes nematodes by causing hypercontraction of body wall muscles [16]. Use the platinum worm pick to transfer ~10 male animals into the liquid on the agar pad. With practice, all ten males can be picked up at once. Then, place the tip of the pick into the M9 plus levamisole solution on the agar pad, and wait for males to leave the pick (or shake them off gently). Be careful not to tear the agar pad, which results in worms pooling in regions that cannot be imaged. Once animals are in the levamisole solution, wait 8 min with slides uncovered for the anesthetic to fully paralyze animals (*see* **Note 4**).
7. Use the eyelash pick to move remaining bacteria away from animals for better image quality, and line the worms up in a row for convenience in finding them on the compound microscope. Gently place coverslip over the preparation. If possible, avoid trapping bubbles. Image within 30–40 min of mounting, because worms may become less healthy and will certainly become starved (*see* **Note 5**).
8. Mount the slide on the compound microscope stage for imaging.

3.5 Acquiring Streaming Video

Only CEM cilia that reside completely within the plane of focus should be imaged. CEM cilia normally curve slightly outward to reach the outside of the animal's cuticle; this curved morphology should be apparent if the cilium is completely within the plane of focus. Some animals may not have any acceptable CEM cilia completely in focus. Therefore, it is likely that you will not collect data from all of the worms you mount for imaging.

1. Set excitation power to 25–50%; set power higher only if necessary to visualize fluorescent puncta (*see* **Note 6**).
2. The following imaging parameters in Metamorph permit you to capture high-quality images of OSM-3::GFP puncta for the duration of the streaming image acquisition. Open the "Acquire" panel using the "Acquire" dropdown menu in Metamorph. Select "GFP" illumination under the "Acquire" tab; then, select "5 MHz Digitizer" and "1× Gain" in the "Special" tab, and set "Exposure Time" to 100 ms (*see* **Note 7**). Select "100× oil" objective in the "Magnification" setting in the "Device Control" toolbar to store the calibration for pixel size with your images. Save your settings so that you can reload them next time you image the same molecular marker.
3. Click "Full Chip" and "Show Live" to see continuous video from the camera. You may select a region of the image with the rectangle tool in the "Region Tools" toolbar, and then select "Use Active Region" rather than "Full Chip" in the "Acquire"

panel, in order to capture streaming video from a small region. Alternatively, you can acquire using "Full Chip" and simply magnify the image later (e.g., to zoom in to the CEM cilium).

4. Open the "Stream Acquisition" panel from the "Acquire" menu, and set the number of frames you are going to capture—for 100 ms exposure, 100 frames will take 10 s. We select "Stream to Ram" in the "Acquisition Mode" and "Acquire images at frame rate" in the "Camera Parameters" tab. We do not select "Display preview image during acquisition" in the "Camera Parameters" tab because this may slow down acquisition in Metamorph.
5. When you are ready to acquire a video stream, turn on fluorescence excitation, quickly fine-tune focus, and press the "Acquire" button in the "Stream Acquisition" panel.
6. After the acquisition, the new image stack of your video will appear. Check for movement of GFP puncta by dragging the bar on the top of the stack or by pressing "play." You should see movement of GFP puncta in the image stream without significant image processing. If you do not see moving particles using a marker known to move in cilia, either photobleaching or improper acquisition settings are typically to blame (*see* **Note 8**). However, some fluorescently marked molecules may not undergo active transport in cilia. For example, we did not detect moving particles for the TRP channel PKD-2::GFP in CEM cilia [17] or the inversin homolog NPHP-2::GFP in phasmid cilia [18]. To determine if your protein is stationary or mobile, you may need to use another imaging technique such as FRAP (fluorescence recovery after photobleaching [19]).

3.6 Analysis of Kymographs

Because scoring by tracing lines on kymographs includes an element of subjectivity, we typically acquire many streaming videos and then have a lab mate blind the genotypes of the videos before scoring. It is essential to score for particle movement in CEM cilia of several different animals for each genotype to ensure that you account for individual variation or those due to focus issues (*see* **Note 9**).

1. Load your streaming video stack or multidimensional TIFF back into Metamorph, and select "Kymograph" from the "Stack" menu to open the "Kymograph" panel.
2. Click "Calibrate distance" under the "Measure" menu. In the "Calibrate Distances" panel, click on the option that indicates calibration for 100X objective. Click "Apply to all open images" (*see* **Note 10**).
3. Using the "Multisegment Line tool" from the "Region Tools" toolbar, carefully draw a line along the cilium or dendrite in the region where you want to measure particle velocity (Fig. 3). When drawing the line, always start proximally, and extend the

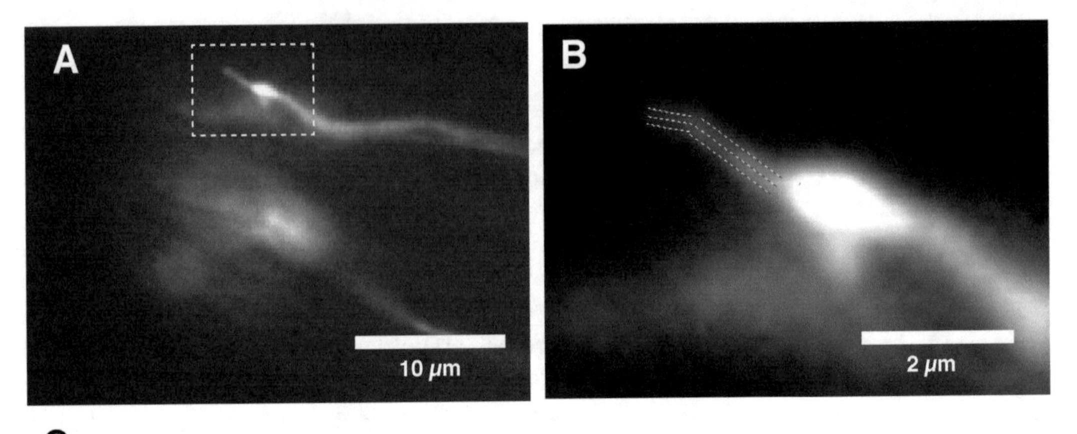

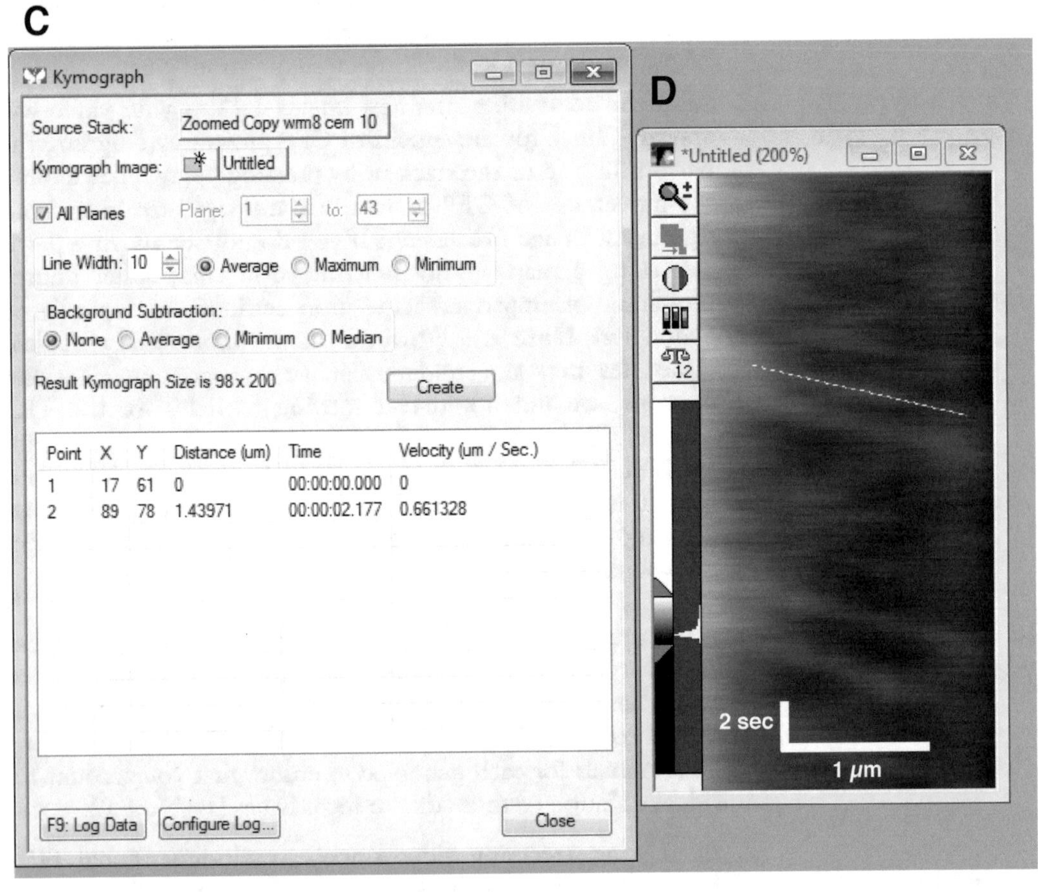

Fig. 3 Analysis of motile particles in streaming images. (**a**) The tip of adult male nose showing OSM-3::GFP fluorescence in a CEM neuron. At the tip of a dendrite, a CEM cilium extends from the bright area at the ciliary base. (**b**) Enlargement of area enclosed by *dashed-line box* in A with a *line* traced down center of the CEM cilium; *outer dashed lines* are set by "Line Width" of 10 pixels, selected in "Kymograph" panel (shown in **c**). This line will be used to generate kymograph. (**c**) "Kymograph" panel with "Line Width" of 10 pixels selected. Clicking "Create" produces the kymograph (shown in **d**). (**d**) Several diagonal lines in the kymograph shown represent motile OSM-3::GFP puncta. The *dotted white line* in the kymograph was traced over a *diagonal line* and automatically analyzed in the kymograph panel (in **c**), showing a velocity of 0.661328 μm/s

line distally so that your kymographs will always have the same orientation and you will be able to distinguish anterograde from retrograde movement.

4. In the "Kymograph" panel (Fig. 3), adjust "Line Width" to include the brightest part of cilium (~5–10 pixels). For "Background subtraction," selecting "None" will not subtract any background from the image, while "Minimum" will subtract the lowest detected value from each pixel in the stack. Either one usually works well.
5. Click "Create" in the "Kymograph" panel to create a kymograph which plots the pixel values for each pixel in the line you drew for each frame of the streaming video, resulting in rectangle with "time" plotted vertically and "distance" plotted horizontally (Fig. 3). You will (hopefully) see diagonal lines, representing particles that have moved during your streaming video (*see* **Note 11**). You can then use the "Straight-line tool" from the "Region Tools" toolbar to trace the diagonal lines in your kymograph; the slope of the line, dD/dT, is the particle velocity (Fig. 3). Because Metamorph stores the pixel and time dimensions for your video, the velocity will be calculated automatically and appear in the "Kymograph" panel (Fig. 3).
6. The data points can be manually logged, or Metamorph can automatically store them in an Excel or text file. To store measurements automatically, click "Open Log" at the bottom of the "Kymograph" panel and follow instructions. Thereafter, to log each data point, click "F9: Log Data" in the "Kymograph" panel or simply press the F9 key.
7. After scoring kymographs from several animals for each genotype, use Graphpad Prism software to perform one-way ANOVA tests to detect differences between genotypes.

4 Notes

1. We often use the *pha-1(e2123ts)* mutation to easily maintain stable transgenic lines [20]. Animals with this mutation develop normally at the permissive temperature of 15 °C, but fail to develop a normal pharynx at the restrictive temperature of 20–25 °C, and therefore do not survive to adulthood [20]. Because the *myEx685* transgene contains DNA encoding the fluorescent marker (*klp-6p::osm-3::gfp*) as well as the plasmid pBX, which contains a wild-type copy of *pha-1*, the extrachromosomal array is conveniently maintained by growing animals at 20–25 °C. Under these conditions, only animals that have inherited the extrachromosomal array, which rescues the lethality of the *pha-1(e2123ts)* mutation, survive to adulthood [20].

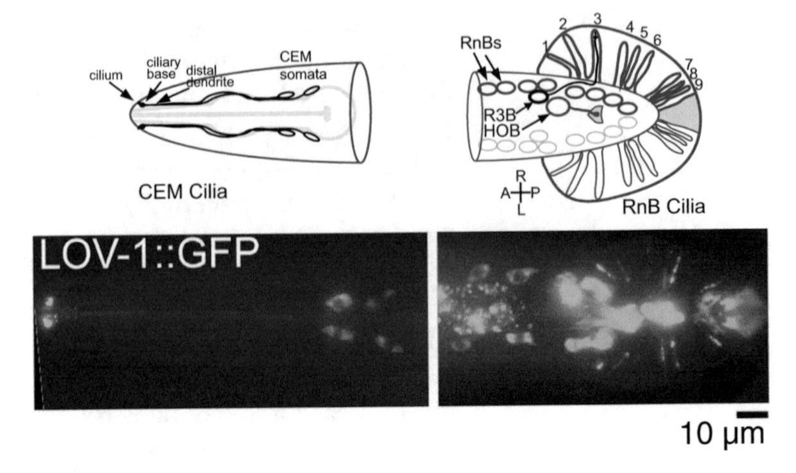

Fig. 4 *C. elegans* male-specific ciliated neurons in diagram (*top*), and in vivo (*bottom*) visualized using a LOV-1::GFP marker. Male-specific neurons with sensory exposed to the environment include the four CEM neurons in the head and the B-type ray neurons ("RnBs") and the HOB neuron in the tail. These neurons express the tagged polycystin LOV-1::GFP (diagrams reproduced from [15] and pictures reproduced from [27]). This LOV-1::GFP reporter does not move in cilia but is released in ciliary extracellular vesicles [28]

The *myEx685 [klp-6p::osm-3::gfp+pBX]* transgene uses the *klp-6* promoter to drive expression of GFP-tagged kinesin-2 motor OSM-3 in only a subset of ciliated sensory neurons (including the male-specific B-type CEM, RnB, and HOB neurons; Fig. 4). In this example, we describe how to measure OSM-3::GFP motility in male-specific CEM cilia, which coexpress the polycystins LOV-1 and PKD-2 (homologs of mammalian PKD1 and PKD2, respectively [21, 22]).

2. Worms should be raised in incubators to provide a stable temperature for growth. The example transgenic strain PT2098 of genotype *pha-1(e2123ts) III; him-5(e1490) V; myEx685 [klp-6p::osm-3::gfp+pBX]* must be grown at 20–25 °C to maintain the transgene. Typically, animals are raised from egg to adulthood in about 3 days at 20 °C. Exposure to temperatures higher than 25 °C is not healthy for *C. elegans.*
3. To avoid possible variability of data caused by differences in the age and health of the animals, image only healthy young adult males. Take care to avoid transferring hermaphrodites along with your L4 males to eliminate differences in neuronal activity due to sexual experience of hermaphrodites. In this way, each male will have never encountered a potential mate during his adulthood, when his CEM neurons become mature. Therefore, any effects of neuronal stimulation on motor transport will be avoided.

4. Adult males are more active than hermaphrodites and larvae. Therefore, a lower concentration of levamisole may be used when adapting this method for imaging adult hermaphrodites or larvae. If your worms burst on the slide, you may be inadvertently using levamisole at a higher concentration.
5. If you find you are wasting too much time refocusing the stereodissecting scope between picking up animals off NGM plates and mounting on agar pads, you can use a "booster stage" comprising stacked and taped glass slides to hold slides in the same focal plane as the NGM plates.
6. Take care to minimize photobleaching. Since time-lapse imaging will expose samples to excitation illumination for a significant amount of time, the sample will bleach during acquisition. Note also that you may be bleaching your sample before you even begin imaging. Make sure you turn off fluorescent excitation and use Nomarski optics or simply transmitted white light for all manipulations prior to imaging.
7. If using another fluorescent reporter, adjust the exposure time (typically 100–300 ms), Gain, and/or EM Gain (in the "Special" Tab of the "Acquire" panel), and intensity of the excitation light source until you see an acceptable Live image (*see* Table 1, e.g., settings). The process of finding the right combination of excitation illumination, microscope, and camera settings is iterative. After determining the best parameters for a particular reporter, use the same parameters for all experiments with this marker.

Table 1
Example molecules tagged for kymography analysis in *C. elegans* CEM male-specific-ciliated sensory neurons

Protein name (*C. elegans*/mammal)	Class	Location	Exposure/gain settings	Reference
PKD-2/PKD2	Polycystin 2	CEM dendrite CEM cilia	300 ms/5 MHz digitizer (Standard gain) 1×	[17, 26]
OSM-3/KIF17	Homodimeric kinesin-2	CEM cilium	100 ms/5 MHz digitizer (standard gain) 1×	[12, 15]
KAP-1/KAP	Heterotrimeric kinesin-2 subunit	CEM cilium	150 ms/5 MHz digitizer (standard gain) 4×	[12]
KLP-6	Kinesin-3	CEM cilium	150 ms/20 MHz digitizer (standard gain) 1×	[12]
OSM-6/IFT152	IFT B complex	CEM cilium	200 ms/5 MHz digitizer (standard gain) 4×	[15, 23]

Location indicates where motile particles have been analyzed in *C. elegans* male-specific-ciliated sensory neurons. PKD-2::GFP puncta move in CEM dendrites, but active transport was not detectable in CEM cilia [17]

8. Your exposure may be too long or too short to detect movement. Check that the combination of your camera resolution and exposure time is likely to detect movement of the velocity expected of your marker. For example, for our Photometrics Cascade 512B EMCCD camera, the distance calibration for the 100× objective is 0.15997 μm/pixel. CEM cilia are approximately 3 μm (or 19 pixels) long. OSM-3::GFP puncta in wild type move at approximately 0.7 μm/s (or 4.4 pixels/s), so for each frame video acquired with 100 ms exposure time (or approximately ten video frames per second), we would expect to see 0.44 pixel displacement of a moving OSM-3::GFP particle in each frame, and each particle could travel from ciliary base to tip in 4.3 s. In contrast, if the exposure were set to 1000 ms, a given particle might appear to "jump" about one third of the way down the cilium and might not be readily detectable as a single moving particle.
9. Anomalous velocities can result from analyzing video of a cilium that does not reside completely in the plane of focus, causing the cilium to appear shorter than it is in reality. This leads to an underestimate of the velocity.
10. We have found applying "Calibrate Distances" each time before analysis helps to avoid mistakes in the original calibration (i.e., in case you forgot to select the 100X objective in Metamorph when acquiring the stream).
11. Vertical lines indicate stationary puncta. Some stationary puncta are expected. However, if your marker should be moving and is not, several explanations exist. Problems could arise due to overexpression of particular transgenes on the cell or cilium. For example, we have found that some GFP-tagged IFT reporters produce a synthetic ciliogenesis defect depending on the genetic background [23]. Alternatively, problems in transport could arise from poor health of mounted worms, owing to either problems in mounting or culture conditions. For an excellent overview of *C. elegans* biology and methodology, the reader is directed to the Worm Methods section of *Wormbook* (http://wormbook.org/toc_wormmethods.html). Your reporter may be too highly expressed, leading to intense fluorescence obscuring the detail necessary for kymography. Extrachromosomal arrays containing a high copy number of the fluorescent marker genes can lead to excessively high expression or overexpression phenotypes [24]. In this case, you may generate a dimmer reporter by injecting your transgene at a lower concentration and looking at more transgenic lines to select one with appropriate fluorescence expression. In the near future, genomic engineering techniques such as CRISPR-mediated genome editing [25] will be used to GFP-tag the endogenous gene of interest to ensure markers are expressed at native levels.

Acknowledgments

The authors were supported by NJCSCR Grant CSCR15IRG014 (R.O.) and NIH Grants DK059418 and DK074746 (M.B.).

References

1. Kozminski KG, Johnson KA, Forscher P, Rosenbaum JL (1993) A motility in the eukaryotic flagellum unrelated to flagellar beating. Proc Natl Acad Sci U S A 90(12):5519–5523
2. Pedersen LB, Geimer S, Rosenbaum JL (2006) Dissecting the molecular mechanisms of intraflagellar transport in Chlamydomonas. Curr Biol 16(5):450–459. doi:10.1016/j.cub.2006.02.020
3. Cole DG, Diener DR, Himelblau AL, Beech PL, Fuster JC, Rosenbaum JL (1998) Chlamydomonas kinesin-II-dependent intraflagellar transport (IFT): IFT particles contain proteins required for ciliary assembly in *Caenorhabditis elegans* sensory neurons. J Cell Biol 141(4):993–1008
4. Scholey JM (2008) Intraflagellar transport motors in cilia: moving along the cell's antenna. J Cell Biol 180(1):23–29
5. Inglis PN, Ou G, Leroux MR, Scholey JM (2006) The sensory cilia of *Caenorhabditis elegans*. WormBook:1–22
6. Ou G, Blacque OE, Snow JJ, Leroux MR, Scholey JM (2005) Functional coordination of intraflagellar transport motors. Nature 436(7050):583–587. doi:10.1038/nature03818
7. Snow JJ, Ou G, Gunnarson AL, Walker MR, Zhou HM, Brust-Mascher I, Scholey JM (2004) Two anterograde intraflagellar transport motors cooperate to build sensory cilia on *C. elegans* neurons. Nat Cell Biol 6(11):1109–1113. doi:10.1038/ncb1186
8. Jenkins PM, Hurd TW, Zhang L, McEwen DP, Brown RL, Margolis B, Verhey KJ, Martens JR (2006) Ciliary targeting of olfactory CNG channels requires the CNGB1b subunit and the kinesin-2 motor protein, KIF17. Curr Biol 16(12):1211–1216. doi:10.1016/j.cub.2006.04.034
9. Insinna C, Humby M, Sedmak T, Wolfrum U, Besharse JC (2009) Different roles for KIF17 and kinesin II in photoreceptor development and maintenance. Dev Dyn 238(9):2211–2222
10. Jiang L, Tam BM, Ying G, Wu S, Hauswirth WW, Frederick JM, Moritz OL, Baehr W (2015) Kinesin family 17 (osmotic avoidance abnormal-3) is dispensable for photoreceptor morphology and function. Faseb J. doi:10.1096/fj.15-275677
11. Miki H, Okada Y, Hirokawa N (2005) Analysis of the kinesin superfamily: insights into structure and function. Trends Cell Biol 15(9):467–476. doi:10.1016/j.tcb.2005.07.006
12. Morsci NS, Barr MM (2011) Kinesin-3 KLP-6 regulates intraflagellar transport in male-specific cilia of *Caenorhabditis elegans*. Curr Biol 21(14):1239–1244. doi:10.1016/j.cub.2011.06.027
13. Perkins LA, Hedgecock EM, Thomson JN, Culotti JG (1986) Mutant sensory cilia in the nematode *Caenorhabditis elegans*. Dev Biol 117(2):456–487
14. Mukhopadhyay S, Lu Y, Qin H, Lanjuin A, Shaham S, Sengupta P (2007) Distinct IFT mechanisms contribute to the generation of ciliary structural diversity in *C. elegans*. Embo J 26(12):2966–2980
15. O'Hagan R, Piasecki BP, Silva M, Phirke P, Nguyen KC, Hall DH, Swoboda P, Barr MM (2011) The tubulin deglutamylase CCPP-1 regulates the function and stability of sensory cilia in *C. elegans*. Curr Biol 21(20):1685–1694. doi:10.1016/j.cub.2011.08.049
16. Rand JB (2007) Acetylcholine. WormBook:1–21. doi:10.1895/wormbook.1.131.1
17. Qin H, Burnette DT, Bae YK, Forscher P, Barr MM, Rosenbaum JL (2005) Intraflagellar transport is required for the vectorial movement of TRPV channels in the ciliary membrane. Curr Biol 15(18):1695–1699
18. Warburton-Pitt SR, Silva M, Nguyen KC, Hall DH, Barr MM (2014) The *nphp-2* and *arl-13* genetic modules interact to regulate ciliogenesis and ciliary microtubule patterning in *C. elegans*. PLoS Genet 10(12), e1004866. doi:10.1371/journal.pgen.1004866
19. Cevik S, Sanders AA, Van Wijk E, Boldt K, Clarke L, van Reeuwijk J, Hori Y, Horn N, Hetterschijt L, Wdowicz A, Mullins A, Kida K, Kaplan OI, van Beersum SE, Man Wu K, Letteboer SJ, Mans DA, Katada T, Kontani K, Ueffing M, Roepman R, Kremer H, Blacque OE (2013) Active

transport and diffusion barriers restrict Joubert Syndrome-associated ARL13B/ARL-13 to an Inv-like ciliary membrane subdomain. PLoS Genet 9(12), e1003977. doi:10.1371/journal.pgen.1003977

20. Granato M, Schnabel H, Schnabel R (1994) pha-1, a selectable marker for gene transfer in *C. elegans*. Nucleic Acids Res 22(9): 1762–1763
21. Barr MM, Sternberg PW (1999) A polycystic kidney-disease gene homologue required for male mating behaviour in *C. elegans*. Nature 401(6751):386–389
22. Barr MM, DeModena J, Braun D, Nguyen CQ, Hall DH, Sternberg PW (2001) The *C. elegans* autosomal dominant polycystic kidney disease gene homologs *lov-1* and *pkd-2* act in the same pathway. Curr Biol 11(17): 1341–1346
23. Jauregui AR, Nguyen KC, Hall DH, Barr MM (2008) The *C. elegans* nephrocystins act as global modifiers of cilium structure. J Cell Biol 180(5):973–988
24. Prelich G (2012) Gene overexpression: uses, mechanisms, and interpretation. Genetics 190(3):841–854. doi:10.1534/genetics.111.136911
25. Frokjaer-Jensen C (2013) Exciting prospects for precise engineering of *C. elegans* genomes with CRISPR/Cas9. Genetics 195(3):635–642. doi:10.1534/genetics.113.156521
26. Bae YK, Qin H, Knobel KM, Hu J, Rosenbaum JL, Barr MM (2006) General and cell-type specific mechanisms target TRPP2/PKD-2 to cilia. Development 133(19):3859–3870
27. O'Hagan R, Wang J, Barr MM (2014) Mating behavior, male sensory cilia, and polycystins in *C. elegans*. Seminars in Cell & Developmental Biology 33:25–33. doi:10.1016/j.semcdb.2014.06.001
28. Wang J, Silva M, Haas LA, Morsci NS, Nguyen KC, Hall DH, Barr MM (2014) *C. elegans* ciliated sensory neurons release extracellular vesicles that function in animal communication. Curr Biol 24(5):519–525. doi:10.1016/j.cub.2014.01.002

Chapter 9

Visualization and Manipulation of Cilia and Intraciliary Calcium in the Zebrafish Left–Right Organizer

Shiaulou Yuan and Martina Brueckner

Abstract

Cilia play a key role in the determination of the left–right axis in vertebrates by generating and sensing flow of extraembryonic fluid at the left–right organizer (LRO). Perception of cilia-directed flow triggers a calcium signaling cascade which originates within the cilium itself and then is relayed into the surrounding mesendodermal tissue, thereby directing organ situs via the nodal pathway. Two types of cilia, motile and immotile, function simultaneously to coordinate and direct asymmetric intraciliary calcium signaling cues in the LRO. Here, we describe tools, reagents, and methodologies for the visualization and manipulation of both cilia types as well as intraciliary calcium signaling in the LRO of zebrafish.

Key words Cilium, Nodal cilia, Kupffer's vesicle, GECI

1 Introduction

Congenital heart disease (CHD) encompasses a wide spectrum of structural cardiac defects that are present from birth and is the most common birth defect, affecting an estimated one in 130 live births [1–5]. Although there is strong evidence to support a genetic contribution to CHD, the etiology of the disease has remained poorly understood [6–10]. Heterotaxy (Htx) is a rare congenital defect in which numerous visceral organs, such as the heart, are abnormally distributed due to disruptions in normal left–right (LR) axis patterning during embryogenesis. Htx is tightly correlated with complex CHD, as over 90% of Htx patients have CHD and display high mortality rates despite surgical intervention [6, 11–13]. Mutations affecting at least 12 genes affecting cilia structure and function have been identified in patients with Htx and other CHD [14].

Cilia are hairlike organelles found on the surface of most vertebrate cell types and serve a multitude of functions, including signaling, extracellular fluid propulsion, and cell cycle control [15–18]. The cilium is central to normal LR development, and

Peter Satir and Søren Tvorup Christensen (eds.), *Cilia: Methods and Protocols*, Methods in Molecular Biology, vol. 1454,
DOI 10.1007/978-1-4939-3789-9_9, © Springer Science+Business Media New York 2016

mutations affecting cilia structure and/or function result in abnormal LR development in humans, mice, and zebrafish [19]. LR asymmetry is initiated during the early somite stage at an evolutionarily conserved ciliated organ of asymmetry, the left–right organizer (LRO; also referred to as the "embryonic node" in mouse, "gastrocoel roof plate" in *Xenopus*, and "Kupffer's vesicle" in zebrafish). In the LRO, two populations of cilia coordinate to break left–right symmetry: motile cilia, which serve as motors to generate leftward flow of extraembryonic fluid [20], and immotile cilia which function as sensors to perceive flow [21]; both cilia and flow are required for LR development [22]. Flow has been shown to initiate the asymmetric expression of the *Nodal* signaling pathway to break bilateral symmetry and induce organ situs, such as heart looping [20, 23].

The ciliary localization of Pkd2, a voltage-sensitive, calcium-permeable cation channel, and other cation channels have suggested that the primary cilium itself may serve as a discrete calcium signaling compartment. However, fluorescent calcium dyes cannot be easily targeted to the cilium, making visualization of intraciliary calcium very challenging. This has recently been made possible by the development of genetically encoded calcium indicators (GECIs) [24]. We and others [25–29] have successfully targeted and imaged GECIs in cilia by fusing them to ciliary proteins such as the small ciliary GTPase Arl13b [30, 31]. In addition, by fusing Arl13b to the calcium-binding protein (CBP) parvalbumin, we have developed a novel tool to suppress intraciliary calcium [28]. Most importantly, these new tools have opened the door to observing and manipulating intraciliary calcium and understanding its role in cilia-mediated signaling. Utilizing these novel tools, we and others have revealed that high levels of calcium are detectable in the cilium of epithelial cells relative to the cytoplasm at resting state, confirming that the cilium is a discrete subcellular calcium compartment that is separated from the cytoplasm [25–29]. Further, we have revealed that dynamic oscillations in calcium levels occur in cilia of the LRO that are driven by Pkd2 and fluid flow generated by motile cilia.

In this chapter, we provide protocols and tools that have been recently developed by our research group to visualize and manipulate the motility and physiology of cilia in the LRO of zebrafish.

2 Materials

Embryos and Mounting

1. 1.0% Low-melt agarose.
2. 1× E3 embryo medium (5 mM NaCl, 0.17 mM KCl, 0.33 mM $CaCl_2$, 0.33 mM $MgSO_4$).

3. Watchmaker's forceps (Dumont, no. 5 or 55).
4. 2 mg/mL Pronase (Roche, #10165921001).
5. 60 mm glass-bottom petri dish.
6. Press-to-Seal Silicone Isolator (Sigma-Aldrich; #664301 nonadhesive).
7. Sharp tungsten needle (Roboz Surgical Instruments; #RS-6063).
8. Fine micro scalpel (Fine Science Tools; #10316-14).
9. Cover glass-bottom petri dish (MatTek, 20 mm cover glass, #P35G-1.5-20-C).
10. 25 × 75 mm glass microscope slides.
11. 24 × 60 mm cover glass (no. 1.5 thickness; 0.17 mm).
12. Water bath at 50 °C.
13. Zebrafish morpholino oligos (Sequences are shown in Table 1; custom synthesized by Genetools).
14. In vitro-synthesized mRNA for *arl13b-EGFP* [31, 36, 37] or transgenic *Tg(arl13b-EGFP)* zebrafish line [38] (Protocols for in vitro synthesis of mRNA for zebrafish microinjection has been previously reviewed [36, 37]).
15. In vitro-synthesized mRNA for *arl13b-mCherry* [39].
16. In vitro-synthesized mRNA for *arl13b-GCaMP6s* [28].
17. In vitro-synthesized mRNA for *arl13b-Pvalb-mCherry* [28].
18. Fluorescence stereoscope (e.g., Zeiss Lumar V12).

Imaging and Analysis

1. Zeiss Observer Z1 microscope or similar.
2. Zeiss 63× C-Apochromat water objective (1.2 numerical aperture) or similar.

Table 1
Morpholinos for paralyzing cilia motility in the LRO

Gene	Morpholino type	Morpholino oligo sequence	Effect on cilia motility	Recommended dose[a] (ng)	Reference
c21orf59	AUG	5′-CGCGTGTTATAT AAATATGAGACTT-3′	Paralysis	16	[42]
	Splice	5′-CACGTAAATATCA GTCATACCAGTG-3′	Paralysis	20	[42]
dnah9 (previously referred to as "*lrdr1*")	AUG	5′-GCGGTTCCTGCTCC TCCATCGCGCC-3′	Paralysis or partial paralysis	12	[43]
	Splice	5′-ACTCCAAGCCCTCA CCTCTTATCTA-3′	Paralysis or partial paralysis	16	[43]

This list represents morpholinos that were tested in our laboratory for the purpose of paralyzing motile cilia in the zebrafish LRO via high-speed DIC analysis. All morpholinos were manufactured by Genetools
[a]The suggested dose is for 1-cell embryo microinjections

3. MotionPro Y4 high-speed camera (Integrated Design Tools) or similar.
4. ORCA Flash4.0 v2 high-speed camera (Hamamatsu) or similar.
5. Inverted 710 DUO/NLO (Zeiss) confocal microscope with a 7-LIVE scanner (multiple high-speed CCD line detectors), a 488 nm laser, and a 561 nm laser or similar. Upright microscope systems are also suitable.
6. Stage-mounted thermoplate (Tokai Hit Corporation; #TP-ST2) or similar.
7. Motion Studio (Integrated Design Tools) acquisition software.
8. Definite Focus system (Zeiss) or similar.
9. Zen (Zeiss) acquisition software.
10. NIH ImageJ software (http://imagej.nih.gov/ij).
11. "R" graphing software (http://www.r-project.org) or Matlab (Mathlabs).

3 Methods

3.1 Mounting Approaches for Analyzing the Zebrafish LRO

Unlike the mouse LRO, which is located on the ventral surface of the developing embryo, the zebrafish LRO is buried within the tail bud near the interface of the yolk and embryo proper (Fig. 1). Thus, we and others have developed various approaches to precisely orient the tail bud directly toward the microscope objective to minimize the working distance between the LRO and objective [40]. Further, the optical clarity and ex vivo development of zebrafish further facilitate live imaging approaches.

3.1.1 Inverted Microscopes

With this mounting scheme and an inverted microscope (Fig. 2a, b), the zebrafish LRO can be viewed from the dorsal side of the embryo as a frontal section. The main objective of this mounting approach is to minimize the distance between the LRO and imaging objective in order to capitalize on the low relative working distance of high magnification and numerical aperture objectives (40×, 63×, 100×, etc.) to achieve maximal photon collection and spatial resolution of LRO cilia. Of note, we recommend utilizing long working distance water-immersion objectives, as they are better suited for imaging live whole embryos in an aqueous medium. All zebrafish stages and times indicated are following a standard incubation temperature of 28.5 °C.

1. Prepare a solution of 1.0% low-melt agarose in standard 1× E3 embryo medium without methylene blue (*see* **Note 1**). Once the low-melt agarose is dissolved, maintain the solution at 50 °C in an incubator or water bath.

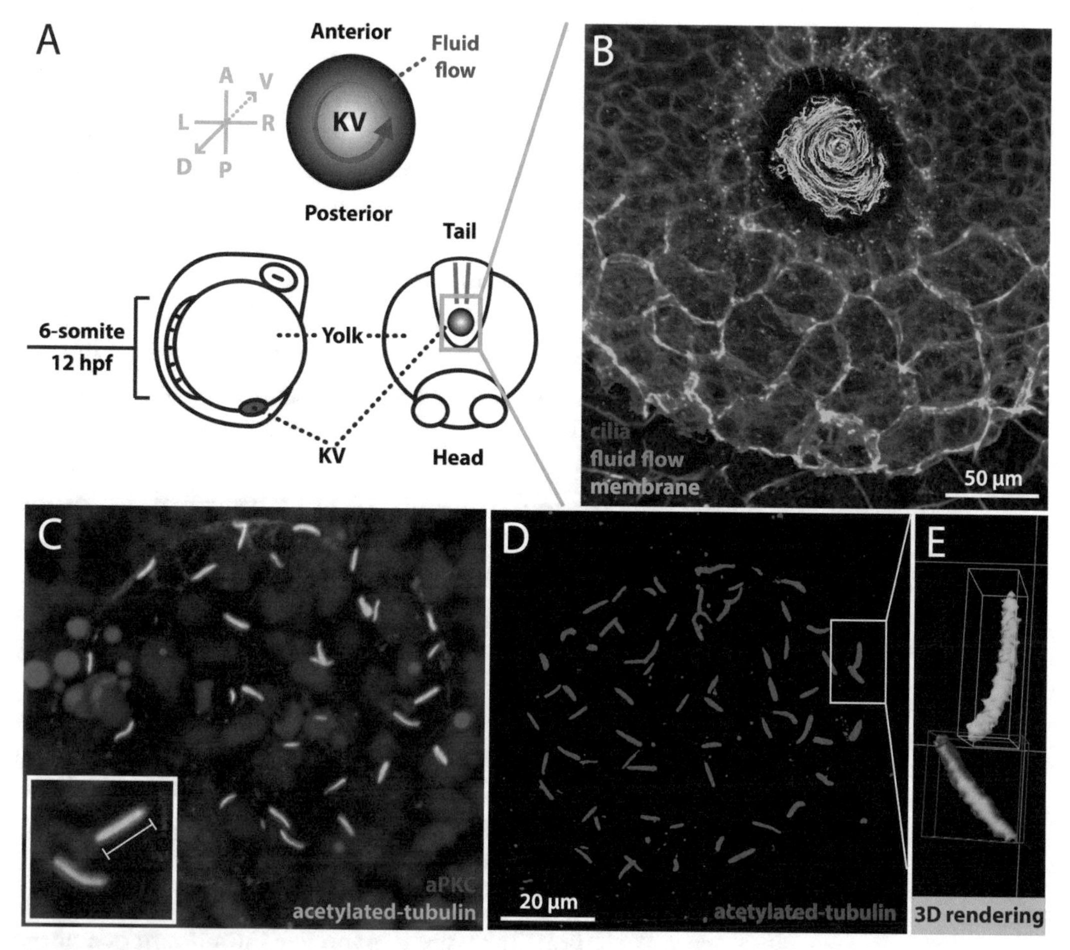

Fig. 1 Cilia in the zebrafish left–right organizer. (**a**) Schematic illustrating the morphology and positioning of the zebrafish LRO at the 6-somite stage. (**b**) Fluorescence montage of the zebrafish LRO labeled to visualize the plasma membrane (*cyan*) and cilia (*red*). Cilia-driven fluid flow is visualized by tracking the trajectory of fluorescent beads that were microinjected into the lumen of the organ at 8-somite stage. Scalebar = 50 μm. (**c**) Immunofluorescence image of the zebrafish LRO at the 8-somite stage, stained with antibodies against atypical PKC (*red*; apical membrane) and acetylated tubulin (*green*, cilia). (**d**, **e**) Immunofluorescence image (**d**) of cilia (*magenta*) in the zebrafish LRO at the 6-somite stage and a 3D volumetric rendering of two cilia along the periphery (**e**). Scalebar = 20 μm. Panels (**a**, **c**) are reproduced from Yuan 2012 with permission from PNAS

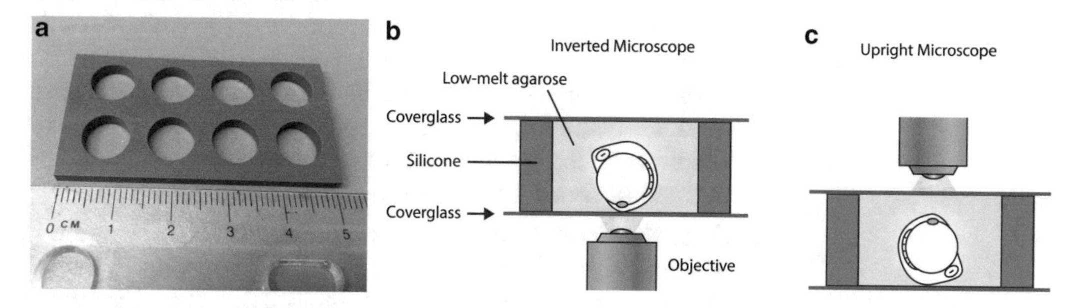

Fig. 2 Mounting embryos for LRO imaging. (**a**) Photograph of the Press-to-Seal Silicone Isolator that is utilized to form an imaging chamber. Each well has a diameter of 9 mm and a depth of 2.5 mm. (**b**, **c**) Schematic of an imaging chamber intended for inverted (**b**) or upright (**c**) imaging approaches utilizing low-melt agarose. Panel (**b**) is modified from Yuan 2013 with permission from Elsevier Inc

2. Collect fertilized embryos between bud (10 hpf) and 16-somite (16 hpf) stage. Detection of the zebrafish LRO is heavily stage dependent as this organ is transient: it forms at the end gastrulation and dissipates by mid-somitogenesis (*see* **Note 2**).
3. For analysis of cilia motility and cilia-directed fluid flow, collect embryos between 6- to 12-somite stage (12–15 hpf).
4. For detection of intraciliary calcium oscillations, collect embryos between bud and 5-somite stage (10.0–11.6 hpf).
5. For visualization of asymmetry in mesendodermal calcium levels around the LRO, collect embryos between 13 and 16 somite stages (15.5–17.0 hpf).
6. Manually dechorionate embryos using watchmaker's forceps or by enzymatic assistance using Pronase. For further details on dechorionation techniques, please *see* [41].
7. After dechorionation, handle the embryos using glass pipettes and glass-bottom petri dishes as dechorionated embryos may adhere to plastic.
8. Fill a 60 mm glass-bottom petri dish with 5 mL of the low-melt agarose solution. To prevent the agarose from cooling and solidifying, place the petri dish on a 50 °C heat block.
9. Place a "Press-to-Seal Silicone Isolator" (Fig. 2a) onto a 24 mm × 60 mm cover glass (*see* **Note 3**).
10. Transfer one embryo using a glass Pasteur pipette into the agarose solution. Take care to minimize the amount of embryo medium transferred into the agarose, and ensure the embryos are completely immersed in agarose. Young zebrafish embryos will disintegrate when exposed to air.
11. Once fully immersed, immediately transfer the embryo along with approximately 200 μL of agarose solution into a single well formed by the Press-to-Seal Silicone Isolator and 24 × 60 mm cover glass.
12. Under a dissecting microscope and using watchmaker's forceps or a sharp tungsten needle, quickly position the embryos at the bottom of the agarose-filled well, and then orient the embryo so that the dorsal side of the tail bud is touching the cover glass. In this orientation and developmental stage, the embryonic head will be facing upward toward the eyepiece, while the tail bud will be facing downward toward the stage (Fig. 2b).
13. Once the desired orientation is achieved, hold the embryo in position with the forceps or needle while the agarose polymerizes. Positioning should be accomplished within 15–20 s as the low-melt agarose will polymerize rapidly. Alternatively, to decrease the polymerization rate, mounting can be performed on a stage-mounted thermoplate.

14. Let the agarose further solidify and stabilize at RT for 5 min.
15. To prevent desiccation, add a small drop of E3 embryo medium to the surface of the agarose pad.
16. If desired, repeat **steps 10** through **15** to embed more embryos in additional wells (up to eight wells per silicone isolator).
17. After mounting all embryos, add a 24 × 50 mm cover glass on top of the silicone isolator to enclose the wells.
18. To find the zebrafish LRO, first locate the tail bud and then the notochord. The LRO will form at the posterior end of the notochord in the tail bud and will appear as a bubble (Fig. 1). The LRO morphology is that of an elliptical sphere that is highly enriched with cilia that are approximately 5–7 μm in length, while the surrounding mesendodermal tissue will have much shorter cilia. Typically, the LRO diameter ranges from 40 to 60 μm at the 6- to 8-somite stage.

3.1.2 Upright Microscopes

Compared to an inverted microscope, an upright microscope is advantageous for the analysis and manipulation of cilia-driven fluid flow in the LRO as the mounting scheme facilitates microinjection approaches (Fig. 2c) [40]. Further, water dipping objectives, which exhibit extremely long working distances, are available for most upright setups and may facilitate LRO work.

1. Follow **steps 1** through **11** from Subheading 3.1.1, but replace the 24 mm × 60 mm cover glass with a standard 25 mm × 75 mm glass microscope slide.
2. Under a dissecting microscope and using watchmaker's forceps or a sharp tungsten needle, quickly position the embryos at the bottom of the agarose-filled well, and then orient the embryo so that the dorsal side of the embryo's head is touching the cover glass. In this orientation and developmental stage, the tail bud will be facing upward toward the eyepiece, while the embryonic head will be facing downward toward the stage.
3. Follow **steps 13** through **18** from Subheading 3.1.1.

3.2 Recovery of Mounted Embryos for Analysis of Downstream Phenotypes

After mounting and all desired imaging is accomplished, each embryo may be unmounted and raised to a later embryonic stage of interest or to adulthood. This can be accomplished because the whole embryo is kept alive and intact during all mounting and imaging steps. In addition, low-melt agarose permits exchange of gases such as oxygen and is not inherently toxic to the embryo at the low concentration used. However, it should be noted that long-term culturing of embryos in low-melt agarose may stunt growth of the embryo and inhibit normal tail extension. Thus, after imaging the LRO, it is recommended to unmount the embryo from the agarose at approximately the 16- to 18-somite stage.

1. Remove the cover glass and Press-to-Seal Silicone Isolator using watchmaker's forceps. With the silicone spacer remover, the agarose-filled well will resemble the shape of a hockey puck and will commonly remain adherent to the bottom cover glass or slide.
2. Immerse the agarose pad and cover glass/slide into a standard 100 mm glass petri dish filled with E3 embryo medium.
3. Using watchmaker's forceps, gently dislodge and free the agarose pad from the cover glass/slide.
4. Hold the agarose pad with the forceps, and then gently slice away the agarose using a fine micro scalpel until the embryo falls out of the agarose pad. A single slice in the agarose along a plane that coincides with the embryo will often quickly free the embryo with ease.
5. Place each embryo in a single well of a 24-well plastic plate filled with E3 embryo medium and label accordingly. As young dechorionated embryos may adhere to plastic surfaces, it is recommended to pre-coat the bottom of each well with 1% agarose.
6. Culture the embryos normally at 28.5 °C.
7. For analysis of downstream LR readouts, please *see* Subheading 3.10.

3.3 Direct Visualization of Cilia Motility in the LRO by DIC Videomicroscopy

Due to the unique morphology and transparent characteristics of the cilium, differential interference contrast (DIC) microscopy is useful for enhancing the contrast of the organelle without the need for staining or labeling. Further, DIC imaging offers outstanding spatial resolution of both cilia and LRO (Fig. 3a) and affords the opportunity to capture the movement of motile cilia at extremely high acquisition rates.

The provided protocol utilizes a MotionPro Y4 camera on a Zeiss Observer Z1 with a 63× C-Apochromat water objective (1.2 numerical aperture) to achieve ultra high-speed DIC videomicroscopy capture rates. Notably, this particular camera has 16GB of onboard RAM memory, which hastens acquisition speeds (*see* **Note 4**).

1. To ensure an even field of illumination and the highest possible sample contrast, optimize Kohler illumination following your microscope manufacturer's recommended steps. Briefly, partially close the field diaphragm, focus the condenser lens and diaphragm, and finally open the field diaphragm to just beyond the field of view.
2. At the 6- to 8-somite stage (12–13 hpf at 28.5 °C), find the LRO (detailed in Subheading 3.1.1, **step 18**) and image using DIC illumination on a microscope with a 63× C-Apochromat water-immersion objective (or similar), a MotionPro Y4 high-speed camera, and Motion Studio acquisition software.

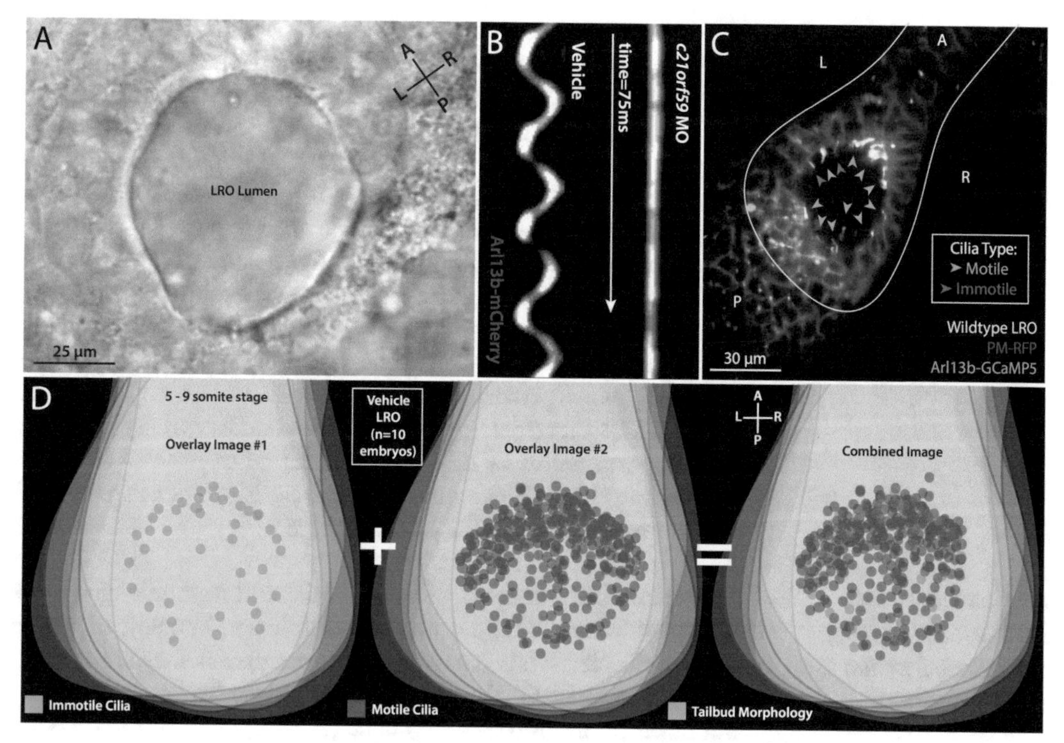

Fig. 3 In vivo visualization of cilia in the zebrafish LRO. (**a**) DIC micrograph of a live LRO at the 4-somite stage. Scalebar = 25 μm. (**b**) Kymographs of motile LRO cilia wave form from wild-type vehicle (*left*) or *c21orf59* morphant (*right*) embryos. *c21orf59* morphants display paralyzed cilia. Cilia were visualized by expression of *arl13b-mCherry* in the zebrafish LRO. (**c**) Live fluorescence image from a movie demonstrating both motile and immotile cilia in a LRO co-expressing *arl13b-GCaMP5* (*cyan*) and *membrane-RFP* (*magenta*). Morphology of the LRO and notochord is outlined in *yellow*. Scalebar = 30 μm. (**d**) Schematic illustrating the additive mapping approach for identifying the spatial positioning of motile and immotile cilia in the zebrafish LRO. Panels (**b**, **d**) are modified from Yuan 2015 with permission from Cell Press

3. Several recordings should be performed at various focal planes to accurately capture the *Z*-axis of the LRO as cilia will be distributed along the entirety of the apical membrane. All time series should be collected at a rate of 1000–2000 frames per second (fps) with a 499 to 999 μs exposure time or less and a 1016 × 1016 pixel resolution in Motion Studio.
4. Once the camera's internal memory bank is full, export the data as a TIFF image sequence.
5. Import the image sequence for a single recording into ImageJ (NIH) using the File menu command. The recordings can be analyzed or converted into a Quicktime movie in ImageJ (*see* **Note 5**).
6. Cilia beating dynamics may be analyzed and quantified by generating kymographs of an individual motile cilium from a reconstructed time series in MetaMorph (Molecular Devices) using the kymograph tool or in ImageJ using the reslice function (Fig. 3b). In both applications, use the draw line tool along

the maximal beat path of a single LRO cilium [40], and then generate a kymograph using the aforementioned functions. With a kymograph, the beat frequency of an individual cilium can be scored by counting the number of sinusoidal peaks.

3.4 Paralysis of Cilia Motility in the LRO

To paralyze LRO cilia motility, a morpholino to knockdown *c21orf59*, a gene implicated in human primary ciliary dyskinesia, or *dnah9*, which encodes a component of the axonemal dynein motor complex, may be utilized. Indeed, we have found that *c21orf59* knockdown results in total paralysis of LRO cilia, while *dnah9* results in partial paralysis of LRO cilia [28] (Fig. 3b). The sequences for these oligos are provided in the table below (Table 1). Standard techniques for microinjecting morpholinos and in vitro-synthesized mRNA for expression of transgenes in zebrafish have been previously reviewed [37]. Mechanical cilia paralysis provides an alternative to the use of morpholinos (*see* **Note 6**).

3.5 Simultaneous Fluorescence Imaging of Motile and Immotile LRO Cilia

The zebrafish LRO consists of both motile and immotile cilia types [28, 38]. Although DIC permits high-resolution imaging of motile cilia, immotile cilia can be challenging to visualize as they are stationary. Alternatively, LRO cilia may be visualized via fluorescence by targeting fluorophores into the organelle (Fig. 3c). Our research group has tested numerous ciliary proteins which successfully function as vectors for localizing fluorophores into the cilium. Among these, proteins that localize to the ciliary membrane, such as the small GTPase Arl13b [44], tend to be the most robust (Table 2 (Ciliary proteins suitable for fluorescence live imaging of cilia)) and are recommended. The provided protocol utilizes *arl13b-EGFP* [31] for high-speed fluorescence live imaging of zebrafish LRO cilia, followed by spatial mapping. The provided protocol utilizes a Hamamatsu ORCA Flash4.0 v2 sCMOS camera on a Zeiss Observer Z1 with a 63× C-Apochromat water objective (1.2 numerical aperture) to achieve

Table 2
Ciliary proteins suitable for fluorescence live imaging of cilia

Gene	Ciliary component	Fluorophore(s)	Reference
Arl13b	Small GTPase (membrane)	EGFP, mCherry, mApple, mRuby2	[30, 31]
Sstr3	GPCR (transmembrane)	EGFP	[32]
Htr6	GPCR (transmembrane)	EGFP	[32]
Mchr1	GPCR (transmembrane)	EGFP	[32]
Smo	GPCR (transmembrane)	EGFP, YFP	[33, 34]
Pkd2	Cation channel (transmembrane)	EGFP, Venus	[21, 35]

This list represents ciliary proteins that were tested in our laboratory for the purpose of live fluorescence visualization of cilia in the zebrafish LRO. Each ciliary protein was fused and tested with the indicated fluorophores high-speed fluorescence microscopy acquisitions. Please *see* **Note 7** for selection of proteins for cilia localization.

1. Microinject fertilized zebrafish embryos at the 1-cell stage with 100 pg of in vitro-synthesized *arl13b-EGFP* mRNA [31]. Protocols for microinjection of mRNA and morpholinos in zebrafish and preparation of in vitro-synthesized mRNA have been previously described [37]. Alternatively, embryos may be collected from mating a *Tg(arl13b:EGFP)* transgenic zebrafish line [36].
2. Culture and mount the embryos according to Subheading 3.1.
3. From the 6- to 12-somite stage, find the LRO (detailed in Subheading **3.1.1**, step **18**), and image using a 488 nm excitation wavelength on an Observer Z1 microscope with a 63× C-Apochromat water-immersion objective, an ORCA Flash4.0 v2 high-speed camera, and Zen acquisition software.
4. Several recordings should be performed at various focal planes to accurately capture the *Z*-axis of the LRO as cilia will be distributed along the entirety of the apical membrane [45]. All time series should be collected at a rate of 100–200 frames per second (fps) with a 5 ms exposure time and a 2048 × 2048 or 1024 × 1024 pixel resolution in Zen.
5. Once acquired, export the data as a TIFF image sequence.
6. Import the image sequence for a single recording into ImageJ using the File menu command. The recordings can be analyzed or converted into a Quicktime movie in ImageJ. Alternatively, for higher quality compression codecs, *see* **Note 5**.
7. Kymographic analysis may also be performed utilizing fluorescently labeled cilia (please *see* Subheading 3.3, **step 6**).

3.6 Spatiotemporal Mapping of Motile and Immotile Cilia in the LRO

To analyze the spatial distribution of motile and immotile cilia in the LRO, additive analysis can be applied to the LRO of multiple embryos to achieve a heat map-like distribution of LRO cilia (Fig. 3d) [28, 39].

1. In ImageJ, open movies containing live high-speed acquisitions of zebrafish LROs from bud to 16-somite stage expressing *arl13b-EGFP*.
2. Utilizing the time scale bar and multipoint tool, movies can be slowed down, played back, and scored for motile and immotile cilia in ImageJ onto two overlaying images: one for motile and the second for immotile cilia.
3. Convert each overlaying image into a binary image and then convert a second time into a false-colored 8-bit image.
4. Merge both images to generate a two-color map of cilia distribution for an individual LRO. For each experimental condition, numerous two-color maps from several embryos can be

assembled into an image series and processed into a single composite image using the maximal Z-project stack function in ImageJ. The tail bud morphology surrounding each LRO may be traced from corresponding DIC images in ImageJ as a third overlaying image.

3.7 Altering the Ratio of Motile and Immotile Cilia in the LRO

Recent studies have indicated a role for notch signaling in determining the ratio of motile and immotile cilia in the LRO [39]. Activation of notch signaling decreases the ratio of motile to immotile cilia, while inhibition of notch increases this ratio. Thus, manipulation of notch signaling may be utilized to precisely determine the patterning of the two cilia types in the LRO. The provided protocol utilizes a Hamamatsu ORCA Flash4.0 v2 sCMOS camera on a Zeiss Observer Z1 with a 63× C-Apochromat water objective (1.2 numerical aperture) to achieve high-speed fluorescence microscopy acquisitions (*see* **Note 4**).

1. To activate notch signaling, microinject 12.5 pg of *Nicd* (Notch1 intracellular domain) mRNA at the 1-cell stage [39]. In contrast, to inhibit notch signaling, microinject 3 ng of a translational blocking morpholino against *deltaD* (5′-AAACA GCTATCATTAGTCGTCCCAT-3′), a ligand of the Notch1 receptor at the 1-cell stage [46].
2. To fluorescently label LRO cilia, co-inject these embryos with 100 pg of *arl13b-eGFP* mRNA.
3. Culture, mount, and image the embryos according to Subheadings 3.1 and 3.5.
4. Immotile and motile cilia types may also be mapped using the protocol provided in Subheading 3.6.

3.8 Spatiotemporal Mapping of Intraciliary Calcium Oscillations in LRO

For visualization of intraciliary calcium oscillations during LRO development, we have developed a ratiometric fluorescence approach utilizing a cilia-targeted genetically encoded calcium indicator (GECI) [28]. We have previously found that calcium oscillations in cilia of the zebrafish LRO are most abundant on the left-side of the organ and that they require Pkd2 and cilia-driven fluid flow. Here, we provide a protocol using Arl13b-GCaMP6s in combination with mCherry as a ratiometric biosensor for detecting these oscillations in the cilium (Fig. 4a). Please *see* **Note 8** for a

Fig. 4 Detection and analysis of intraciliary calcium oscillations in the LRO. (**a**) Schematic illustrating the ratiometric time-lapse imaging approach utilizing cilia-targeted calcium biosensors in the zebrafish LRO to detect calcium dynamics. (**b**) Graph of quantified fluorescence intensity over time from a representative control LRO cilium expressing *arl13b-GCaMP6s* (*cyan*) and *arl13b-mCherry* (*magenta*). Large repetitive peaks correspond to intraciliary calcium oscillations. (**c**) Ratiometric representation of the same data shown in graph (**b**). *Black line* across the *Y*-axis indicates threshold for identifying oscillation peaks (ICOs). Panels (**b**, **c**) are reproduced from Yuan 2015 with permission from Cell Press

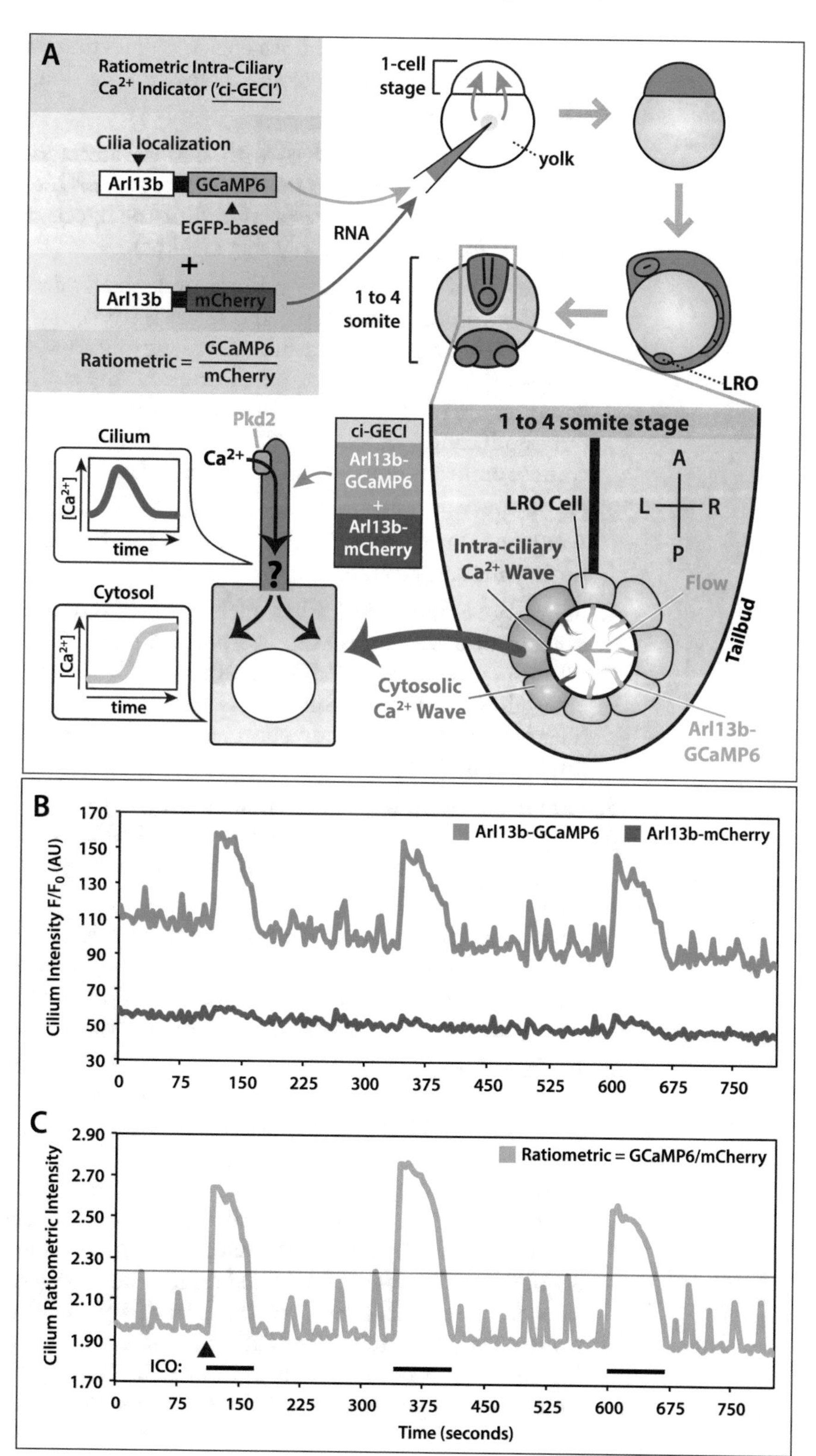
A
Ratiometric Intra-Ciliary
Ca2+ Indicator ('ci-GECI')
Cilia localization
Arl13b
GCaMP6
EGFP-based
+
Arl13b
mCherry
Ratiometric = GCaMP6 / mCherry
1-cell stage
yolk
RNA
1 to 4 somite
LRO
1 to 4 somite stage
A
L
R
P
LRO Cell
Intra-ciliary Ca2+ Wave
Flow
Tailbud
Cytosolic Ca2+ Wave
Arl13b-GCaMP6
Pkd2
Ca2+
ci-GECI
Arl13b-GCaMP6
+
Arl13b-mCherry
?
Cilium
[Ca2+]
time
Cytosol
[Ca2+]
time
B
Arl13b-GCaMP6
Arl13b-mCherry
Cilium Intensity F/F0 (AU)
170
150
130
110
90
70
50
30
0
75
150
225
300
375
450
525
600
675
750
C
Ratiometric = GCaMP6/mCherry
Cilium Ratiometric Intensity
2.90
2.70
2.50
2.30
2.10
1.90
1.70
ICO:
0
75
150
225
300
375
450
525
600
675
750
Time (seconds)

discussion of suitable confocal systems for detecting intraciliary calcium oscillations. Further, please *see* **Note 9** regarding selection of calcium biosensors.

1. Co-inject wild-type embryos with sterile water and an equimolar concentration of *arl13b-GCaMP6s* and *arl13b-mCherry* mRNA (78 and 60 pg, respectively) at the 1-cell stage, and culture until the 90% epiboly stage (9 hpf).
2. At late epiboly stages, sort embryos under a standard fluorescence dissection microscope, and select for imaging based on the following criteria: (1) normal gross morphology, (2) global positive expression of reporters throughout the entire embryo based on the mCherry signal (does not exhibit mosaicism), (3) not significantly developmentally delayed (±1 h) relative to control embryos, and (4) intact and not damaged yolk sac.
3. Dechorionate embryos using watchmaker's forceps or with 5 mg/mL Pronase for 5 min, and then thoroughly wash five times with embryo medium.
4. Submerge embryos in 1% low-melt agarose, and mount at the bottom of a 35 mm cover glass-bottom petri dish (MatTek, 20 mm cover glass, #P35G-1.5-20-C), with the tail bud oriented down toward the cover glass. Alternatively, the Press-to-Silicone Isolator mounting approach may be also utilized (*see* Subheading 3.1).
5. Commence long-term time-lapse imaging of the LRO dorsally starting at bud stage (10 hpf), when cilia begin to form in the zebrafish LRO, through 16-somite stage (~16 hpf), when the LRO begins to dissipate.
6. To accurately stage embryos under acquisition, take stage-matched sibling embryos, mounted in a similar fashion, and place near the microscope stage. Periodically score for somite progression under a standard dissection microscope.
7. Perform imaging on an inverted confocal microscope with a 63× C-Apochromat water objective as described in the equipment section. Using an incubated stage heater, maintain conditions at 28.5 °C, the optimal temperature for zebrafish embryonic development. Capture simultaneous two-channel recordings at an acquisition rate of 0.2 Hz (or more frequently) at a 512×512 pixel resolution with no binning. The acquisition rate was chosen based on the duration (~40 s) and slow frequency (2–4 min) of the intraciliary calcium oscillations at the 1–4 somite stage [28] but at the same time balancing this with a need to limit phototoxicity during prolonged time-lapse imaging sessions. Of note, recordings of intraciliary calcium oscillations obtained at 1 Hz revealed no significant

differences in frequency, duration, amplitude, and total number relative to 0.2 Hz; however, photobleaching became a limiting factor for long-term imaging. Acquire optical sections (5 μm thick) at a plane of focus within the zebrafish LRO that is located along the equator of the organ and which visualizes the maximal number of cilia along the LR axis.

8. Acquire time-lapse recordings spanning the entire course of LRO development, from bud to 16-somite stage (~6 h in duration). It is recommended to monitor the acquisition in order to correct for sample, focal, and thermal drift. To assist with thermal drift, a Definite Focus system (Zeiss) may be utilized at a frequency of once per minute.
9. After acquiring data, intraciliary calcium oscillations may be quantified in Zen or ImageJ (Fig. 4b). To perform ratiometric fluorescence analysis, divide the GCaMP6 signal by mCherry using the ratio tool in Zen or the "Image Calculator" tool in ImageJ (Fig. 4c). To false color the data according to intensity (and thus concentration of calcium), a rainbow-colored intensometric look-up table ("Rainbow RGB") may be applied in either Zen (Zeiss) ("Channel Color" tool) or ImageJ (NIH) ("LUT" tool). Define a region of interest in the cilium using the appropriate tool in either software.
10. To analyze the dynamics of intraciliary calcium oscillations over time, a baseline reading of steady-state intraciliary calcium levels for 5–30 s may be used to divide the experimental data (F/F_0 or $\Delta F/F_0$; this type of fluorescence analysis is standard in the calcium field and has been previously reviewed [47]).
11. To facilitate classification of calcium dynamics in the LRO [28], we defined a ratiometric threshold value of >2.25 for positive calcium oscillations and waves based on the following observations: (1) wild-type intraciliary oscillations, which were leftward biased at the LRO, consistently displayed a peak amplitude value that ranged from approximately 2.3 to 3, a duration of at least 30 s, a frequency of at least 2 min, and a high signal-to-noise ratio in the GCaMP channel and (2) smaller fluctuations in intraciliary calcium were commonly detected that exhibited an amplitude of approximately 2.1–2.2 and a brief duration (~5 s) yet were irregular in frequency and distribution and exhibited a low signal-to-noise ratio in the GCaMP channel. Cytosolic calcium activity was defined utilizing similar criteria.
12. To perform spatiotemporal mapping of intraciliary calcium oscillations in the LRO, partition the entire acquisition by embryonic stage. We previously detected significant differences in the abundance and frequency of intraciliary

oscillations in the following four stages: bud, 1–4 somite, 5–9 somite, and 10–16 somite [28]. Overlay a transparent image on top of the time-lapse acquisition in ImageJ. Play back the partitioned acquisition slowly and manually mark where intraciliary oscillations occur in the LRO onto the transparent layer using the multipoint tool.

13. Pinpoint the center of the LRO lumen using the 'Angle' tool in ImageJ, and quantify the spatial position of each oscillation around the center as an angle, with the anterior–posterior and left–right axis as the *Y*- and *X*-axis, respectively. Quantify the percentage of cilia with intraciliary oscillations by region in the LRO (treating the lumen of the LRO as a circle), and present as a rose diagram using standard statistical software such as R (open source software; http://www.r-project.org) or Matlab.

3.9 Suppression of Intraciliary Calcium Oscillations in LRO

In order to demonstrate the functional relevance of intraciliary calcium signaling to embryonic development, we devised a tool to suppress calcium elevations in the cilium by targeting a small 11 kDa calcium-binding protein, parvalbumin (Pvalb), into the cilium via fusion to Arl13b [28, 48]. In addition, we attached a mCherry fluorophore to the C-terminus of this fusion protein to facilitate ratiometric calcium imaging in combination with Arl13b-GCaMP6s (Fig. 5a). Co-expression of *arl13b-Pvalb-mCherry* with *arl13b-GCaMP6s* in the zebrafish resulted in the suppression of intraciliary calcium oscillations and prevalance of cardiac looping defects (Fig. 5b-e). Here we provide a protocol outlining this experiment.

1. Microinject fertilized zebrafish embryos at the 1-cell stage with 138 pg of *arl13b-Pvalb-mCherry* mRNA and 78 pg of *arl13b-GCaMP6* mRNA.
2. Follow Subheading 3.8 for mounting, imaging, and analysis of intraciliary calcium oscillations in response to Arl13b-Pvalb-mCherry.
3. For downstream analysis of phenotype LR defects in the embryos, follow Subheading 3.2 for recovery of embryos and then the following section (Subheading 3.10) for specific assays to determine LR defects.

3.10 Assessing for Left–Right Phenotypic Defects in Zebrafish

A robust and easy to assess readout for dysfunctional ciliary signaling at the LRO is the establishment of cardiac laterality. Cardiac laterality may be analyzed by gross morphological inspection of heart morphology and placement under a stereoscope using basic transmitted light. Heart jogging is the positioning of the primitive heart tube across the left–right axis (Fig. 5c–e), while heart looping is the morphogenetic process in which the primitive heart loops asymmetrically across the left–right axis.

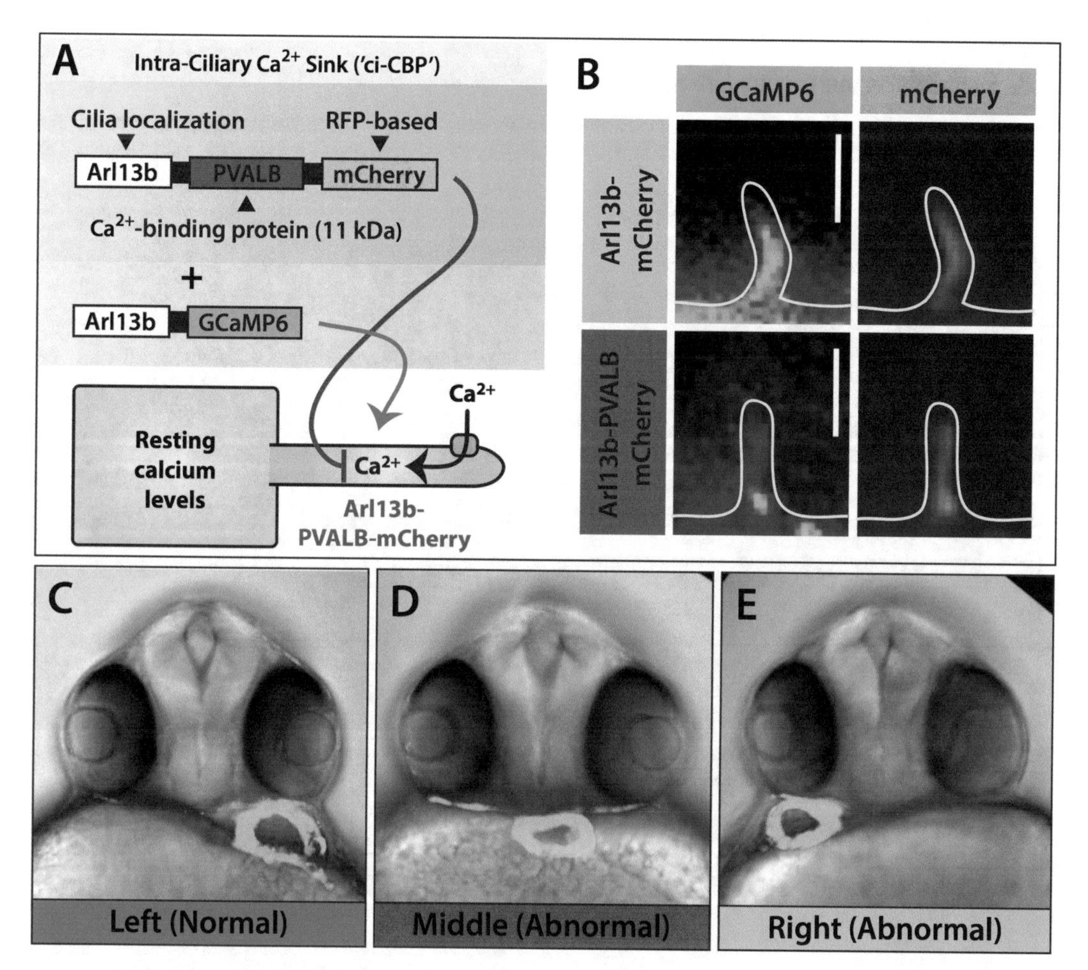

Fig. 5 Suppression of intraciliary calcium oscillations in the LRO. (**a**) Schematic illustrating fusion protein design for an intraciliary calcium sink that enables simultaneous suppression and visualization of intraciliary calcium oscillations. (**b**) Representative fluorescence image from a live LRO expressing intraciliary calcium sink, with *arl13b-PVALB-mCherry* signal in magenta and *arl13b-GCaMP6* signal in blue-green. Scalebar = 5 μm. (**c–e**) Merged fluorescence and bright-field image of cardiac jogging in *Tg(cmlc2:EGFP)* transgenic zebrafish (stage: 20 hpf), demonstrating normal left (**c**) or abnormal middle (**d**) and abnormal right-positioned (**d**) heart tubes. Panels (**a–e**) are reproduced from Yuan 2015 with permission from Cell Press

In zebrafish, heart jogging can be easily assessed at 24 h postfertilization, while heart jogging is best assessed at 36 h postfertilization and onward.

Alternatively, in situ hybridization for well-characterized molecular markers of LR asymmetric patterning may also be assessed. We provide a list of LR markers that are commonly utilized in zebrafish below (Table 3). Further, in situ hybridization protocols for analysis of LR markers in zebrafish have also been previously described [36, 54].

Table 3
In situ hybridization markers for left–right patterning

Step of the LR cascade	Gene	Stage to analyze	Tissue	Wild-type pattern	Reference
Early	*dand5* (also referred to as "*charon*")	8- to 10-somite	Peripheral region of LRO	Right sided	[46, 49]
	southpaw	18- to 20-somite	Lateral plate mesoderm	Left sided	[50]
	lefty2	22-somite	Lateral plate mesoderm	Left sided	[51]
	pitx2	22- to 25-somite	Dorsal diencephalon, developing gut and heart fields, lateral plate mesoderm	Left sided	[52]
	cmlc2	26 hpf	Heart	Left sided or D looped	[53]
Late					

This list represents well-established molecular markers for various components of the left–right signaling cascade in zebrafish that can be assayed by in situ hybridization

4 Notes

1. *Embryo medium without methylene blue.* Methylene blue is commonly used as a fungicide in E3 embryo medium; however, it is a potent cationic dye that absorbs light particularly in the red and far-red range.
2. *Embryo timing.* As zebrafish typically mate in the mornings in traditional zebrafish laboratory vivariums, these stages occur at night. To alleviate staging inconvenience, animals of interest can be maintained and mated under a nontraditional light–dark cycle. Another solution is to incubate the fertilized embryos at nontraditional temperatures (between 21 and 33 °C) to alter the rate of development. Due to temperature variations in incubators, quantification of somites is a more robust method for staging embryos rather than time (hpf) alone.
3. *Mounting chamber.* The silicone isolator is an autoclavable and reusable hydrophobic gasket that functions as a spacer and creates a 2.5 mm deep well for the isolation of live specimens. The silicone isolator readily adheres to the surface of glass and forms a hydrophobic seal.
4. *Selection of cameras.* In order to perform high-speed DIC videomicroscopy of LRO cilia which may beat at a rate greater than 40 Hz [55], an appropriate scientific digital camera that

is capable of high-speed acquisitions must be interfaced with the microscope. In the past, suitable cameras were few and costly. Currently, there is a wide selection of commercially available cameras across a range of costs depending on the technology utilized. Below, we list examples that we have verified for live cilia work in our research group (Table 4).

5. *Movie processing.* For higher quality compression codecs and container options, the image sequence can be imported into Quicktime Pro 7 (Apple), Compressor (Apple), Final Cut X (Apple), Premiere CC (Adobe), or an equivalent application and converted into a plethora of movie formats. We suggest the following codec and movie container format for viewing and distribution: H.264 and Quicktime MOV.
6. *Mechanical cilia paralysis.* Cilia motility may be inhibited by direct microinjection of 2% methylcellulose, a clear yet highly viscous solution, into the lumen of the LRO. Methylcellulose methods for the LRO have been previously reviewed [43].
7. *Selection of proteins for cilia localization.* Recent proteomic studies have identified over 300 proteins in the primary cilium [56, 58–60]. Thus, it is not surprising that several of these ciliary proteins have been successfully utilized as effective vectors for trafficking fluorophores, epitope tags, and biosensors into the cilium. Among these, proteins that localize to the ciliary membrane have been successfully adapted for fluorescence live imaging of primary cilia. In the zebrafish LRO, we have utilized the following ciliary membrane-related fusion proteins for the purpose of in vivo imaging of both motile and immotile cilia (Table 2).
8. *Confocal systems for detection of intraciliary calcium oscillations.* Although our research group has documented the existence of intraciliary calcium oscillations in the zebrafish LRO with an inverted 710 DUO/NLO (Zeiss) confocal microscope with a 63× C-Apochromat water objective (1.2 numerical aperture) and a 7-LIVE scanner, numerous other commercial confocal systems are suitable. Notably, the most important factor is the capability for high-speed confocal fluorescence acquisition. We have tested and recommend the following suitable microscopes: Leica SP8 (with a resonant scanner), Nikon A1R (with a resonant scanner), or Bruker Opterra (with a EMCCD camera from Table 4).
9. *Selection of calcium biosensors.* Due to significant advances in genetically encoded calcium indicators (GECIs) in recent years, there is now a wide choice of biosensors that may be adapted for probing the ionic composition of the cilioplasm [57]. Below we provide a table of calcium biosensors that we have tested in the zebrafish LRO and report their ability to localize to cilia when fused with Arl13b and their capability to detect

Table 4
Digital cameras for in vivo imaging of cilia

Sensor type	Camera model	Manufacturer	Pixel size (μm)	Resolution (pixels)	Speed at full resolution (FPS)	Quantum efficiency (%)	Suitable for cilia DIC?	Suitable for cilia fluorescence?	Cost
Cooled CCD	Coolsnap HQ2[a]	Photometrics	6.5	1392 × 1040	11	>60	Yes	Yes	$
Cooled CCD	Retiga-2000R[a]	Q Imaging	7.4	1600 × 1200	7.5	55	Yes	Yes	$
sCMOS	ORCA Flash4.0 v2	Hamamatsu	6.5	2048 × 2048	100	82	Yes	Yes	$$
sCMOS	Zyla 5.5	Andor	6.5	2560 × 2160	100	60	Yes	Yes	$$
CMOS	MotionPro Y4	IDT	13.9	1024 × 1024	7000	NA	Yes	No	$$
EMCCD	iXon Ultra 897	Andor	16	512 × 512	56	>90	Yes	Yes	$$$
EMCCD	ImagEM *X*2	Hamamatsu	16	512 × 512	70	>90	Yes	Yes	$$$

This list represents cameras that were tested in our laboratory for the purpose of visualizing LRO cilia by DIC and fluorescence. The manufacturers provided technical specifications

[a]These two cameras are only suitable for accurately capturing LRO cilia beating only if part of the sensor is utilized (via a region of interest) or if binning is applied

Table 5
Calcium biosensors for detection of intraciliary calcium dynamics

Reporter name	Calcium affinity K_d (nM)[a]	Fluorophore	Biosensor type	Detects calcium oscillations in LRO?	Localizes to LRO cilia (via Arl13b)?	Reference
GCaMP3	540	EGFP	Single wavelength	No	Yes	[61]
GCaMP5g	460	EGFP	Single wavelength	No	Yes	[24]
GCaMP6s	144	EGFP	Single wavelength	Yes	Yes	[62]
GCaMP6f	375	EGFP	Single wavelength	No	Yes	[62]
GCaMP7a	243	EGFP	Single wavelength	Yes	Yes	[63, 64]
GCaMP8	200	EGFP	Single wavelength	Yes	Yes	[64]
RCaMP1h	1300	mRuby	Single wavelength	No	Yes	[65]
G-GECO1.0	750	EGFP	Single wavelength	No	Yes	[66]
G-GECO1.1	620	EGFP	Single wavelength	No	Yes	[66]
G-GECO1.2	1150	EGFP	Single wavelength	No	Yes	[66]
R-GECO1.0	480	mApple	Single wavelength	No	Yes	[66]
B-GECO1.0	480	BFP	Single wavelength	No	Yes	[66]
YCnano15	15	CFP-YFP	FRET	No	Yes	[67]
YCnano30	30	CFP-YFP	FRET	No	Yes	[67]
YCnano50	50	CFP-YFP	FRET	No	Yes	[67]
YCnano65	65	CFP-YFP	FRET	No	Yes	[67]
YCnano140	140	CFP-YFP	FRET	No	Yes	[67]

This list represents calcium bionsensors that were tested in our laboratory for the purpose of live fluorescence visualization of intraciliary calcium dynamics in the zebrafish LRO. Each biosensor was trafficked to the cilium by fusing to Arl13b

[a]Calcium K_d values were derived by the original authors in the corresponding references; conditions and method of measurement may vary

intraciliary calcium oscillations (Table 5). Notably, we have experienced the most success with single wavelength biosensors with a high affinity for calcium (<200 nM K_d) and rapid on–off calcium ion-binding kinetics. For the targeted purpose of detecting intraciliary calcium oscillations in the zebrafish LRO, we specifically recommend GCaMP6s, 7a, or 8.

Acknowledgments

This work was funded by NIH grant 1R01HL125885-01A1 (to M.B.) and a Hartwell Foundation postdoctoral fellowship (to S.Y.).

References

1. Dolk H, Loane M, Garne E (2010) The prevalence of congenital anomalies in Europe. Adv Exp Med Biol 686:349–364. doi:10.1007/978-90-481-9485-8_20
2. Dolk H, Loane M, Garne E (2011) Congenital heart defects in Europe: prevalence and perinatal mortality, 2000 to 2005. Circulation 123(8):841–849. doi:10.1161/CIRCULATIONAHA.110.958405
3. Heron M, Hoyert DL, Murphy SL, Xu J, Kochanek KD, Tejada-Vera B (2009) Deaths: final data for 2006. National vital statistics reports: from the Centers for Disease Control and Prevention, National Center for Health Statistics, National Vital Statistics System 57(14):1–134
4. Cuneo BF (2006) Outcome of fetal cardiac defects. Curr Opin Pediatr 18(5):490–496
5. van der Linde D, Konings EE, Slager MA, Witsenburg M, Helbing WA, Takkenberg JJ, Roos-Hesselink JW (2011) Birth prevalence of congenital heart disease worldwide: a systematic review and meta-analysis. J Am Coll Cardiol 58(21):2241–2247. doi:10.1016/j.jacc.2011.08.025
6. Zhu L, Belmont JW, Ware SM (2006) Genetics of human heterotaxias. Eur J Hum Genet 14(1):17–25
7. Pierpont ME, Basson CT, Benson DW Jr, Gelb BD, Giglia TM, Goldmuntz E, McGee G, Sable CA, Srivastava D, Webb CL (2007) Genetic basis for congenital heart defects: current knowledge: a scientific statement from the American Heart Association Congenital Cardiac Defects Committee, Council on Cardiovascular Disease in the Young: endorsed by the American Academy of Pediatrics. Circulation 115(23):3015–3038
8. Ransom J, Srivastava D (2007) The genetics of cardiac birth defects. Semin Cell Dev Biol 18(1):132–139
9. Weismann CG, Gelb BD (2007) The genetics of congenital heart disease: a review of recent developments. Curr Opin Cardiol 22(3):200–206
10. Bruneau BG (2008) The developmental genetics of congenital heart disease. Nature 451(7181):943–948. doi:10.1038/nature06801
11. Belmont JW, Mohapatra B, Towbin JA, Ware SM (2004) Molecular genetics of heterotaxy syndromes. Curr Opin Cardiol 19(3):216–220
12. Kennedy MP, Omran H, Leigh MW, Dell S, Morgan L, Molina PL, Robinson BV, Minnix SL, Olbrich H, Severin T, Ahrens P, Lange L, Morillas HN, Noone PG, Zariwala MA, Knowles MR (2007) Congenital heart disease and other heterotaxic defects in a large cohort of patients with primary ciliary dyskinesia. Circulation 115(22):2814–2821
13. Sutherland MJ, Ware SM (2009) Disorders of left-right asymmetry: heterotaxy and situs inversus. Am J Med Genet C Semin Med Genet 151C(4):307–317
14. Yuan S, Zaidi S, Brueckner M (2013) Congenital heart disease: emerging themes linking genetics and development. Curr Opin Genet Dev 23(3):352–359. doi:10.1016/j.gde.2013.05.004
15. Goetz SC, Anderson KV (2010) The primary cilium: a signalling centre during vertebrate development. Nat Rev Genet 11(5):331–344. doi:10.1038/nrg2774
16. Satir P, Christensen ST (2007) Overview of structure and function of mammalian cilia. Annu Rev Physiol 69:377–400. doi:10.1146/annurev.physiol.69.040705.141236

17. Satir P, Pedersen LB, Christensen ST (2010) The primary cilium at a glance. J Cell Sci 123(Pt 4):499–503. doi:10.1242/jcs.050377
18. Pedersen LB, Rosenbaum JL (2008) Intraflagellar transport (IFT) role in ciliary assembly, resorption and signalling. Curr Top Dev Biol 85:23–61. doi:10.1016/S0070-2153(08)00802-8
19. Basu B, Brueckner M (2008) Cilia multifunctional organelles at the center of vertebrate left-right asymmetry. Curr Top Dev Biol 85:151–174. doi:10.1016/S0070-2153(08)00806-5
20. Nonaka S, Tanaka Y, Okada Y, Takeda S, Harada A, Kanai Y, Kido M, Hirokawa N (1998) Randomization of left-right asymmetry due to loss of nodal cilia generating leftward flow of extraembryonic fluid in mice lacking KIF3B motor protein [published erratum appears in Cell 1999 Oct 1;99(1):117]. Cell 95(6):829–837
21. Yoshiba S, Shiratori H, Kuo IY, Kawasumi A, Shinohara K, Nonaka S, Asai Y, Sasaki G, Belo JA, Sasaki H, Nakai J, Dworniczak B, Ehrlich BE, Pennekamp P, Hamada H (2012) Cilia at the node of mouse embryos sense fluid flow for left-right determination via Pkd2. Science 338(6104):226–231. doi:10.1126/science.1222538
22. Yoshiba S, Hamada H (2014) Roles of cilia, fluid flow, and Ca2+ signaling in breaking of left-right symmetry. Trends Genet 30(1):10–17. doi:10.1016/j.tig.2013.09.001
23. Nakamura T, Hamada H (2012) Left-right patterning: conserved and divergent mechanisms. Development 139(18):3257–3262. doi:10.1242/dev.061606
24. Akerboom J, Chen TW, Wardill TJ, Tian L, Marvin JS, Mutlu S, Calderon NC, Esposti F, Borghuis BG, Sun XR, Gordus A, Orger MB, Portugues R, Engert F, Macklin JJ, Filosa A, Aggarwal A, Kerr RA, Takagi R, Kracun S, Shigetomi E, Khakh BS, Baier H, Lagnado L, Wang SS, Bargmann CI, Kimmel BE, Jayaraman V, Svoboda K, Kim DS, Schreiter ER, Looger LL (2012) Optimization of a GCaMP calcium indicator for neural activity imaging. J Neurosci 32(40):13819–13840. doi:10.1523/JNEUROSCI.2601-12.2012
25. Delling M, DeCaen PG, Doerner JF, Febvay S, Clapham DE (2013) Primary cilia are specialized calcium signalling organelles. Nature 504(7479):311–314. doi:10.1038/nature12833
26. Jin X, Mohieldin AM, Muntean BS, Green JA, Shah JV, Mykytyn K, Nauli SM (2013) Cilioplasm is a cellular compartment for calcium signaling in response to mechanical and chemical stimuli. Cell Mol Life Sci 71(11):2165–2178. doi:10.1007/s00018-013-1483-1
27. Su S, Phua SC, Derose R, Chiba S, Narita K, Kalugin PN, Katada T, Kontani K, Takeda S, Inoue T (2013) Genetically encoded calcium indicator illuminates calcium dynamics in primary cilia. Nat Methods. doi:10.1038/nmeth.2647
28. Yuan S, Zhao L, Brueckner M, Sun Z (2015) Intraciliary calcium oscillations initiate vertebrate left-right asymmetry. Curr Biol 25(5):556–567. doi:10.1016/j.cub.2014.12.051
29. DeCaen PG, Delling M, Vien TN, Clapham DE (2013) Direct recording and molecular identification of the calcium channel of primary cilia. Nature 504(7479):315–318. doi:10.1038/nature12832
30. Caspary T, Larkins CE, Anderson KV (2007) The graded response to Sonic Hedgehog depends on cilia architecture. Dev Cell 12(5):767–778
31. Duldulao NA, Lee S, Sun Z (2009) Cilia localization is essential for in vivo functions of the Joubert syndrome protein Arl13b/Scorpion. Development 136(23):4033–4042. doi:10.1242/dev.036350
32. Berbari NF, Johnson AD, Lewis JS, Askwith CC, Mykytyn K (2008) Identification of ciliary localization sequences within the third intracellular loop of G protein-coupled receptors. Mol Biol Cell 19(4):1540–1547. doi:10.1091/mbc.E07-09-0942
33. Chen JK, Taipale J, Cooper MK, Beachy PA (2002) Inhibition of Hedgehog signaling by direct binding of cyclopamine to Smoothened. Genes Dev 16(21):2743–2748. doi:10.1101/gad.1025302
34. Corbit KC, Aanstad P, Singla V, Norman AR, Stainier DY, Reiter JF (2005) Vertebrate smoothened functions at the primary cilium. Nature 437(7061):1018–1021. doi:10.1038/nature04117
35. McGrath J, Somlo S, Makova S, Tian X, Brueckner M (2003) Two populations of node monocilia initiate left-right asymmetry in the mouse. Cell 114(1):61–73
36. Malicki J, Avanesov A, Li J, Yuan S, Sun Z (2011) Analysis of cilia structure and function in zebrafish. Methods Cell Biol 101:39–74. doi:10.1016/B978-0-12-387036-0.00003-7
37. Yuan S, Sun Z (2009) Microinjection of mRNA and morpholino antisense oligonucleotides in zebrafish embryos. J Vis Exp 27:pii:1113. doi:10.3791/1113
38. Borovina A, Superina S, Voskas D, Ciruna B (2010) Vangl2 directs the posterior tilting and asymmetric localization of motile primary cilia.

Nat Cell Biol 12(4):407–412. doi:10.1038/ncb2042

39. Boskovski MT, Yuan S, Pedersen NB, Goth CK, Makova S, Clausen H, Brueckner M, Khokha MK (2013) The heterotaxy gene GALNT11 glycosylates Notch to orchestrate cilia type and laterality. Nature 504(7480):456–459. doi:10.1038/nature12723
40. Yuan S, Zhao L, Sun Z (2013) Dissecting the functional interplay between the TOR pathway and the cilium in zebrafish. Methods Enzymol 525:159–189. doi:10.1016/B978-0-12-397944-5.00009-2
41. Westerfield M (1993) The zebrafish book: a guide for the laboratory use of zebrafish (Brachydanio rerio). M. Westerfield, Eugene, OR
42. Austin-Tse C, Halbritter J, Zariwala MA, Gilberti RM, Gee HY, Hellman N, Pathak N, Liu Y, Panizzi JR, Patel-King RS, Tritschler D, Bower R, O'Toole E, Porath JD, Hurd TW, Chaki M, Diaz KA, Kohl S, Lovric S, Hwang DY, Braun DA, Schueler M, Airik R, Otto EA, Leigh MW, Noone PG, Carson JL, Davis SD, Pittman JE, Ferkol TW, Atkinson JJ, Olivier KN, Sagel SD, Dell SD, Rosenfeld M, Milla CE, Loges NT, Omran H, Porter ME, King SM, Knowles MR, Drummond IA, Hildebrandt F (2013) Zebrafish ciliopathy screen plus human mutational analysis identifies C21orf59 and CCDC65 defects as causing primary ciliary dyskinesia. Am J Hum Genet 93(4):672–686. doi:10.1016/j.ajhg.2013.08.015
43. Essner JJ, Amack JD, Nyholm MK, Harris EB, Yost HJ (2005) Kupffer's vesicle is a ciliated organ of asymmetry in the zebrafish embryo that initiates left-right development of the brain, heart and gut. Development 132(6):1247–1260. doi:10.1242/dev.01663
44. Sun Z, Amsterdam A, Pazour GJ, Cole DG, Miller MS, Hopkins N (2004) A genetic screen in zebrafish identifies cilia genes as a principal cause of cystic kidney. Development 131(16):4085–4093
45. Kreiling JA, Prabhat, Williams G, Creton R (2007) Analysis of Kupffer's vesicle in zebrafish embryos using a cave automated virtual environment. Dev Dyn 236(7):1963–1969. doi:10.1002/dvdy.21191
46. Lopes SS, Lourenco R, Pacheco L, Moreno N, Kreiling J, Saude L (2010) Notch signalling regulates left-right asymmetry through ciliary length control. Development 137(21):3625–3632. doi:10.1242/dev.054452
47. Perez Koldenkova V, Nagai T (2013) Genetically encoded Ca(2+) indicators: properties and evaluation. Biochim Biophys Acta 1833(7):1787–1797. doi:10.1016/j.bbamcr.2013.01.011
48. Pusl T, Wu JJ, Zimmerman TL, Zhang L, Ehrlich BE, Berchtold MW, Hoek JB, Karpen SJ, Nathanson MH, Bennett AM (2002) Epidermal growth factor-mediated activation of the ETS domain transcription factor Elk-1 requires nuclear calcium. J Biol Chem 277(30):27517–27527. doi:10.1074/jbc.M203002200
49. Hashimoto H, Rebagliati M, Ahmad N, Muraoka O, Kurokawa T, Hibi M, Suzuki T (2004) The Cerberus/Dan-family protein Charon is a negative regulator of nodal signaling during left-right patterning in zebrafish. Development 131(8):1741–1753. doi:10.1242/dev.01070
50. Long S, Ahmad N, Rebagliati M (2003) The zebrafish nodal-related gene southpaw is required for visceral and diencephalic left-right asymmetry. Development 130(11):2303–2316
51. Bisgrove BW, Essner JJ, Yost HJ (1999) Regulation of midline development by antagonism of lefty and nodal signaling. Development 126(14):3253–3262
52. Essner JJ, Branford WW, Zhang J, Yost HJ (2000) Mesendoderm and left-right brain, heart and gut development are differentially regulated by pitx2 isoforms. Development 127(5):1081–1093
53. Chen JN, van Eeden FJ, Warren KS, Chin A, Nusslein-Volhard C, Haffter P, Fishman MC (1997) Left-right pattern of cardiac BMP4 may drive asymmetry of the heart in zebrafish. Development 124(21):4373–4382
54. Thisse C, Thisse B (2008) High-resolution in situ hybridization to whole-mount zebrafish embryos. Nat Protoc 3(1):59–69. doi:10.1038/nprot.2007.514
55. Yuan S, Li J, Diener DR, Choma MA, Rosenbaum JL, Sun Z (2012) Target-of-rapamycin complex 1 (Torc1) signaling modulates cilia size and function through protein synthesis regulation. Proc Natl Acad Sci U S A 109(6):2021–2026. doi:10.1073/pnas.1112834109
56. Ishikawa H, Thompson J, Yates JR 3rd, Marshall WF (2012) Proteomic analysis of mammalian primary cilia. Curr Biol 22(5):414–419. doi:10.1016/j.cub.2012.01.031
57. Tian L, Hires SA, Looger LL (2012) Imaging neuronal activity with genetically encoded calcium indicators. Cold Spring Harb Protoc 2012(6):647–656. doi:10.1101/pdb.top069609
58. Narita K, Kozuka-Hata H, Nonami Y, Ao-Kondo H, Suzuki T, Nakamura H, Yamakawa K, Oyama M, Inoue T, Takeda S (2012) Proteomic analysis of multiple primary cilia reveals a novel mode of ciliary development in mammals. Biol Open 1(8):815–825. doi:10.1242/bio.20121081

59. Gherman A, Davis EE, Katsanis N (2006) The ciliary proteome database: an integrated community resource for the genetic and functional dissection of cilia. Nat Genet 38(9):961–962. doi:10.1038/ng0906-961
60. Arnaiz O, Malinowska A, Klotz C, Sperling L, Dadlez M, Koll F, Cohen J (2009) Cildb: a knowledgebase for centrosomes and cilia. Database (Oxford) 2009:22. doi:10.1093/database/bap022
61. Tian L, Hires SA, Mao T, Huber D, Chiappe ME, Chalasani SH, Petreanu L, Akerboom J, McKinney SA, Schreiter ER, Bargmann CI, Jayaraman V, Svoboda K, Looger LL (2009) Imaging neural activity in worms, flies and mice with improved GCaMP calcium indicators. Nat Methods 6(12):875–881. doi:10.1038/nmeth.1398
62. Chen TW, Wardill TJ, Sun Y, Pulver SR, Renninger SL, Baohan A, Schreiter ER, Kerr RA, Orger MB, Jayaraman V, Looger LL, Svoboda K, Kim DS (2013) Ultrasensitive fluorescent proteins for imaging neuronal activity. Nature 499(7458):295–300. doi:10.1038/nature12354
63. Muto A, Ohkura M, Abe G, Nakai J, Kawakami K (2013) Real-time visualization of neuronal activity during perception. Curr Biol 23(4):307–311. doi:10.1016/j.cub.2012.12.040
64. Ohkura M, Sasaki T, Sadakari J, Gengyo-Ando K, Kagawa-Nagamura Y, Kobayashi C, Ikegaya Y, Nakai J (2012) Genetically encoded green fluorescent Ca2+ indicators with improved detectability for neuronal Ca2+ signals. PLoS One 7(12), e51286. doi:10.1371/journal.pone.0051286
65. Akerboom J, Carreras Calderon N, Tian L, Wabnig S, Prigge M, Tolo J, Gordus A, Orger MB, Severi KE, Macklin JJ, Patel R, Pulver SR, Wardill TJ, Fischer E, Schuler C, Chen TW, Sarkisyan KS, Marvin JS, Bargmann CI, Kim DS, Kugler S, Lagnado L, Hegemann P, Gottschalk A, Schreiter ER, Looger LL (2013) Genetically encoded calcium indicators for multi-color neural activity imaging and combination with optogenetics. Front Mol Neurosci 6:2. doi:10.3389/fnmol.2013.00002
66. Zhao Y, Araki S, Wu J, Teramoto T, Chang YF, Nakano M, Abdelfattah AS, Fujiwara M, Ishihara T, Nagai T, Campbell RE (2011) An expanded palette of genetically encoded Ca(2)(+) indicators. Science 333(6051):1888–1891. doi:10.1126/science.1208592
67. Horikawa K, Yamada Y, Matsuda T, Kobayashi K, Hashimoto M, Matsu-ura T, Miyawaki A, Michikawa T, Mikoshiba K, Nagai T (2010) Spontaneous network activity visualized by ultrasensitive Ca(2+) indicators, yellow Cameleon-Nano. Nat Methods 7(9):729–732. doi:10.1038/nmeth.1488

Chapter 10

Methods for Studying Ciliary-Mediated Chemoresponse in *Paramecium*

Megan Smith Valentine and Judith L. Van Houten

Abstract

Paramecium is a useful model organism for the study of ciliary-mediated chemical sensing and response. Here we describe ways to take advantage of *Paramecium* to study chemoresponse.

Unicellular organisms like the ciliated protozoan *Paramecium* sense and respond to chemicals in their environment (Van Houten, Ann Rev Physiol 54:639–663, 1992; Van Houten, Trends Neurosci 17:62–71, 1994). A thousand or more cilia that cover *Paramecium* cells serve as antennae for chemical signals, similar to ciliary function in a large variety of metazoan cell types that have primary or motile cilia (Berbari et al., Curr Biol 19(13):R526–R535, 2009; Singla V, Reiter J, Science 313:629–633, 2006). The *Paramecium* cilia also produce the motor output of the detection of chemical cues by controlling swimming behavior. Therefore, in *Paramecium* the cilia serve multiple roles of detection and response.

We present this chapter in three sections to describe the methods for (1) assaying populations of cells for their behavioral responses to chemicals (attraction and repulsion), (2) characterization of the chemoreceptors and associated channels of the cilia using proteomics and binding assays, and (3) electrophysiological analysis of individual cells' responses to chemicals. These methods are applied to wild type cells, mutants, transformed cells that express tagged proteins, and cells depleted of gene products by RNA Interference (RNAi).

Key words *Paramecium*, T-Maze, Chemoresponse, Receptors, Ciliary membrane, Electrophysiology, Deciliation

1 Introduction

A very simple and reliable method of determining whether a chemical causes attraction or repulsion of *P. tetraurelia* is the T-maze (Fig. 1). Paramecium, like many other ciliated cells, rely on receptors in their cilia and on the cell surface to respond to their environment [1, 2]. Behavioral assays using a T-maze are useful for characterizing chemosensory receptors in *Paramecium* and receptor and ion defects in mutant or RNAi treated cells. The cilia on Paramecium cells serve as antenae, similar to other cell types [3, 4], relaying information to the cell. Behavioral and electrophysiological studies have shown that extracellular small molecules, such as cyclic

Peter Satir and Søren Tvorup Christensen (eds.), *Cilia: Methods and Protocols*, Methods in Molecular Biology, vol. 1454, DOI 10.1007/978-1-4939-3789-9_10, © Springer Science+Business Media New York 2016

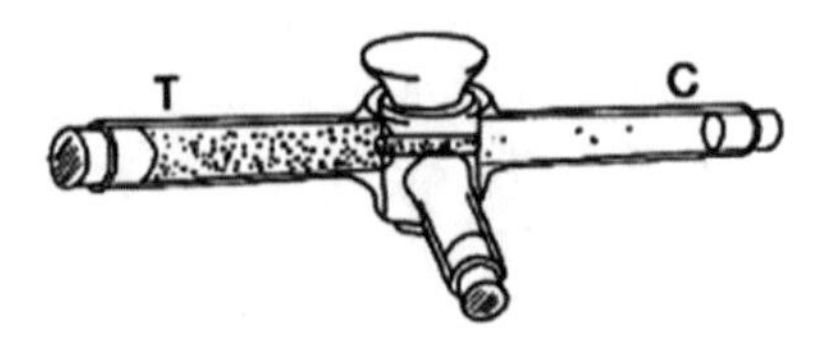

Fig. 1 T-maze apparatus [25] is modified from [56] in that a two-way plug is substituted for the original three-way plug. Test solution is in the T arm; control solution in the C arm; cells are loaded through the remaining arm. Cells are allowed to move through the T-maze for 30 min after the plug is turned to allow them access to the C and T arms. After 30 min, the plug is turned, arms emptied and cells counted

AMP (cAMP), folate, glutamate, and biotin cause a hyperpolarization of the cell and fast, smooth forward swimming [5–11]. Attractants fall into three different signal transduction pathways: Ammonium is expected to cross the membrane as ammonia, causing a change in intracellular pH without binding to a surface receptor [9]. Acetate, glutamate, cAMP, folate and biotin on the other hand, are receptor-mediated. The binding of these attractants to the cells causes a membrane hyperpolarization initiated by a K conductance and sustained by electrogenic calcium pumps [6, 7, 10, 12]. Among the stimuli, only glutamate has been shown to mediate a very rapid increase in intracellular cAMP as a second messenger [11]. Cyclic AMP appears to activate protein kinases that continue the signal transduction because kinase inhibitors specifically interfere with the chemoresponse of paramecia to glutamate and to no other attractant tested [11]. (*See* ref. 12 for a helpful figure.)

Paramecium affords a source of pure cilia that are easily separated from the cell body for protein analysis. *P. tetraurelia* cilia have been used for comparison of proteins in whole cilia with those in ciliary membranes. The proteomic studies most relevant for chemoresponse focus on the chemoreceptors and also on channels and ion pumps that help govern the ciliary beat, i.e., the motor output following the detection of chemical signals. Specifically, the electrical properties of the ciliary membrane govern the direction and frequency of the ciliary beat [13]. Attractants hyperpolarize the cells sending them swimming fast forward; repellents depolarize the cells which slows the cells and increases turning [8]. The hyperpolarization in attractants is due to activation of various K channels and Ca pumps [12]; repellents depolarize the cells to the point of initiating an action potential by activating voltage-gated Ca channels [14, 15]. These Ca channels are found *exclusively* in the cilia [16] and are especially difficult to isolate from cilia because of their low abundance among the highly abundant glycosylphosphatidyl inositol (GPI)-anchored proteins that coat the cell surface and cilia and the abundant axonemal cytoskeletal components [17, 18]. The peripheral GPI-anchored proteins comprise the majority of ciliary proteins in *Paramecium*, almost two-thirds by mass

[17–20]. Other transmembrane channels (Transient Receptor Protein channel, voltage-gated and Ca-activated K channels), adenylyl cyclase that is involved in chemo-signal transduction for glutamate, pumps and transporters are also found in the cilia and their proteomic analysis benefits from enrichment as well. (*See* ref. 21 for a review of channels of the cilia.)

Proteomics provides an unbiased way to inventory the ciliary proteins involved in chemoreception and motor response. The challenge is in identifying the transmembrane proteins of interest which are difficult to solubilize because of their lipid environment and difficult to enrich because they are in low abundance [22]. One solution to the enrichment of ciliary transmembrane proteins is phase separation, i.e., the partitioning of proteins into Triton X-114 detergent and aqueous phases as described in the protocol below. Most GPI-anchored proteins are found in the aqueous phase because their anchor has been cleaved [17, 18], and axonemal and other cytoskeletal elements of the cilia are less soluble in the non-ionic detergent, leaving many transmembrane proteins in the detergent phase.

The T-mazes can be done efficiently and quantitatively, but there are times that a more rapid and non-quantitative assay is useful, as when screening for mutants [23, 24]. Capillary tubes filled with test and control buffers can be propped in depression slides to determine whether more cells enter the tubes with test solution as judged by visual inspection using a dissecting microscope. Small mazes cut into plexiglass and mounted on slides can also be used to visually determine whether more cells appear to enter the arm with test solution or not [25].

Chemoreceptors come from several protein families including transmembrane and GPI-anchored proteins [26–30]. Binding assays have been useful to characterize the receptors for their numbers, binding affinity, and location (cilia vs cell body membrane). Binding assays help to locate the membrane binding sites that could correlate with receptors and determine their affinity and number. The receptors might be on the cilia, cell body membrane or both. To have confidence that the binding site represents the receptor, the binding affinity can be compared with the kinetics of behavior in assays such as the T-maze (e.g., [31]). There are reasons why the two might not match completely, such as a rightward shift of concentration plots of behavior indicating that most receptors must be occupied before the cells respond. Competition for binding of ^{3}[H] ligands should match the results of competition studies in T-mazes if the binding sites are truly the receptors.

Binding studies can be carried out on isolated cilia and on whole cells. For example, these two kinds of assays showed that L-glutamate binding sites are in the cilia and also likely to be on cell somas [11, 32]. Recent measurements of the hyperpolarization of cells in L-glutamate showed that deciliated cells do not hyperpolarize in L-glutamate, implicating the ciliary binding sites as the location of functional receptors (Valentine, unpublished results). (*See* Electrophysiology below.) below). Binding assays using ^{3}[H] compounds on isolated cilia have also been useful for the study of the folate receptor [27, 33] and the cAMP receptor in *Paramecium* [31, 34].

Binding assays can be combined with molecular approaches to down regulate putative receptors and behavioral assays to confirm whether the binding sites function as receptors for the ligand of interest. When the GPI-anchored folate receptor (GSPATG00025147001) or a *Paramecium* glutamate receptor (pGluR1; GSPATG00000858001) were reduced using RNAi, the depleted cells no longer showed attraction to folate or L-glutamate, respectively in T-maze assays [26–28]. In contrast, RNAi targeting a potential folate transporter protein (GSPATG00030563001) had no effect on folate chemoresponse [27]. The reduction of the GPI-anchored folate receptor using RNAi caused a 50% reduction in folate binding compared to the control-fed cells using ^{3}[H]-folate binding studies [27]. The ability to perform both RNAi and binding assays on isolated cilia from *Paramecium* facilitates relatively straight-forward studies of ciliary receptors, assuming the appropriate ^{3}[H] compounds are available.

Electrophysiology has long been used to study membrane potential and ion channels in *Paramecium*. These cells, 100–200 μm in length, are quite amenable to electrophysiological studies. The electrical properties of the membrane help to govern the swimming speed and direction of the cell. The membrane is more permeable to some ions, such as K^+, Na^+, and Mg^{2+}, while it is very selective concerning the passage of ions such as Ca^{2+} and Cl^- [35–39].

Resting membrane potential of *Paramecium* is mostly governed by the cells' permeability to K^+ ions, allowing them to act as a K^+ electrodes [38]. In general, attractant stimuli hyperpolarize the cells while repellents depolarize [8]. These changes in membrane potential are measurable using recording electrodes. For example, a *Paramecium* cell will establish a resting membrane potential in 5 mM KCl in buffer of about −40 mV. If the bath solution is changed to a 5 mM K-acetate in buffer solution, a new more hyperpolarized membrane potential will be established (about −7 mV more negative) [10]. Changes in potential for other attractants (2.5 mM K_2-folate, 5 mM K-L- or K-D-glutamate) have been established relative to comparable resting buffers [8, 10]. Membrane potential recordings could therefore be used on cells to test new compounds for their effects on potential and the concentration studies compared with behavioral responses [31]. Membrane potential changes from cells depleted of a protein by RNAi or antisense are useful to establish whether the protein is a potential receptor [27]. Numerous studies have used electrophysiology to study the hyperpolarization of the membrane in attractants [8, 10, 15, 32]. Two electrode voltage clamp studies have been used to characterize the conductances and localization of chemoreceptive properties on the cell surface [10, 40].

Researchers have also used ciliated and deciliated cells to demonstrate that there is no difference in the amount the membrane hyperpolarizes in the presence of some attractants, suggesting the receptors for the attractant are also present on the cell body. For example, a deciliated *Paramecium* does not show changes in the magnitude of the hyperpolarization in the presence of acetate or folate [10]. Paramecia are easily deciliated using gentle trituration of the cells in 4 or 5 mM KCl solution in buffer with 5% (v/v) ethanol. The deciliation process does not change the resting membrane potential of the cells in 5 or 1 mM KCl [41, 42]. As with cells depleted of receptors using RNAi or antisense, deciliation can also help to tease apart the location of a receptor, whether it is found in the cilia, the cell body, or both.

Combining mutants with electrophysiology and behavior assays can also be used establish the conductances that are involved in the attraction hyperpolarization. For example, the on-response for the behavior in acetate and biotin are attributed to the conductance from $I_{K(Ca)}$ and the off response for biotin only to I_{Ca} through the use of a battery of mutants, each of which has a different single site mutation, in behavior and electrophysiology studies [15]. Other examples are included in these refs. 1, 43, 44.

2 Materials

2.1 General Equipment and Solutions for Concentrating and Working with *P. tetraurelia*

1. IEC clinical continuous flow centrifuge to harvest liters of culture.
2. IEC HNS-II Clinical centrifuge to concentrate cells for many purposes; cells collect in the stem of pear shaped tubes (Kimax, 100 mL) when centrifuged at about 500 × *g*.
3. Dryl's solution: 1 mM NaH_2PO_4, 1 mM Na_2HPO_4, 2 mM sodium citrate, 1.5 mM $CaCl_2$, pH 6.80.
4. STEN Buffer: 0.5 M sucrose, 20 mM tris base, 2 mM EDTA, 6 mM NaCl, pH 7.5.
5. Cilia shock buffer: 180 mM KCl, 60 mM $CaCl_2$.
6. 10× Solution: 10 mM Calcium citrate. This is the base buffer for most other solutions for T-mazes and electrophysiology.
7. Dissecting binocular microscope for observation of cells, 20× and 40× objective.
8. Glass depression slides for counting cells and growing small quantities.
9. Refrigerated centrifuge and tubes, capable of spinning at 29,000 × *g* at 4 °C (Beckman J2-21 model).

2.2 Materials for T-Maze Assays

1. Cultures of *Aerobacter aerogenes.*
2. Culture tubes (pear shaped or Kolmer tubes).
3. Glass 3-way stop cock fitted with 2 way plug for T-maze construction (*see* **Note 1**).
4. Silicone high vacuum grease.
5. Corks for T-mazes.
6. Pasteur pipettes.
7. T-Maze buffers as desired (*see* **Note 2**).
8. Test attractant, repellent or control as desired (e.g., 2 mM Na-acetate for test attractant and 2 mM NaCl for control).

2.3 Materials for Proteomics

1. 1% Triton X-114 (i.e., GBiosciences, cat# DG509).
2. Tris buffered saline (TBS): 20 mM tris, 150 mM NaCl, pH 7.5.
3. 6% sucrose in TBS.
4. 1 mM Tris-EDTA, pH 8.3 (1 mM Tris, 0.1 mM EDTA, pH 8.3, made fresh).
5. Coomassie blue (i.e., Coomassie Brilliant Blue G-250, GBiosciences, cat# 786-497) or silver stain (i.e., FOCUS™FASTsilver™, GBiosciences, cat #786-240).
6. Sterile razor blades.
7. Enzyme for protein digestion, such as sequencing grade modified trypsin (Promega, cat # V5111).

2.4 Materials for Filter Binding Assays

1. Binding filters: 25 mm 0.45 μm nitrocellulose filters (Millipore, cat#: HAWP 025 00).
2. Powdered cAMP salt (Na or K) or other attractant. If using cAMP, 1 mM isobutylmethylxanthine (IBMX) may also be considered to prevent degradation of cAMP by phosphodiesterases.
3. Radioisotope for binding assays: ^{3}[H]-cAMP (ICN 27 Ci/mM, 1 μCi/μL). ^{3}[H]-glutamate or ^{3}[H]-folate is also available. Contact your institution's radiation safety office for ordering information and necessary training and clearance.
4. Binding buffers and wash buffer: All binding and wash buffers should contain the basic buffer described for T-mazes (*see* **Note 2**). Salt concentrations of binding and wash buffers should all match that of the binding buffer containing the radioisotopes.
5. Filter apparatus with coarse glass sintered disks or wire mesh supported platform to fit filter paper.
6. Scintillation counter, scintillation fluid (Aqualyte or Filter-Count™, PerkinElmer), and vials.

2.5 Materials for Electrophysiology

1. Glass electrode puller (Narishige) with 3 in. 1.0 mm 1/0.58 OD/ID standard glass capillary tubes, with filament (World Precision Instruments, cat # 1B100F-3).

2. Inverted light microscope with micromanipulator(s), lens (40× objective) preferred. Dual stands are required for chamber maintenance.
3. Computer and Intracellular recording equipment, Warner Intracellular Electrometer IE-251A processed through AD Instruments PowerLab 4/35.
4. Glass Pasteur pipettes (6″ or 9″) for pulling micropipettes for transferring cells and for making bridges (*see* **Notes 3** and **4**).
5. Faraday cage and necessary grounding equipment.
6. Cell chamber with a small reference well for the agar bridge. Chamber should be glued securely to a glass slide. Reference chamber must contain an Ag/AgCl wire. Agar bridge positioned using Vaseline should connect the reference chamber to the recording bath.
7. Agar bridges: 3% agar made in 3 M KCl (*see* **Note 4**).
8. Polystaltic pump and tubing is required for controlled fluid flow across the cell in the recording chamber. Flow rate should be ~1 mL/min.
9. KCl solutions: 3 M KCl (bridges and reference well of chamber); 500 mM KCl (back-filling electrodes and electrode holder).
10. Recording solutions (*see* **Note 2**): All solutions should be made from 100 mM stock solutions in base buffer. Resting buffers are typically 1 mM KCl or 5 mM KCl in base buffer. pH solutions to 7.03–7.06 using 100 mM tris base.

3 Methods

3.1 Assays of Attraction and Repulsion

1. Cells are grown in standard wheat grass culture with *A. aerogenes* bacteria [45] and generally harvested at late log phase (*see* **Note 5**).
2. Solutions should be brought to room temperature. While the center stopcock is perpendicular to the two long arms (Fig. 1), the control solution should be placed in the C arm, the test solution in the T arm (*see* **Note 6**). Corks should be gently but securely placed into the arms to contain the solutions but not to force them into the other arms when the T-mazes are opened.
3. Ensure no air bubbles are blocking the arms when the plug is opened. If so, bang the T-maze on your hand to loosen them.
4. Collect cells using a Pasteur pipette from the stems of pear-shaped tubes or Kolmer tubes spun in a table top clinical or IEC HNS-II centrifuge at about 400 ×*g*.

5. Resuspend the harvested cells in the control buffer (*see* **Note 5**).
6. The resuspended cells are transferred with a Pasteur pipette into the plug through the arm that is at a right angle to the T-maze arms. Cells should be loaded quickly and T-mazes should be tapped with a hand to loosen any air bubbles.
7. Once the T-mazes have cells, the plug should be turned at an angle such that the cells are trapped.
8. The T-maze should be set in a stand so that the arms are level.
9. Once all the T-mazes are loaded with solutions and cells, the plug is turned so that the cells are free to swim between the arms for 30 min (*see* **Note 7**).
10. The plug is returned to its original position.
11. The contents of the arms are poured into separate tubes. Two or more aliquots (usually 0.5 mL) from each tube are removed for counting cells using a dissecting microscope (*see* **Note 8**).
12. T-mazes and corks should be rinsed very well with hot water and purified water after use (*see* **Note 9**). Allow them to sit at room temperature to dry.
13. The ratio of cells in T arm vs C arm is constructed. A ratio of 0.5 means the two solutions behaved identically. Attractant solutions will give higher cell counts on the attractant side; repellent solutions will give lower cell counts on the repellent side (*see* **Note 10**).

3.2 Protocol for Proteomic Analysis (Adapted from [46])

1. Prepare cilia, axonemal and membrane fractions as described below (Subheading 3.3), or as described previously [46, 47]. It is crucial that the cilia are separated from the cell bodies.
2. Pure ciliary membrane (1–5 mg) should be solubilized at 4 °C while rocking in 1 mL of 1% Triton X-114 (i.e., GBiosciences, cat# DG509) in tris buffered saline (20 mM tris, 150 mM NaCl, pH 7.5) for 2 h (*see* **Note 11**).
3. Centrifuge the suspension at 14,000 ×*g* for 10 min at 4 °C. Resuspend resulting pellet in TBS (20 mM tris base, 150 mM NaCl, pH 7.5).
4. Overlay resuspended pellet over 100 μL 6% sucrose in TBS, incubate at 37 °C for 4 min.
5. Centrifuge sample at 8000 ×*g* for 2 min at room temperature to separate proteins into the aqueous phase (upper layer) and detergent phase (lower layer) (*see* **Note 12**).
6. Precipitate proteins in the different phases using cold acetone at 4 °C for 30 min and centrifuge at 14,000 ×*g* for 15 min at 4 °C.
7. Wash acetone-precipitated pellet three times in 1 mL cold 90% acetone and allow to air dry before resuspending in 80 μL TBS or other buffer.

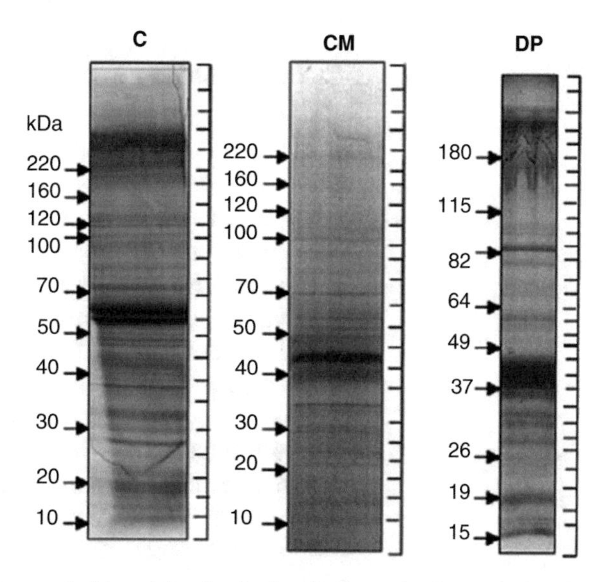

Fig. 2 Coomassie blue stained gels showing samples from whole cilia (C), ciliary membrane (CM), and ciliary membranes partitioned into the Triton X-114 detergent phase (DP) on 7–18 % gels. *Numbers* with *arrows* on the *left* of each gel indicate approximate molecular masses and lines on the *right* of each gel indicate where the gels were cut for processing using trypsin before LC-MS/MS analysis

8. Resuspended proteins should be analyzed for protein content (i.e., Pierce BCA Assay) and ~200 μg of protein should be separated by SDS-PAGE using standard methods on a well-aged gradient gel (i.e., 7–18 %).
9. Gel should be stained using Coomassie blue (i.e., Coomassie Brilliant Blue G-250, GBiosciences, cat# 786-497) or silver stain (i.e., FOCUS™FASTsilver™, GBiosciences, cat #786-240) (Fig. 2) (*see* **Note 13**).
10. Once the gel has been stained and destained, it should be imaged. The lane containing the proteins desired for analysis should be cut into strips and then small cubes (~1 mm × 1 mm) using a sterile razor blade on a sterile glass plate and processed for analysis. A typical lane of a "mini gel" should be cut into 20–35 slices and placed in sterile 1.5 mL Eppendorf tubes for processing (*see* **Note 14**).
11. Gel slices should be processed to further remove the stain (water for Coomassie Blue or Silver D-Stain™, GBiosciences, cat# 786-199) followed by buffering and in-gel digestion with an enzyme such as sequencing grade modified trypsin (Promega, cat # V5111) (*see* **Note 15**).
12. The facility where the samples are processed will assist with the proteomic identification and analysis. Samples are typically dried in a SpeedVac before being sent to the facility. Before LC-MS/MS analysis, samples will be resuspended in the appropriate buffer, 2.5 % formic acid and 2.5 % acetonitrile.

13. When analyzing data, both forward (target) and reverse (decoy) proteome databases should be analyzed using SEQUEST, or other analysis software. A precursor tolerance of 2 Da should be used along with a fragment ion tolerance of 0.5 Da (*see* **Note 16**).
14. Search results should be filtered, previous settings used a dCn score of 0.2 and Xcorr values of 1.5, 2.0, and 2.5 for singly, doubly, and triply charged ions, respectively [46]. Although these values can be adjusted, the false-positive discovery rate should be kept below 1 % for a reliable analysis and a minimum of two unique peptides is necessary for a positive identification.

3.3 Isolating Cilia

1. Filter 4.5 L late log phase cells through a piece of cheese cloth sandwiched between two large Kimwipes™ to remove debris. Concentrate cells in a continuous flow centrifuge (speed 5, ~500 × *g*) (*see* **Note 17**).
2. Centrifuge cells in pear shaped tubes in an IEC HNS-II clinical centrifuge at ¾ speed for 2 min (~350 × *g*) and using a Pasteur pipette, collect cells and resuspend in 200 mL room temperature Dryl's solution.
3. Collect cells using the pear shaped tubes and remove and discard the trichocyst layer (white fluffy layer above the cells). Remove cells using a Pasteur pipette and again resuspend in 200 mL room temperature Dryl's solution.
4. Repeat the washing in Dryl's solution so the cells have been washed three times in 200 mL of Dryl's solution.
5. Resuspend the cells 50 mL room temperature Dryl's solution and add 50 mL ice cold STEN buffer. Swirl and rest cells on ice for 10 min. Check cells periodically for motion.
6. After 10 min, when cells are mostly motionless, add 20 mL of cilia shock buffer to deciliate the cells. Swirl, keep cells on ice and observe a small sample closely under a microscope for 5 min (*see* **Note 18**).
7. Centrifuge the cells in clean pear shaped tubes at full speed (IEC HNS-II centrifuge, 1350 × *g*) for 2 min to pellet the cell bodies.
8. Collect the supernatant by pouring it into new pear shaped tubes and repeat centrifugation. Supernatant should appear cloudy.
9. Centrifuge the supernatant at 29,000 × *g* for 20 min at 4 °C to pellet the cilia.
10. Discard the supernatant and resuspend the pellet (cilia) in 1 mL of the desired buffer, such as 1 mM Tris-EDTA (1 mM Tris, 0.1 mM EDTA, pH 8.3).
11. Follow protocols described refs. 46–48 to subfractionate cilia into membrane and axoneme fractions, if desired.

3.4 Binding Assay Using Cilia (Adapted from [34]), for cAMP

1. All filters should be pre-soaked in 1 mM cAMP for a minimum of 10 min.
2. Wash and incubation solutions should contain the same concentrations of cAMP (1 mM to 1 μM) and 1 mM IBMX (*see* **Note 19**). The incubation solution should also contain 1 μL/mL ^{3}[H]-cAMP (*see* **Note 20**).
3. The nonspecific binding of the isotope must also be tested. To do so, duplicate samples should be run containing the radioisotopes but no cilia. These samples should be treated the same and receive the same washes.
4. At this point, a protein assay is necessary to later normalize ligand binding to mg of protein (*see* **Note 21**).
5. Remove 0.5 mL of cilia suspension and add to the incubation solution containing the ^{3}[H]-cAMP. At varying time points (15 s to 180 min), 1 mL duplicate aliquots should be removed, filtered through the presoaked filters, and washed with 40 mL of the corresponding wash solution (*see* **Note 22**).
6. To terminate, remove the filters and place in 3 mL of scintillation fluid, such as Aqualyte or Fluid-Count™ scintillation fluid (PerkinElmer) for 18 h (overnight).
7. Determine the counts per minute (cpm) associated with the cilia by subtracting the cpm of the nonspecific binding to filters (control) from the cpm of the filters of the test samples containing cilia. Knowing the specific activity, convert cpm to moles of ligand per mg protein.

3.5 Whole Cell Binding Assay

A centrifugation method using Pasteur pipettes can be used for whole cell binding [31, 33, 48].

1. Paramecia cells should be grown to a density of ~4000 cells/mL and have cleared their culture fluid.
2. Cells should be cleaned with an antibiotic wash (*see* **Note 17**).
3. Pellet the antibiotic-treated cells using the IEC HNS-II clinical centrifuge for 2 min, ¾ speed (~350 × *g*). Move the pellet of cells using a Pasteur pipette to the incubation mixture (8 mL buffer with K_2-folate at desired concentration and 10 μL ^{3}[H]-folate (1 μCi/mL)) (*see* **Note 23**). Before adding the cells, remove a small amount of the incubation buffer to test the specific activity.
4. During the incubation, at four different time points between 30 s and 5 min, gently mix the cells and remove duplicate 1 mL aliquots.
5. In two 18.5 cm *sealed* Pasteur pipettes, layer the aliquots over 1.5 mL wash buffer (same salt and K_2-folate concentration as the binding buffer, but no ^{3}[H]-folate) containing 1 % sucrose. (Give the Pasteur pipette centrifuge tubes a quick spin prior to

use to move the sucrose solution to the tip of the pipet. You may need to modify the centrifuge buckets for the 100 mL pear shaped tubes by inserting a #10 black rubber stopper with hole drilled in the middle to accept the Pasteur pipette centrifuge tubes.)

6. Centrifuge the cells for 2 min in the IEC HNS-II centrifuge, ¾ speed (~350 × *g*) using adapted holders to hold the Pasteur pipettes.
7. Collect the cell pellets by breaking the pipette tips into scintillation vials and crushing the tips. Add scintillation fluid and assay the radioactivity using a liquid scintillation counter.
8. Remaining cell suspension should be used to determine protein concentrations using a BCA protein assay (typically 0.2–0.3 mg/mL).
9. The results of each experiment can be expressed as moles of folate associated per mg of cell protein. Data can be used to extrapolate results to time zero and determine instantaneous binding (*see* **Note 24**).
10. In and binding assay, it is possible to include other compounds in the binding assay to probe for competition between molecules, such as L-glutamate and nucleoside monophosphates which could synergize with the attractant [49]. Competition studies for binding can be compared to studies of behavior over comparable concentrations of attractant and competitors.

3.6 Electrophysiology Procedure (Adapted from [10, 42])

Here we describe only the measurement of membrane potential that can be relatively easily established in standard laboratories.

1. Grow *Paramecium* cells until they reach early log phase. Cells should be large and not starved.
2. Place 10–20 cells in 5 mM KCl in buffer using a pulled micropipette (*see* **Note 3**).
3. Backfill recording electrode (*see* **Note 25**) with 500 mM KCl and place in the holder, which should also be filled with 500 mM KCl.
4. Position the agar bridge (*see* **Note 4**) between the recording bath and the reference bath using Vaseline to prevent fluid from flowing between the two chambers. The reference bath should be filled with 3 M KCl. Bridges should be changed each day, but can be cleaned and reused (*see* **Note 26**).
5. Properly position recording electrode in the field of view using the micromanipulator. Test electrode resistance in 5 mM KCl in buffer in the recording chamber. Resistance should be between 40 and 100 MΩ. Remove the 5 mM KCl from the chamber.

6. Remove one *Paramecium* cell and draw off the fluid around the cell to minimal levels using a micropipette (*see* **Note 27**). Place the cell in the chamber and continue to draw off buffer until the cell becomes still.
7. Carefully impale cell with recording electrode in the center of the cell, gently flood the chamber with 5 mM KCl in buffer, and turn on the polystaltic pump to move 5 mM KCl through the recording chamber.
8. Turn on recording device and begin recording.
9. To change the solution in the bath, turn off the pump and move the "in" tube to a new beaker containing the next solution and turn the pump back on. Cells should spend a minimum of 4 min in each solution.
10. Return cell to 5 or 1 mM KCl solution between test solutions to ensure cell returns to baseline.
11. When complete, remove cell from the recording electrode by lifting the electrode out of the bath and placing it back in the chamber bath to check the drift in potential (*see* **Note 28**).
12. Changes in membrane potential (ΔV_m) can be calculated by subtracting the steady-state difference between the membrane potential in the control solution from the membrane potential in the test solution.

4 Notes

1. T-Maze construction: Glass three-way stopcocks (e.g., 4 mm Ace Glass three-way stopcock part #8145-08) are fitted with two-way plugs (4 mm part #8179-09). Assembling the T-mazes requires some care in that the plugs must be lubricated with Dow Corning™ silicone high vacuum grease but not to the extent that the grease gets into the arms of the T-maze or plug. Too little grease will allow the sintered glass of the plug and stopcock to seize up.
2. Resting buffer is 5 mM KCl, 1 mM $Ca(OH_2)$, 1 mM citric acid, 1.3 mM tris base, adjusted to pH 7 with tris base. Other buffers use the same recipe except that other salts, e.g., K-acetate, are substituted for KCl. Buffers should not be made by adding the salt of interest to an already made buffer because all cations should be controlled and kept the same from test to control buffers. Attractants or repellents to be tested are added to buffer solutions at desired concentrations. Acetate, folate, glutamate, ammonium, biotin, and lactate are all excellent attractants. The attractant and control salts are added from stock solutions before the pH is adjusted and the solution brought to volume.

The final pH is brought to 7 with tris base, making the final tris concentration a bit higher than 1.3 mM. A common mistake is to adjust the buffer pH with NaOH or some other base or acid. This changes the ionic composition of the buffer and greatly affects the behavior of the cells. Since the accumulation or repulsion of the cells in the T-maze is an indirect measure of their swimming behavior, this kind of mistake can make it impossible to reproduce results. Na^+ or K^+ salts can generally be used for attractants, but they must be balanced. For example, 2 mM Na-acetate buffer is used with 2 mM NaCl for control and not KCl. If there must be a choice between balancing Na^+ or K^+ salts vs keeping ionic strength the same, the salts must be balanced. For example, 2 mM NaCl control is used for 1 mM Na_2-folate because folate is a divalent anion.

3. For pulling micropipettes to isolate individual cells: Heat a glass Pasteur pipette in the center of the wider part of the pipette. This can be the same kind of pipette as used to make the bridges. Rotate the pipette at a position 2/3 of the way up the large part of the pipette in a flame until the glass becomes soft and malleable. While removing the pipette from the heat, pull the pipette from both ends making the glass stretch out. Once it has cooled, heat the thinned pipette near where the glass widens briefly. Once the glass is hot, pull again while removing quickly from the flame to create a much finer pipette.
4. For making bridges: Heat the narrow end of a glass Pasteur pipette until it makes a bend, becoming U-shaped. Using a glass cutting pencil, scar the glass to make the bridge ~3–4 mm on one side of the bend and ~5–6 mm on the other. To fill the bridges, bring 3% agar in 10 mL of 3 M KCl to a gentle boil to dissolve the agar. Pour in sterile bridges and stir with a Pasteur pipette. Allow agar to harden, pour 3 M KCl over the surface. Cover with Parafilm and store at 4 °C.
5. Cells in all growth conditions can be used in the T-mazes, but late log cells are most responsive. They are also the most dense which means that very little of the cell pellet harvested by centrifugation is needed for resuspension in the control buffer and, therefore, little residual culture fluid is transferred with the cells. It is critical to dilute away the culture medium as it contains many bacterial metabolites that are attractants for paramecia. (If necessary, the cells can be suspended in the control solution and centrifuged again, ensuring that little culture fluid remains with the cells. If cultures are very sparse, as sometimes happens with RNAi treated cultures, the cells can be harvested by centrifugation and allowed to swim to the top of a tube of control buffer. There they concentrate themselves and can be collected with a pipet.)

6. A common mistake is to fail to make the solutions with sterile stocks and water, or to otherwise let the buffers become contaminated with bacteria. Buffers should be streaked on nutrient agar plates regularly or made fresh each day. The attractants for *P. tetraurelia* are products of bacterial metabolism, which means that contaminated buffers already have unknown amounts of undefined stimuli that will confound the assay. Shorter time periods yield similar results, and longer time periods risk the diffusion of the solutions between arms [25].
7. The bore size of the stopcock arms and plug are also critical to a successful assay. We have established that the cells move in and out of the arms during the assay, which requires that they not become trapped in arms with very small bores. Larger species of *Paramecium*, such as *P. caudatum*, and small ciliates would need the apparatus customized.
8. Ideally there will be about 30–50 cells per aliquot in the test arm. It takes a practiced eye to dilute the cells in control buffer before loading the cells into the T-maze to the point that there are not too many or too few to count for an accurate calculation of response.
9. Corks are kept clean by washing in reverse osmosis water and air-drying thoroughly. The glass T-mazes must be kept clean with vigorous rinsing in hot water between uses. If the results of wild type cells with known attractants become variable, the T-mazes should be disassembled and cleaned with KOH saturated ethanol to remove organic materials from the arms and grease from the plug. Thorough rinsing is critical before the T-mazes are reassembled.
10. Inexperienced investigators typically learn this assay in 2 days and get results such as: 0.75 ± 0.05 (SD) for three T-mazes testing 2 mM Na-acetate in buffer vs 2 mM NaCl in buffer (average cells in control aliquots/total average cells in both control and test aliquots).
11. Separating cilia into membrane and axoneme is followed by phase separation in which GPI anchored proteins enter the aqueous phase and transmembrane proteins, even those from lipid rafts, partition in the detergent phase [46, 50]. This is relatively easily done using established and implemented protocols.
12. Solubilizing the pure ciliary membrane samples in Triton X-114 followed by heating and centrifugation enriches for the transmembrane proteins in the detergent phase, leaving many soluble and non-membrane proteins in the aqueous phase [46, 51]. This approach of combining the ciliary isolation established by Adoutte and colleagues [47] with a Triton X-114 phase separation of the isolated ciliary membranes resulted in identification

by liquid chromatography tandem mass spectrometry (LC/MS/MS) of many transmembrane proteins that had not been identified previously [46]. These proteins included the ciliary voltage-gated Ca-channels and the Mg^{2+}-specific channel like protein, XntA [52, 53], among others [46].

13. The proteomics facility where the samples are analyzed typically will have protocols for staining and processing the gels. It is recommended that their protocols be followed.
14. The gel is cut into slices in order to prevent any one protein from dominating the samples. Cutting the gel into slices allows for a more complete analysis of the proteins present.
15. The proteomics facility where your samples are to be processed will have a detailed protocol as to how to complete the in-gel digestion.
16. Commonly, when using SDS-PAGE and in-gel digestion, cysteine residues are identified with an acrylamide adduction. Therefore, searches may consider a static increase of 71 Da for acrylamide adduction [46].
17. Before beginning binding assay experiments, consider treating collected paramecia with antibiotics to kill any remaining bacteria, both on the surface of the cells as well as in food vacuoles. Treat cells for 1 h in 2 mM NaCl in buffer containing 2000 U/mL penicillin and 2 mg/mL streptomycin, pH 7.02 before beginning deciliation process [34]. Alternatively, pellet the cells and incubate them in 500 μg/mL gentamycin solution in a standard buffer for 30 min.
18. If cells begin to lyse, begin centrifugation before the 5 min are up. Cells that lyse will contaminate the isolated cilia with cell membrane and other cellular components. If too many cells lyse, cilia collection should be abandoned.
19. Caution when using IBMX, cells can begin to swell after 30 min [31].
20. For other ^{3}[H] isotopes, the recommended starting concentrations are 40–50 nM ^{3}[H]-glutamate [29] or 5 nM ^{3}[H]-folate spiked into 25 μM carrier Na_2-folate [27, 33, 48].
21. A BCA Protein assay (i.e., ThermoFisher BCA protein assay kit, cat# 23227) or similar kit should be considered to determine protein concentration of the cilia in the reactions. This will allow for the results to be normalized to the protein concentration of each sample, allowing for multiple experiments to be combined.
22. Instantaneous binding assays can also be done. In this case, the cilia are incubated for only short periods of time with the ^{3}[H] isotope before the reaction is stopped. The cpm can then be plotted as a function of time and a straight line can be fit to extrapolate time zero [34, 48].

23. Binding studies should first determine the time to establish equilibrium binding (e.g., 60 min for ^{3}[H]-biotin [7]). Binding over a range of concentrations is then used to establish numbers of sites and affinity of those sites. If equilibrium cannot be established because counts continue to rise, instantaneous binding can be measured. As in the protocol for folate whole cell binding, cells incubated in various specific activities of the ^{3}[H]-attractant ligand (i.e., concentrations of cold attractant and with the same amount of ^{3}[H]-ligand) are sampled over time. Once the data are plotted over time, the binding can be extrapolated to zero time. The values at zero time are then plotted as a function of ligand concentration. In all binding analyses, the nonspecific binding should be subtracted from the final binding curve. Isomers that are not active as attractants generally are useful. (*See* refs. 33, 48 for a discussion.) The resulting binding curve should saturate and be useful for kinetic analysis.

24. Another approach to kinetic analysis of binding is to plot binding of ^{3}[H]-ligand as a function of cold carrier ligand [31]. The same kinetic parameters as from association curves should result. It is important that the amount of ligand (cold and hot) bound be small compared to the total ligand. Also, it is important that a sufficient number of concentrations be used to accurately calculate the dissociation constant K_d—several concentrations lower than and higher than the K_d are critical [54].

    The standard kinetic analyses for ligand binding are very well described in ref. 55 for equilibrium binding, competitive binding, affinity and number of different sites through the use of Scatchard, Hill, and other plots. *See* refs. 33, 48 for an example of analysis of Kd and also reference to caveats for binding assays [54].

25. Clean capillary tubes (3 in. 1.0 mm 1/0.58 OD/ID standard glass capillary tubes, with filament (World Precision Instruments, cat # 1B100F-3)). Wipe tubes with 95% ethanol and allow to dry before pulling them. Use tweezers to remove pulled pipettes from puller and from that point on if possible to avoid transferring oil and grease from your hands. Resistance should be between 40 and 100 MΩ [10, 42].

26. Bridges should be used 1 day for recording, but they can be cleaned and reused. To do so, first wipe the glass clean of Vaseline after use. Once ~12–20 used bridges have been collected, boil bridges in distilled water, then remove and boil briefly in 95% ethanol. Remove bridges to a Kimwipes and dry. Repeat process if necessary. Store wrapped in a Kimwipes or sealed beaker to protect from dust.

27. When using deciliated cells, cells should be deciliated fresh for each recording as cilia will regenerate over time. The work by

Machemer and Ogura provides excellent tips and methods for deciliation of paramecia cells using 5% ethanol (v/v) in a KCl solution in buffer along with trituration [41]. Cells should be moved to 5 mM KCl in buffer after deciliation to remove as much ethanol as possible.

28. After recording from deciliated cells, high potassium (30 mM in buffer) should be added to the bath to ensure no action potentials are produced, signifying the regrowth of the cilia and the return of the voltage-dependent calcium channels [16, 41].

References

1. Van Houten JL (1992) Chemoresponse in microorganisms. Ann Rev Physiol 54:639–663
2. Van Houten JL (1994) Chemoreception in microorganisms: trends for neuroscience? Trends Neurosci 17:62–71
3. Berbari NF, Oconnor AK, Haycraft CJ, Yoder BK (2009) The primary cilium as a complex signaling center. Curr Biol 19(13):R526–R535
4. Singla V, Reiter J (2006) The primary cilium as the cell's antenna: signaling at a sensory organelle. Science 313:629–633
5. Van Houten J, Preston RR (1987) Chemoreception: *Paramecium* as a receptor cell. Adv Exp Med Biol 221:375–384
6. Van Houten J (1998) Chemosensory transduction in *Paramecium*. Eur J Protistol 34:301–307
7. Bell WE, Karstens W, Sun Y, Van Houten JL (1998) Biotin chemoresponse in *Paramecium*. J Comp Physiol A 183(3):361–366
8. Van Houten J (1979) Membrane potential changes during chemokinesis in *Paramecium*. Science 204(4397):1100–1103
9. Davis DP, Fiekers J, Van HOuten J (1998) Intracellular pH and chemoresponse to NH^{4+} in *Paramecium*. Cell Motil Cytoskeleton 40:107–118
10. Preston RR, Van Houten JL (1987) Chemoreception in *Paramecium tetraurelia*: acetate and folate-induced membrane hyperpolarization. J Comp Physiol A 160(4):525–535
11. Yang WQ, Braun C, Plattner H, Purvee J, Van Houten JL (1997) Cyclic nucleotides in glutamate chemosensory signal transduction of *Paramecium*. J Cell Sci 110(Pt 20):2567–2572
12. Van Houten JL (2000) Chemoreception in microorganisms. In: Finger T, Silver W, Restrepo D (eds) The neurobiology of taste and smell. Wiley-Liss, New York, NY, pp 11–40
13. Machemer H (1974) Frequency and directional responses of cilia to membrane potential changes in *Paramecium*. J Comp Physiol 92:293–316
14. Preston RR, Usherwood PN (1988) L-Glutamate-induced membrane hyperpolarization and behavioural responses in *Paramecium tetraurelia*. J Comp Physiol A 164(1):75–82
15. Bell WE, Preston RR, Yano J, Van Houten JL (2007) Genetic dissection of attractant-induced conductances in *Paramecium*. J Exp Biol 210(Pt 2):357–365
16. Dunlap K (1977) Localization of calcium channels in *Paramecium caudatum*. J Physiol 271(1):119–133
17. Capdeville Y, Benwakrim A (1996) The major ciliary membrane proteins in *Paramecium primaurelia* are all glycosylphosphatidylinositol-anchored proteins. Eur J Cell Biol 70(4): 339–346
18. Paquette CA, Rakochy V, Bush A, Van Houten JL (2001) Glycophosphatidylinositol-anchored proteins in *Paramecium tetraurelia*: possible role in chemoresponse. J Exp Biol 204(Pt 16):2899–2910
19. Yano J, Rachochy V, Van Houten JL (2003) Glycosyl phosphatidylinositol-anchored proteins in chemosensory signaling: antisense manipulation of *Paramecium tetraurelia* PIG-A gene expression. Eukaryot Cell 2(6): 1211–1219
20. Merkel SJ, Kaneshiro ES, Gruenstein EI (1981) Characterization of the cilia and ciliary membrane proteins of wild-type *Paramecium tetraurelia* and a pawn mutant. J Cell Biol 89(2):206–215
21. Kleene S, Van Houten JL (2014) Electrical signaling in motile and primary cilia. BioScience 64:1092–1102
22. Barrera NP, Robinson CV (2011) Advances in the mass spectrometry of membrane proteins: from individual proteins to intact complexes. Annu Rev Biochem 80:247–271

23. Van Houten JL (1977) A mutant of *Paramecium* defective in chemotaxis. Science 198:746–749
24. DiNallo MC, Wohlford M, Van Houten J (1982) Mutants of *Paramecium* defective in chemokinesis to folate. Genetics 102(2):149–158
25. Van Houten J, Martel E, Kasch T (1982) Kinetic analysis of chemokinesis of *Paramecium*. J Protozool 29(2):226–230
26. Romanovitch M (2012) The L-glutamate receptor in *Paramecium tetraurelia*. M.S. thesis, Department of Biology, University of Vermont, Burlington, VT
27. Weeraratne SD (2007) GPI receptors in folate chemosensor transduction in *Paramecium tetraurelia*. Ph.D. thesis, Department of Biology, University of Vermont, Burlington, VT
28. Jacobs CL (2007) NMDA receptor associated protein in *Paramecium* and it involvement in glutamate chemoresponse. M.S. thesis, Department of Biology, University of Vermont, Burlington, VT
29. Preston RR, Usherwood PNR (1988) Characterization of a specific L-[H-3]glutamic acid binding-site on cilia isolated from *Paramecium tetraurelia*. J Comp Physiol B 158(3):345–351
30. Czapla H (2012) Cyclic adenosine monophosphate receptors in *Paramecium tetraurelia*. M.S. thesis, Department of Biology, University of Vermont, Burlington, VT
31. Smith R, Preston RR, Schulz S, Gagnon ML, Van Houten J (1987) Correlations between cyclic AMP binding and chemoreception in *Paramecium*. Biochim Biophys Acta 928(2): 171–178
32. Van Houten JL, Yang WQ, Bergeron A (2000) Chemosensory signal transduction in *Paramecium*. J Nutr 130(4S Suppl): 946S–949S
33. Schulz S, Denaro M, Xypolyta-Bulloch A, Van Houten J (1984) The relationship of folate binding to chemoreception in *Paramecium*. J Comp Physiol A 155:113–119
34. Smith RA (1987) The association between external binding of cyclic AMP chemoattraction in *Paramecium tetraurelia*. M.S. thesis, Department of Biology, University of Vermont, Burlington, VT
35. Hansma HG (1979) Sodium uptake and membrane excitation in *Paramecium*. J Cell Biol 81(2):374–381
36. Machemer-Rohnisch S, Machemer H (1989) A Ca paradox: electric and behavioural responses of *Paramecium* following changes in external ion concentration. Eur J Protistol 25(1):45–59
37. Naitoh Y (1968) Ionic control of the reversal response of cilia in *Paramecium caudatum*. A calcium hypothesis. J Gen Physiol 51(1): 85–103
38. Naitoh Y, Eckert R, Friedman K (1972) A regenerative calcium response in *Paramecium*. J Exp Biol 56(3):667–681
39. Preston RR (1990) A magnesium current in *Paramecium*. Science 250(4978):285–288
40. Preston RR, Van Houten JL (1987) Localization of the chemoreceptive properties of the surface membrane of *Paramecium tetraurelia*. J Comp Physiol A 160(4):537–541
41. Machemer H, Ogura A (1979) Ionic conductances of membranes in ciliated and deciliated *Paramecium*. J Physiol 296:49–60
42. Valentine MS (2015) Polycystin-2 (*PKD2*), Eccentric (*XNTA*), and Meckelin (*MKS3*) in the ciliated model organism *Paramecium tetraurelia*. Ph.D. dissertation, Department of Biology, University of Vermont, Burlington, VT
43. Oami K (1996) Distribution of chemoreceptors to quinine of the cell surface of *Paramecium caudatum*. J Comp Physiol A 179:345–352
44. Oami K (1998) Membrane potential response of *Paramecium caudatum* to bitter substances: existence of multiple pathways for bitter responses. J Exp Biol 201:13–20
45. Sasner J, Van Houten JL (1989) Evidence for a *Paramecium* folate chemoreceptor. Chem Senses 14:587–595
46. Yano J, Rajendran A, Valentine MS, Saha M, Ballif BA, Van Houten JL (2013) Proteomic analysis of the cilia membrane of *Paramecium tetraurelia*. J Proteomics 78:113–122
47. Adoutte A, Ramanathan R, Lewis RM, Dute RR, Ling KY, Kung C, Nelson DL (1980) Biochemical studies of the excitable membrane of *Paramecium tetraurelia*. III. Proteins of cilia and ciliary membranes. J Cell Biol 84(3): 717–738
48. Schulz S, Denaro M, Xypolytabulloch A, Vanhouten J (1984) Relationship of folate binding to chemoreception in *Paramecium*. J Comp Physiol 155(1):113–119
49. Van Houten J, Yang W, Bergeron A (2000) Glutamate chemosensory signal transduction in *Paramecium*. J Nutr 130:946S–949S
50. Arnaiz O, Malinowska A, Klotz C, Sperling L, Dadlez M, Koll F, Cohen J (2009) Cildb: a knowledgebase for centrosomes and cilia. Database (Oxford) 2009:bap22
51. Qoronfleh MW, Benton B, Ignacio R, Kaboord B (2003) Selective enrichment of membrane proteins by partition phase separation for proteomic studies. J Biomed Biotechnol 2003(4): 249–255

52. Preston RR, Kung C (1994) Inhibition of Mg^{2+} current by single-gene mutation in *Paramecium*. J Membr Biol 139(3):203–213
53. Haynes WJ, Kung C, Saimi Y, Preston RR (2002) An exchanger-like protein underlies the large Mg^{2+} current in *Paramecium*. Proc Natl Acad Sci U S A 99(24):15717–15722
54. Klotz I (1982) Number of receptor sites from Scatchard graphs: facts and fantasies. Science 217:1247–1249
55. Motulsky H, Christopoulos A (2005) Fitting models to biological data using linear and nonlinear regression: a practical guide to curve fitting. Oxford University Press, San Diego, CA
56. Van Houten J, Hansma H, Kung C (1975) Two quantitative assays for chemotaxis in *Paramecium*. J Comp Physiol 104: 211–223

Chapter 11

STED and STORM Superresolution Imaging of Primary Cilia

T. Tony Yang*, Weng Man Chong*, and Jung-Chi Liao

Abstract

The characteristic lengths of molecular arrangement in primary cilia are below the diffraction limit of light, challenging structural and functional studies of ciliary proteins. Superresolution microscopy can reach up to a 20 nm resolution, significantly improving the ability to map molecules in primary cilia. Here we describe detailed experimental procedure of STED microscopy imaging and dSTORM imaging, two of the most powerful superresolution imaging techniques. Specifically, we emphasize the use of these two methods on imaging proteins in primary cilia.

Key words Primary cilium, Ciliary protein, Superresolution microscopy, Diffraction limit, Fluorophore, STED, STORM

Abbreviations

BaLM	Bleaching/blinking assisted localization microscopy (BaLM)
CW	Continuous wave (CW)
dSTORM	Direct stochastic optical reconstruction microscopy (dSTORM)
FPALM	Fluorescence photoactivation localization microscopy (FPALM)
GSD	Ground state depletion (GSD)
GSDIM	Ground state depletion microscopy followed by individual molecule return (GSDIM)
gSTED	Gated stimulated emission depletion (gSTED)
IFT	Intraflagellar transport (IFT)
PAINT	Point accumulation for imaging in nanoscale topography (PAINT)
PALM	Photoactivated localization microscopy (PALM)
PALMIRA	PALM with independently running acquisition (PALMIRA)
PSF	Point spread function (PSF)
RESOLFT	Reversible saturable optical fluorescence transitions (RESOLFT)
SIM	Structured illumination microscopy (SIM)
SOFI	Superresolution optical fluctuation imaging (SOFI)

*Author contributed equally with all other contributors.

Peter Satir and Søren Tvorup Christensen (eds.), *Cilia: Methods and Protocols*, Methods in Molecular Biology, vol. 1454,
DOI 10.1007/978-1-4939-3789-9_11, © Springer Science+Business Media New York 2016

SSIM	Saturated structured illumination microscopy (SSIM)
STED	Stimulated emission depletion (STED)
STORM	Stochastic optical reconstruction microscopy (STORM)

1 Introduction

Primary cilium is an organelle with a complex structure confined in a small volume. The diameter of the ciliary membrane is approximately 250–300 nm and its length is about 2–10 μm. It is composed of different elements and compartments, including the axoneme, the ciliary membrane, the ciliary pocket, the transition zone, the transition fibers or distal appendages, the inversin region, and the ciliary tip region, each containing numerous structural and functional proteins. The relative localization of ciliary proteins is essential for understanding interactions and active roles of each ciliary protein, as well as their potential association with different ciliopathic disease phenotypes. However, the packed arrangement of these proteins largely challenges the resolution limit of conventional fluorescent microscopy. For example, many transition zone proteins have been identified at the ciliary base [1–5], but which of them forms the Y-links remains unknown. Many intraflagellar transport (IFT) proteins are involved in ciliary cargo transport, but where IFT-A and IFT-B proteins are assembled is unclear. Key elements of distal appendages have recently been identified [6–8], but how they coordinate ciliogenesis initiation remains elusive. All these knowledge gaps require a geometric framework of ciliary proteins to determine their potential interactions. We have previously shown that superresolution microscopy can be a suitable approach to address some of these problems [9, 10]. Others have also used superresolution imaging for primary cilium studies [4, 6, 11–16]. Much more efforts will be needed to reconstruct the interaction map upon the ciliary molecular architecture.

Superresolution microscopy breaks the diffraction limit of light for far-field imaging, which includes widefield and confocal fluorescent microscopy techniques we routinely use for biological sample imaging. The diffraction limit is defined by Abbe's law: $d = \lambda/2\mathrm{NA}$, where d is the resolution, λ is the wavelength, and NA is the numerical aperture of the optical system. For a green fluorescent dye or protein emitting at ~520 nm wavelength and a high-end oil immersion objective with an NA of 1.4, a conventional microscopy can only reach ~190 nm resolution. Various superresolution techniques have been invented in the past two decades to reach a subdiffraction resolution, including STED [17, 18], SIM [19], RESOLFT [20], SSIM [21], PALM [22], STORM [23], FPALM [24], PAINT [25], GSD [26], PALMIRA [27], dSTORM [28], GSDIM [29], SOFI [30], gSTED [31], BaLM [32], Bayesian localization methods [33, 34], and other variations. Among these methods, SIM, STED, and PALM/STORM are three of the most widely used approaches for superresolution imaging of biological samples. For fluorescent imaging of cell cultures, SIM can reach ~110 nm lateral resolution; STED can reach ~50 nm lateral resolution; and PALM/STORM can reach ~20 nm lateral resolution. SIM is the easiest to use, with imaging protocols similar to those of epifluorescent imaging. PALM and STORM both fall into the category of localization microscopy, with PALM used for photoactivatable or photoconvertible fluorescent proteins and STORM used for photoswitchable organic fluorophores. STED is a confocal-based method that can reach a relatively high imaging speed. In this Chapter, we will only focus on the methods of STED and STORM (specifically, dSTORM) for ciliary imaging.

Stimulated emission depletion (STED) microscopy achieves superresolution by directing neighboring molecules to another photon release path, so that the center region emitting fluorescent signals is much smaller than the diffraction limited spot [17, 18]. It is a point-by-point scanning method to achieve subdiffractive imaging one point at a time. For each imaging spot, a diffraction-limited focused laser light (excitation laser) illuminates a volume of the point spread function (PSF) to drive molecules to the excited state. Another laser of a different wavelength (depletion laser) focusing at the same focal spot passes through a special lens (e.g., spiral phase plate) which creates a doughnut-shaped PSF, so that the peripheral molecules are forced to emit photons through stimulated emission at a wavelength distinct from the fluorescent wavelength. By tuning the power of the depletion laser, one can control how large a peripheral region is, and thus adjust how small the volume at the center is where molecules emit photons through fluorescence instead of stimulated emission. Therefore, neighboring molecules that cannot be resolved by conventional microscopy can now be distinguished by different wavelength emissions at the central and peripheral regions of the focal spot. In principle, one can arbitrarily increase the power of the depletion laser to reach an arbitrarily high resolution. In fact, a lateral resolution as high as 5.8 nm has been achieved for nanodiamonds [35]. For imaging organic dyes and fluorescent proteins in biological samples, the signal-to-noise ratio remains as one of the major issues, limiting the lateral resolution to ~50 nm. Deconvolution is often, then, applied to the raw data to improve image quality of STED microscopy.

A variety of STED microscopes are available for different applications. The original STED microscope was implemented with a mode-locked pulsed laser, requiring complex optical manipulation to operate [18]. Later, a continuous wave (CW) laser-based STED microscope was implemented [36]. The CW STED system is much easier to build than the other versions of STED microscopes and is able to achieve a ~60 nm resolution. Dual-color CW STED imaging was also demonstrated [37]. One drawback of this system is its need of high depletion power. Despite its relative early release, the CW STED microscopy system is still considered as a good option when the fluorescent signals of the molecule of interest are dim, such as

MKS1 in the transition zone of primary cilia. Time-gated STED (gSTED) microscopy has later been developed [31], reaching ~50 nm resolution with a much lower laser power, and thus reducing the photodamage of samples. One drawback of gSTED microscopy is that the photon count per image frame may be low due to the time gating if a fluorophore is dim. The other drawback of a gSTED system compared to a CW STED system is its higher cost. Currently gSTED microscopy is considered as the primary STED imaging method. 3D STED imaging has also been achieved [38], as well as two-photon STED imaging [39]. Available commercial STED systems (e.g., Leica and Abberior) include CW STED, gSTED, 3D CW STED, and 3D gSTED systems, all of which are compatible with two-color imaging.

Live-cell STED imaging has been shown [40], where the sampling rate is limited by photon counts and the imaging duration is limited by photobleaching and phototoxicity. For intraflagellar transport imaging of IFT88-EYFP in IMCD cells, we have achieved 4 frames/s to collect enough photons for each frame (unpublished). Only a limit number of frames can be imaged due to photobleaching (<100 frames for CW STED). It is possible that gSTED can achieve a longer imaging duration due to its low depletion laser power, although we have not verified it. Two-color live-cell STED imaging has also been reported [41].

The choice of fluorophores depends on the available laser sources and their STED performance. With a 592-nm depletion laser for a CW STED system, Oregon Green 488 and EYFP as an organic dye and a fluorescent protein, respectively, perform well for single-color STED microscopy. For two-color STED imaging, the combination of Oregon Green 488 and BD Horizon V500 is one of the best pairs of organic dyes. BD Horizon V500 has a large Stokes shift, so that one can use a single depletion laser (e.g., 592 nm laser) for Oregon Green 488 and V500, both excited with different lasers (e.g., 491 nm and 447 nm lasers). One can also use SNAP-tag and HaloTag [42, 43] together with the Oregon Green 488 and BD V500 for live cell two-color STED imaging. For two-color STED imaging of fluorescent protein pairs, ECFP and EYFP can be a reasonable choice for a 592-nm depletion laser, although organic dyes are preferred due to their higher photon output. The new commercial STED systems provide multiple depletion laser sources and a wide spectrum of excitation laser sources (e.g., supercontinuum laser), broadening the possible choices of fluorophores. For example, it is suggested that one can use Oregon Green 488 with a 592-nm depletion laser, TRITC with a 660-nm depletion laser, and STAR635P with a 775-nm depletion laser for three-color STED superresolution imaging using a commercial system (Leica). In this chapter we describe the protocol of using the combination of Oregon Green 488 and BD Horizon V500 for two-color STED microscopy, but the procedure of using other combinations of fluorophores can be similar.

Stochastic optical reconstruction microscopy (STORM) and photoactivated localization microscopy (PALM) achieve superresolution by preventing most of the molecules from emitting light and then determining the intensity center of those isolated molecules that do emit fluorescent signals by fitting [22, 23]. Knowing the intensity distribution of photons from a single molecule forms a Gaussian-like distribution (Airy disk) in lateral dimensions, one can fit the distribution function and find the point of the intensity peak with a resolution limited by the signal-to-noise ratio. The subset of fluorophores that light up in a frame will turn dark and a random set of other molecules will be excited in the next frame. Thus, every frame of images contains a portion of molecules shown in a very high resolution. By overlapping a large number of frames, one can then observe most of the fluorescent molecules in superresolution. The key to result in isolated fluorescent molecules is to make them alternate between bright and dark states randomly. This is achieved by creating an environment to promote photoswitchable behaviors of fluorophores.

Various photoactivatable and photoconvertible fluorescent proteins have been used for PALM superresolution imaging [22]. For single-color PALM imaging, mEos3.2 can be one of the best choices with its high photon output [44]. Recent development of mMaple3 can be a good alternative of mEos3.2 [45]. For two-color PALM imaging, combinations of PAGFP/PAmCherry1 and Dronpa/PAmCherry1 are relatively good in performance [46]. The combination of mEos3.2/PSCFP2 has also been used [47].

A variety of photoswitchable organic fluorophores are available for STORM imaging [23, 48]. Mostly these organic dyes are for antibody staining of fixed cells, although they can also be used for live cell imaging when using with target specific labeling techniques including SNAP-tag and HaloTag [42, 43]. STORM was originally implemented using photoswitching activities of Cy3-Cy5 dye pairs [23]. Many related methods based on photoswitching features of fluorescent dyes were later developed. Among them direct STORM (dSTORM) is currently the most popular method due to its compatibility with conventional organic dyes such as Alexa Fluor and ATTO-dyes [28]. By adding a finite amount of a reducing thiol compound (e.g., mercaptoethylamine) and an oxygen scavenging system of catalase/glucose/glucose oxidase to the sample, molecules are populated to a stable non-fluorescent reduced state, i.e., a dark state. A small population of molecules can return to the ground state by spontaneous or light-induced (405 nm light) oxidation to reenter the fluorescent path. Among all organic dyes, Alexa 647 performs the best for dSTORM superresolution imaging due to its high photon yield per switch and low duty cycle. For two-color dSTORM imaging, many choices have been suggested for the second dye, including Alexa Fluor 488, ATTO 488, Alexa Fluor 568, Cy3B, CF568, and Alexa 750, although none of them has a performance close to that of Alexa 647. In our hands, Alexa 647/Cy3B is the best combination for two-color dSTORM imaging thus far.

Several methods have been developed for 3D STORM imaging: the astigmatism method [49], the biplane method [50], the double-helix point spread function method [51], and the opposing objective method [52]. The astigmatism method may be the most widely used among them so far due to its ease of implementation, although other methods may be more precise in the axial localization. For the astigmatism method, the ellipticity and orientation of the image from a single molecule allow one to quantitatively determine how far the molecule is away from the focal plane, and thus each image spot provides coordinate information in all three dimensions.

Here we describe detailed protocols of CW STED and dSTORM methods for ciliary protein imaging. Many tricks and solutions can also be used for other STED and localization based methods such as gSTED and PALM.

2 Materials

2.1 Cell Culture

Sterilized solutions and reagents should be used for cell culture.

1. Cells: Here we use retinal pigment epithelial cells (hTERT RPE-1, ATCC CRL-4000) as an example. Cells are incubated in an incubator at 37 °C in 5% CO_2. Thin tissue samples (<10 μm, preferably thinner) may also be used.
2. Growth medium: Dulbecco's Modified Eagle Medium: Nutrient Mixture F12 (DMEM/F12, Life Technologies, 11330-032) with 10% fetal bovine serum (FBS, Life Technologies, 10437-028) and 1% penicillin-streptomycin (Life Technologies, 15140-122).
3. 0.25% Trypsin-EDTA (Life Technologies, 25200-056).
4. 10-cm tissue culture plates and other basic cell culture consumables.

2.2 Stock Solutions or Common Reagents for Immunofluorescence Staining

1. Poly-L-lysine solution: 0.1 mg/ml (0.01%) poly-L-lysine hydrobromide (Sigma-Aldrich, P1274) dissolved in water, sterilized with a 0.22 μm filter, and stored at 4 °C.
2. 1 M hydrogen chloride (HCl).
3. #1.5 glass coverslips.
4. Phosphate buffer saline (PBS, pH 7.0, Amresco, K813).
5. Bovine serum albumin (BSA, Sigma-Aldrich, A9647).
6. Permeabilizing buffer/washing buffer: PBST, i.e., 0.1% Triton X-100 (Fisher Scientific, BP151-100) in PBS at pH = 7.0.

7. Blocking buffer: 3% BSA in PBST.
8. 4% paraformaldehyde (PFA): four times diluted solution of the 16% aqueous paraformaldehyde (Electron Microscopy Sciences, 15710).
9. 100% methanol (Sigma-Aldrich, 32213).
10. 0.01% sodium azide.
11. Nail polish.

2.3 Reagents for dSTORM Imaging

1. Primary antibodies for ciliary markers and proteins of interest. For example, for ciliary markers, one may use anti-acetylated tubulin (Abcam, ab24610), anti-IFT88 (Proteintech, 13967-1-AP), or anti-Arl13b (Abcam, ab136648).
2. Secondary antibodies: Alexa Fluor 647 conjugated IgG antibody (Life Technologies) and Cy3B maleimide (GE, PA6313) (conjugation protocol for Cy3B is described below).
3. Whole IgG affinity-purified antibodies (Jackson ImmunoResearch).
4. Borate buffer (0.67 M, Thermo Scientific, 1859833).
5. Dimethylsulfoxide (DMSO, Sigma-Aldrich, D8418).
6. Dimethylformamide (DMF).
7. Purification resin (Thermo Scientific, 1860513).
8. Pierce spin column (Thermo Scientific, 69705).
9. TetraSpeck™ microspheres (Life Technologies, T7279).
10. Chamlide magnetic chamber, 35-mm dish (Live Cell Instrument).
11. Imaging buffer for oxygen scavenging:
    (a) Glucose (Sigma-Aldrich, G5767).
    (b) Glucose oxidase (Sigma-Aldrich, G2133).
    (c) Catalase (Sigma-Aldrich, C40).
    (d) Mercaptoethylamine (MEA or Cysteamine, Sigma-Aldrich, 30070).
    (e) TN buffer : Tris–HCl, pH 8.0 (10 mM), NaCl (150 mM).

2.4 Reagents for STED Imaging

1. Secondary antibodies: Oregon Green 488 conjugated IgG (Life Technologies) and biotin conjugated antibody (Sigma-Aldrich).
2. Primary antibodies (*see* Subheading 2.3).
3. BD V500 streptavidin (BD, 561419): streptavidin conjugated to BD-Horizon V500 fluorescent dye.
4. Prolong Antifade Kit (Life Technologies, P-7481) or glycerol.

2.5 A microscopy System for STORM Imaging

1. A STORM/PALM system equipped with a high NA objective (NA ≥ 1.4), an autofocus system, an activation laser (405 nm), multiple excitation lasers depending on the fluorophores in use (e.g., ~640 nm laser for Alexa Fluor 647), and a high-speed, high-sensitivity EMCCD or sCMOS camera (Fig. 1a). As an example, our STORM/PALM system is composed of four lasers (100 mW 405 nm solid state optically pumped semiconductor laser (OPSL), 150 mW 488 nm solid state OPSL, 150 mW 561 nm pumped diode laser, and 140 mW 637 nm solid state OPSL, all integrated to a laser unit by Spectral Applied Research), a light path integration platform (Diskovery Platform, Spectral Applied Research), an EMCCD camera (Evolve 512 Delta, Photometrics), a microscope (Eclipse Ti-E, Nikon) equipped with an NA 1.49 100× objective (CFI APO TIRF 100× OIL, Nikon) and an autofocus system (Perfect Focus System, Nikon). Zeiss, Nikon, and Leica all have integrated superresolution microscopy systems with equivalent elements convenient for STORM/PALM imaging.
2. A computer with a large data storage system.
3. STORM/PALM image analysis software. Multiple STORM/PALM image analysis software packages to find single molecule centers and to merge multiple images together are available (e.g., QuickPalm [53]). We use the MetaMorph Microscopy Automation & Image Analysis Software with a superresolution module, which uses wavelet fitting to allow real-time localization of single molecules [54]. A computer with CPU/GPU integration is used to expedite wavelet fitting.
4. Vibration isolation system. The vibration isolation system such as a high-performance optical table is important for superresolution imaging. To push the limit of mechanical stability, we use negative stiffness vibration isolators (two 250BM-1, Minus K, and one breadboard, SG-35-2, Newport), which have a better performance than optical tables. The choice of Minus K systems depends on the total weight of the optical setup.

2.6 A Microscopy System for STED Imaging

1. A STED microscopy system equipped with a high NA objective (NA ≥ 1.4), a scanning system, excitation lasers, depletion lasers, an optical system integrating both laser sources, and a highly sensitive photon counter (Fig. 1b). Our STED microscopy system is a system using CW lasers, with two excitation lasers (20 mW 447 nm diode laser, PGL-V-H-447, CNI;

Fig. 1 (continued) depletion laser is converted to a doughnut-shaped distribution through a spiral phase plate. Fluorescent signals from a sample are detected pixel-by-pixel by an avalanche photodiode (APD), a single photon counting module. *SMF* single-mode polarization-maintaining fiber, *MMF* multimode fiber, *DC* dichroic mirror

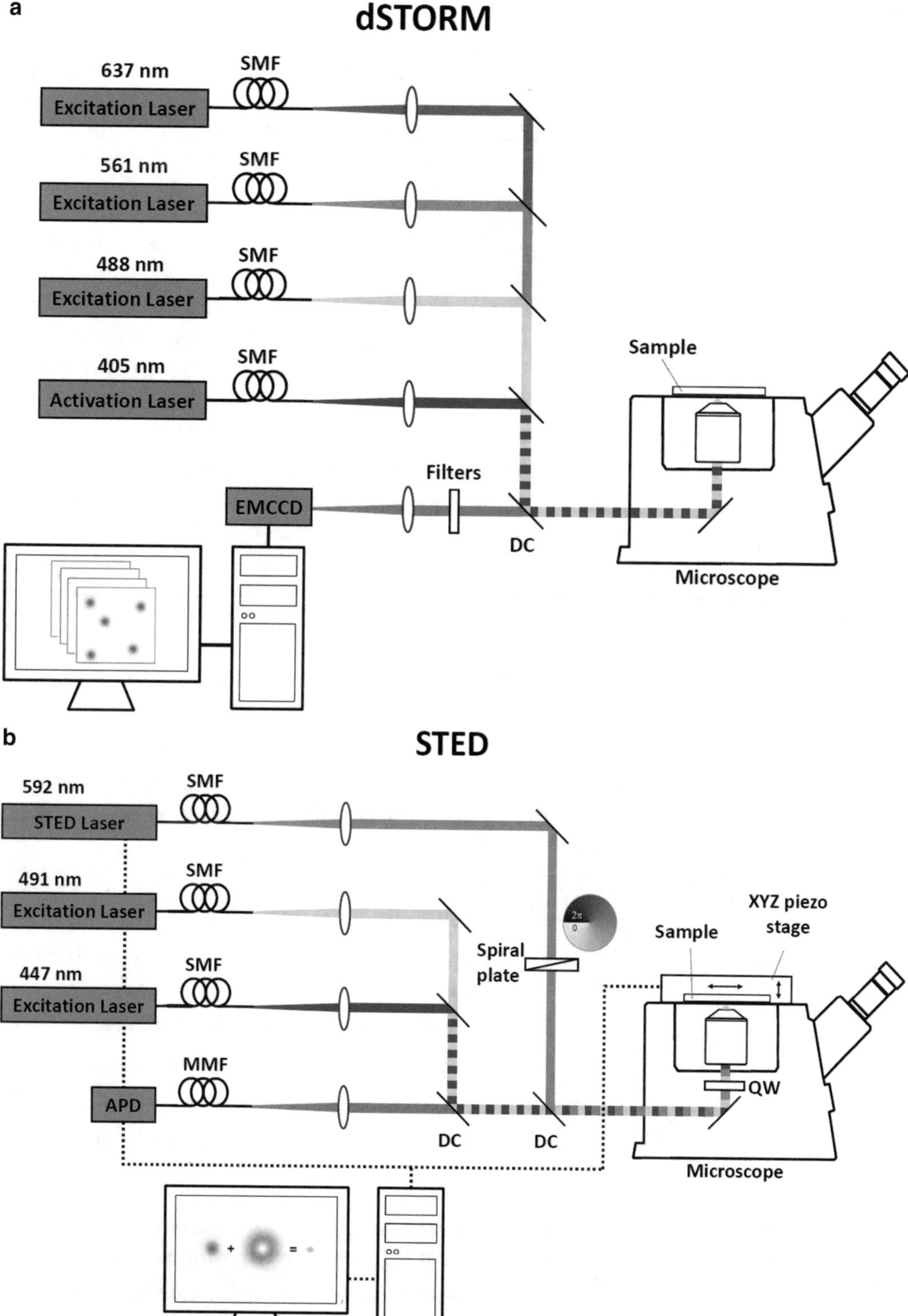

Fig. 1 Schematic diagrams of a dSTORM system and a CW STED system in our lab. (**a**) The dSTORM system consists of three excitation laser lines (637, 561, and 488 nm) and one activation laser (405 nm). A high-speed EMCCD captures single-molecule emission over 5000–20,000 frames to reconstruct a superresolution image. (**b**) In the STED microscope, two excitation lasers (447 and 491 nm) are chosen for Oregon Green 488 and BD V500 whose emission fluorescence can both be depleted with a 592-nm laser. The light pattern of the

25 mW 491-nm DPSS laser, Cobolt AB) and one depletion laser (1 W 592 nm fiber laser, VFL-P-1000-592, MPB Communications). Either a photomultiplier tube (PMT, MP963, PerkinElmer) or an avalanche photodiode single photon counting module (SPCM-AQR-15, PerkinElmer) is used to detect photons from samples held on a three-axis piezo scanner (Nano-PDQ375HS, Mad City Labs). Similar or gSTED versions of STED microscopes are available commercially (Leica and Abberior), some of them including multiple depletion lasers (592, 660, and 775 nm) and a wide spectrum of excitation laser source (supercontinuum laser) for a broad selection of fluorophores.

2. A computer with a large data storage system.
3. Vibration isolation system. To push the limit of mechanical stability, we use negative stiffness vibration isolators (two 350BM-1, Minus K, and one breadboard, SG-46-2, Newport).

3 Methods

3.1 Sample Preparation for dSTORM

3.1.1 Preparation of Poly-L-Lysine Coating Glass Coverslips (See Note 1)

1. Put coverslips in a glass beaker with 1 M HCl and heat to 50–60 °C for more than 4 h or overnight.
2. Rinse off 1 M HCl with ddH_2O for three times.
3. Incubate coverslips in ddH_2O and shake gently using an orbital shaker at low speed for 30 min so as to completely remove the HCl.
4. Wash once with 100% ethanol.
5. Incubate with 100% ethanol for over 1 h.
6. Transfer coverslips into a 10-cm plate inside a biosafety cabinet and air dry overnight.
7. Add 5–10 ml 0.01% poly-L-lysine (0.01%) and shake for 4–6 h.
8. Wash with ddH_2O for three times. Air dry completely overnight.
9. Leave the plate in the biosafety cabinet and sterilize with UV.
10. Wrap the plate containing coated coverslips with aluminum foil and store inside the biosafety cabinet.

3.1.2 Primary Cilium Growth Upon Serum Starvation

1. Place 8–10 poly-L-lysine coated #1.5 glass coverslips into a 10-cm cell culture plate. Avoid overlapping the coverslips.
2. Wash coverslips once with 5 ml PBS. Aspirate PBS completely with a suction to allow adherence of the coverslips to the plate surface.
3. Grow hTERT RPE-1 in the plate with 10 ml growth medium to ~50% confluency.
4. Aspirate medium and wash the cells twice with 5–10 ml PBS.

5. Add 10 ml serum free medium to cells.
6. Incubate cells for 48 h in this serum starved condition to promote cilia formation.

3.1.3 Conjugation of Cy3B Fluorophore to IgG Antibody (Optional, Only Required for 2-Color dSTORM Imaging)

1. Prepare a 10 mg/ml solution of Cy3B monofunctional NHS ester in DMSO/ DMF [1:1] by adding 100 μl DMSO/DMF solution to 1 mg Cy3B mono NHS ester.
2. Prepare IgG solution by adding 8 μl of the Borate Buffer to 100 μl of 1 mg/ml IgG antibody.
3. Add 1.5 μl of the Cy3B solution into the 100 μl IgG solution, pipette up and down to mix.
4. Briefly centrifuge the vial to collect the mixture in the bottom of the tube.
5. Incubate the reaction mixture for 60 min at room temperature protected from light.
6. Place spin column in a microcentrifuge tube.
7. Mix the purification resin to ensure uniform suspension and add 300 μl of the suspension into the spin column.
8. Centrifuge for 1 min at 1000 × *g* to remove the storage solution. Discard the used microcentrifuge tube and place the column into a new tube.
9. Wash column once with PBS and place the column into a new tube.
10. Add the labeling reaction mixture to the column and allow the sample to mix with the resin by briefly vortexing.
11. Centrifuge column for 1 min at 1000 × *g* to collect the purified antibody. Discard the used column.
12. Store the labeled antibody protected from light at 4 °C.

3.1.4 Immuno fluorescence Staining

All steps are performed under room temperature unless otherwise indicated. Solutions or buffers are added to samples at a volume sufficient to cover the glass coverslips.

1. Wash cells by adding 5–10 ml PBS to the 10-cm culture plate.
2. Fix cells with 4% PFA (or ice-cold methanol at −20 °C, *see* **Note 2**) for 10–15 min. Avoid over fixing or otherwise epitope of the protein of interest will be shielded.
3. Wash cells twice with PBS to remove the fixative.
4. Permeabilize cells with PBST for 10 min.
5. Block cells by incubating samples in the blocking buffer for 30–60 min.
6. Prepare primary antibodies in the blocking buffer at an optimized dilution. The optimized dilution of each primary antibody

for superresolution microscopy may be different from that of conventional microscopy and should be found independently. In general, a volume of 50-μl is sufficient to cover one coverslip. For dual color imaging, dilute and mix the two desired primary antibodies at their corresponding optimized dilutions. Incubate cells in the antibody-containing buffer for 1 h.

7. Wash cells five times with the washing buffer.
8. For single color imaging, dilute Alexa Fluor 647 secondary antibody (*see* **Note 3**) in the blocking buffer to an optimized concentration. For dual color imaging, prepare a mixture of Alexa Fluor 647 antibody and Cy3B conjugated IgG antibody at their corresponding optimized dilutions. Add the solution to cells and incubate for 1 h. Protect samples from light.
9. Wash cells with the washing buffer for three times.
10. Dilute TetraSpeck microspheres in a ratio of 1:200 with PBS. Add microspheres to cells and incubate for 30 min (*see* **Note 4**).
11. Wash once with PBS and store samples in PBS at 4 °C. For long-term storage, keep stained samples in PBS with 0.01% sodium azide at 4 °C.

3.2 dSTORM Imaging Procedure

1. The immunostained sample prepared on a glass coverslip is transferred into a custom-built or commercial imaging chamber (for example, imaging chamber, Live Cell Instrument). A custom-built imaging chamber can be made by mounting a coverslip on a carved glass microscope slide and then sealing the edge of the coverslip with nail polish.
2. Prepare the imaging buffer for oxygen scavenging. Imaging buffer consists of 10–100 mM MEA (*see* **Note 5**), 0.5 mg/ml glucose oxidase, 40 μg/ml catalase, and 10% glucose solution in TN buffer at pH 8.0 (*see* **Note 6**). The concentration of MEA depends on the protein density and labeling efficiency and has to be optimized to maintain a condition in which single emitters residing within a diffraction-limited region can be separated.
3. The freshly prepared imaging buffer should be immediately added to the sample and cover the chamber entirely. For the best performance, the imaging chamber is suggested to be sealed with Parafilm to slow down the degradation of the imaging buffer (*see* **Note 7**).
4. All instrument and devices should be turned on at least 30 min before samples are imaged with dSTORM mode to minimize any temperature drift caused by the warming-up of microscope components.
5. The samples should be placed in the warmed-up microscope at least for 15 min after they have been pre-focused (*see* **Note 8**). To stabilize microscope axial position, an electronically feedback control focusing system (for example, Nikon Perfect Focus System) is highly recommended.

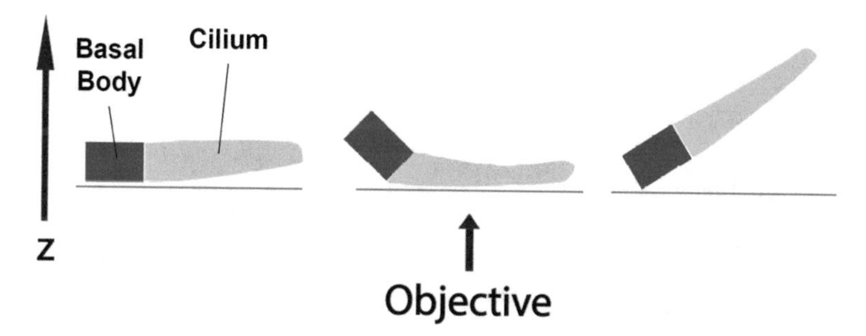

Fig. 2 Cilia grown at different orientations and with different bending characteristics. Most of cilia are bent to a certain degree, so their basal body may not be aligned with the ciliary axoneme (*middle*). Some cilia are not parallel to the surface of coverslip, challenging to focus the entire structure (*right*). One has to search a horizontal cilium (*left*) so that its orientation is ideal for side-view imaging

6. A ciliary marker is required to effectively locate a cilium during imaging. To start imaging, a ciliary marker (e.g., acetylated tubulin) is first observed in the widefield mode to determine whether the orientation of this cilium is suitable. Perfectly horizontal or vertical cilia usually are not easy to find. Most of the cilia are bent to a certain degree (Fig. 2). In practice, a nearly horizontal cilium can still be found with the use of ciliary markers labeling both the cilium compartment and basal body. If both markers of a cilium have the same focus, this cilium is likely horizontal.
7. Once a cilium is found using a marker, the target protein is first imaged with the widefield mode to determine if its signal-to noise level is high enough for dSTORM.
8. Further fine focus adjustment may be needed for the best image contrast of the target protein (*see* **Note 9**). After setting the focus, one should switch the imaging channels between the marker and the target protein back and forth to make sure that both signals are correlated with each other.
9. To successfully achieve superresolution imaging of cilia, fiducial markers (fluorescent beads or gold nanoparticles) should be imaged together with the target protein. Therefore, both signals of proteins of interest and a stable fiducial marker have to appear in the same field of view.
10. The sample is first illuminated at a gradually increasing power (up to 1–4 kW/cm^2) of the excitation laser (e.g., 637 nm laser for Alexa Fluor 647) to drive the dye molecules to a dark state. This step may saturate the EM-CCD camera. To protect CCD, it is suggested to keep the EM gain at a minimal level in the beginning.
11. The dyes immersed in the imaging buffer start to repetitively photoswitch between on and off state (photoblinking). The EM gain of CCD can be changed to the working level for optimal detection without saturation. It usually takes a few

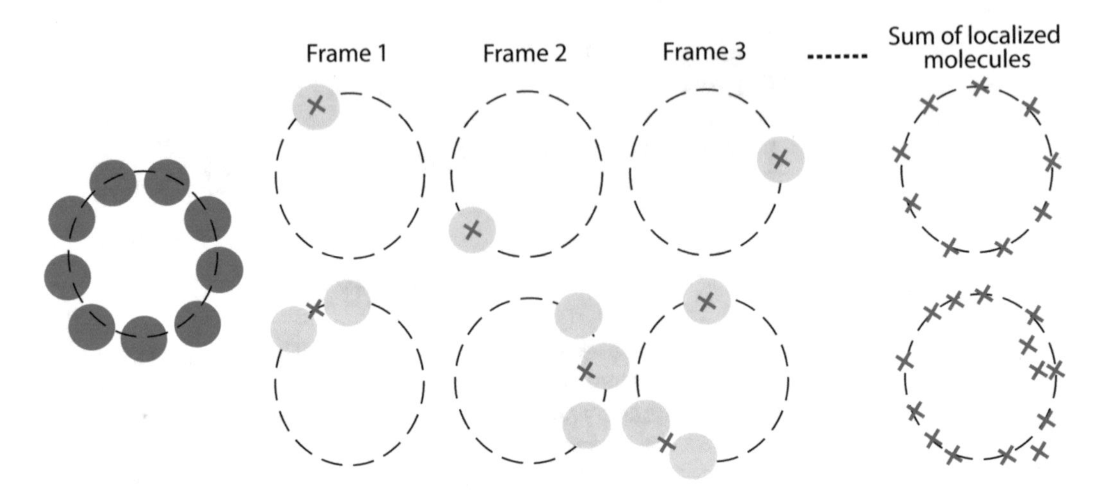

Fig. 3 Achieving isolated single emitters for good dSTORM images. Emitter overlapping should be avoided for a better imaging (*upper*). The overlapping signals from multiple emitters results in mislocalization (*lower*). In this case, the turn-on frequency should be reduced to meet the requirement of separable single-molecule detections per frame

seconds to see an obvious photoswitching mechanism after pumping the intense laser power. However, a good fiducial marker should not show substantial brightness fluctuation and any photoswitching during the imaging procedure.

12. The exposure time should then be adjusted to match the photoswitching rate, preventing the adjacent emitters from overlapping. The exposure time is usually set to 10–50 ms depending on the density of labeled molecules (*see* **Note 10**).
13. The event from single molecule emission can be confirmed by taking several consecutive snapshot images. A good imaging condition is achieved when only isolated single emitters appear in every single snapshot (Fig. 3) (*see* **Note 11**).
14. To estimate the number of single molecule localization per frame, one can select effective single emitters with a brightness threshold to reject undesired noise (*see* **Note 12**).
15. The ciliary feature may not be easily identified while the dyes are photoblinking. A pre-photoblinking snapshot image would be helpful to recognize the feature. In particular, the size of the transition zone and the basal body are too small to identify because their patterns are similar to nonspecific signals.
16. After an appropriate threshold level is determined, the stream acquisition of 5000–20,000 frames is performed for the effective dSTORM imaging.
17. During the acquisition, the total number of single molecule localization should always be monitored to maintain an optimized density of detected localization. The detection number usually decreases with the imaging time (Fig. 4, red). For the

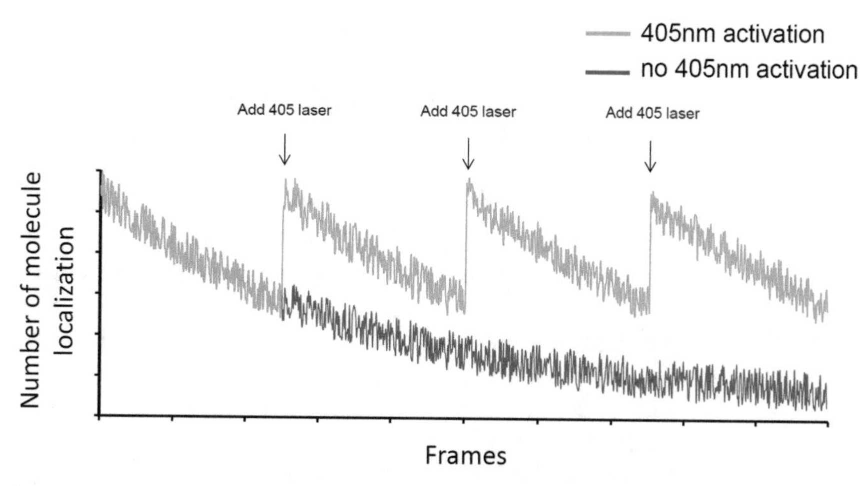

Fig. 4 Tuning the power of the 405-nm laser line to control the activation efficiency of dyes. When seeing the reduction of number of molecules being localized through the imaging process, one should increase of the power of 405-nm laser to maintain an efficient imaging acquisition, minimizing the required number of frames for reconstructing the whole image. Ideally, number of molecule localization should be a constant during the acquisition

best acquisition speed, the detection number has to be maintained approximately constant by adjusting the power of the activation laser (405 nm laser) to activate the transition from the dark state back to the singlet state of a molecule (Fig. 4, green).

18. In addition to registering photon counts of all pixels, real-time snapshot images every 100 frames, if possible, are highly recommended to monitor that the single molecule detection is still maintained in most of the frames.
19. The reconstructed dSTORM image extracted from 20,000 frames should well display a nearly complete pattern of target proteins.
20. For dual-color dSTORM, the first imaging channel is usually assigned to an Alexa Fluor 647-tagged ciliary protein for its good performance. There are multiple options of dyes for the second channel. The working imaging buffer may differ from dye to dye, so switching performance need to be compromised in dual-color imaging. In terms of the brightness and switching properties, the use of red dyes such as Cy3B provides a compromised solution for the second channel. For three-color superresolution imaging, it is possible to have the combination of Alexa Fluor 647/Cy3B/ATTO 488, although the buffer conditions have to be adjusted and the performance is not optimal.
21. Basically, a sequential acquisition is conducted in two-color imaging. The Alexa Fluor 647 dye should be imaged first, and then Cy3B dye is imaged (*see* **Note 13**). For three-color imaging, ATTO 488 is imaged after Cy3B.

22. Similar to the first channel, the laser power and exposure time should be optimized for other channels. In particular, the 405 nm laser is required for Cy3B to maintain a reasonable switching rate while imaging. In general, the second dye does not offer the imaging stability as good as that of the first channel using Alexa Fluor 647 (*see* **Note 14**).
23. Continuous images of 5000–20,000 frames for the second dye are also acquired to reconstruct an image for the second protein.

3.2.1 Image Correction for dSTORM

A successful superresolution image usually cannot be reconstructed directly from the raw streaming images because most of the raw data suffer from non-negligible mechanical drift which causes a localization error far larger than the dSTORM resolution ~20 nm. Moreover, the localization deviation between two image channels is significant at this resolution scale. Several steps for alleviating the above issues will be described in the following.

1. Mechanical drift is induced by an unbalanced sample holding system, immersion oil adhesion force, or temperature change.
2. To reduce the drift caused by holding system, samples should be clamped only with normal force since any shear force may lead to lateral movement of samples.
3. Samples should be placed on the microscope stage at least 15 min before imaging to ensure the immersion oil reaches equilibrium.
4. To actively correct a drifted image, fiducial markers such as fluorescent beads or gold nanoparticles should be immobilized on samples for the tracking of their trajectory during streaming acquisition. All reliable markers should have a nearly identical trajectory (*see* **Note 15**).
5. Based on the pattern correlation analysis of these fiducial markers, drift distance at every single frame can be compensated to generate a drift-free stack image.
6. Mechanical drift can be easily corrected by subtracting the movement of fiducial markers in each frame. However, nonlinear distortion arising from optical issues such as chromatic aberration has to be corrected by either optimizing the optical system or compensating distortion.
7. To correct chromatic aberration between two imaging channels, one can first characterize their distance deviation over a full field of view using fiducial markers (Fig. 5, before) and then create a 2D distance deviation function of lateral position x and y.
8. The error compensation is done by applying this calibration function to a raw image. Therefore, most of fiducial markers of overlap between two channels can be achieved (Fig. 5, after).
9. A good dual-color dSTORM image can be obtained only after two single-color images are correctly aligned.

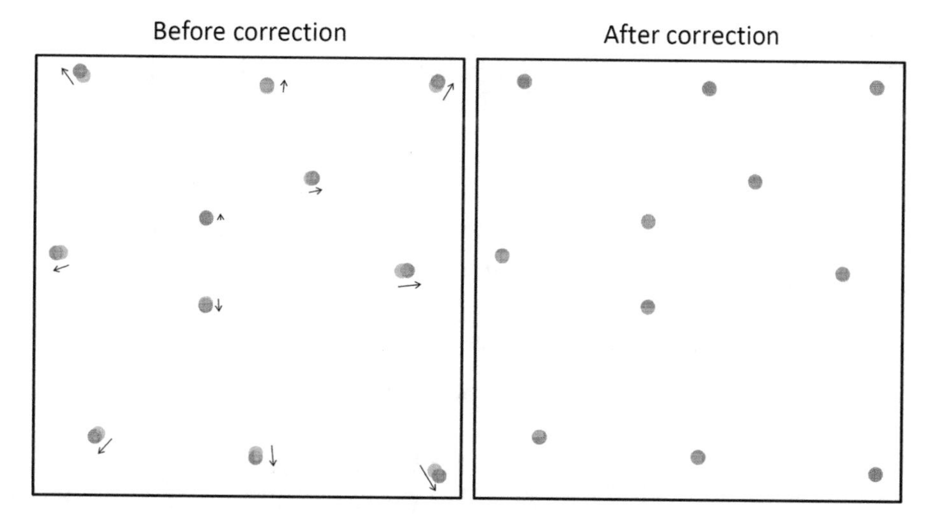

Fig. 5 Chromatic aberration between two imaging channels compensated by calibrating distance deviation of fiducial markers between two channels. The localization error between two channels should be less than optical resolution to ensure an effective superresolution imaging

3.3 Sample Preparation for STED Microscopy

STED microscopy offers a subdiffraction-limited resolution and is a promising tool for ciliary studies even though optical alignment is very challenging when implementing the STED imaging system. Different from confocal fluorescent microscopy, serious care has to be taken to address multiple sample-specific issues of STED imaging. For imaging proteins in cilia, the primary issues are: (1) fluorescent signals from some ciliary proteins are dim; (2) the depletion phenomenon of STED reduces the light being detected from the readily dim signals; (3) the high power of the depletion laser accelerates photodamage; (4) labeling conditions are sample dependent; and (5) the combination of fluorophores for two-color STED imaging have cross-talks or signal leakages between two channels. Thus, in the following sections several tasks of image acquisition and post-processing toward a STED-based superresolution imaging of cilia will be described to address issues of dim ciliary signals caused by inherent properties of fluorophores, the depletion effect, photobleaching, and photodamage. This protocol adapts to a system using two excitation lines, i.e., a 447-nm and a 491-nm line and one 592-nm depletion laser. The setup is suitable for two-color STED imaging for BD Horizon V500 and Oregon Green 488 fluorescent dyes.

3.3.1 Preparation of Poly-L-Lysine Coating Glass Coverslips

See Subheading 3.1.1.

3.3.2 Primary Cilium Growth Upon Serum Starvation

See Subheading 3.1.2.

3.3.3 Immunofluorescence Staining

All steps are performed under room temperature unless otherwise indicated. Solutions or buffers are added to samples at a volume sufficient to cover the glass coverslips.

1. Wash cells by adding 5–10 ml PBS to the 10-cm culture plate.
2. Fix cells with 4% PFA (or ice-cold methanol at –20 °C) for 10–15 min. Avoid over fixing or otherwise epitope of the target protein may be damaged or shielded.
3. Wash cells twice with PBS to remove the fixative.
4. Permeabilize cells with PBST for 10 min.
5. Block cells by incubating samples in the blocking buffer for 30–60 min.
6. Prepare primary antibodies in the blocking buffer at an optimized dilution. The optimized dilution of each primary antibody for superresolution microscopy may be different from that of conventional microscopy and should be found independently. In general, 50-μl blocking buffer is sufficient for one coverslip. For dual color imaging, dilute and mix the two desired primary antibodies at their optimized dilutions. Incubate cells in the antibody-containing buffer for 1 h.
7. Wash cells 5 times with PBST washing buffer.
8. For single color imaging, dilute Oregon Green 488 secondary antibody with the blocking buffer to an optimized concentration. For dual color imaging, prepare a mixture of Oregon Green 488 antibody and biotin-conjugated antibody at their corresponding optimized dilutions. Add the solution to cells and incubate for 1 h. Protect samples from light.
9. Wash cells with the washing buffer for three times.
10. For dual color imaging, incubate cells with BD V500 streptavidin for 30 min, and then wash cells with the washing buffer for three times.
11. Store samples in PBS at 4 °C. For long-term storage, keep stained samples in PBS with 0.01% sodium azide at 4 °C.

3.4 STED Imaging Procedure

1. Mount the stained sample on microscope glass slides with 86% glycerol or Prolong Antifade Kit mounting medium for the best match of refractive index.
2. Seal the sample properly with, for example, nail polish. Let the sample stand for at least 1 h prior to imaging.
3. The sample should be firmly held on the microscope stage. Although STED microscopy suffers less from mechanical drift than dSTORM during the imaging, placing a sample on the stage 15 min before collecting images is still suggested.
4. Search for a target area with ciliary proteins of interest as quickly as possible in the confocal mode. During this step, one should always run the microscope at a high scanning speed, a

large sampling step size, a minimal working laser power, and a minimal number of repetitive scans. When an area of interest is found, the repetitive scanning should be stopped to protect the sample from pre-photobleached.

5. The performance of STED microscopy depends strongly on focusing, because it is based on confocal configuration to possess excellent optical sectioning. Therefore, finding the proteins of interest depends highly on the axial position of the microscope. After a coarse focal adjustment to locate signals of a ciliary marker, one should check layer by layer with a small interval in axial direction (e.g., 100 nm) to identify the proteins of interest (*see* **Note 16**). Imaging an object at different foci may generate different patterns.
6. Before STED imaging, a confocal image is obtained by optimizing the scanning rate, the sampling step size, and the excitation laser power. The scanning speed has to be fast in the beginning to minimize photobleaching. The excitation power may be increased to enhance the signal-to-noise ratio due to high-speed scanning.
7. Once a good confocal image is obtained, STED imaging is then performed first with a low depletion power. The step size of 10–50 nm is used to meet the sampling criterion in STED resolution. When signal-to-noise ratio is high, we use a step size of 10 nm to achieve the resolution as good as possible. Check the improvement in resolution. Next, a moderate STED laser power is used to further improve resolution. As the power increases, the signal-to-noise ratio may be reduced due to signal depletion in the periphery. Use the maximal depletion laser power that remains a reasonable signal-to-noise ratio to maximize the resolution improvement (*see* **Note 17**).
8. For two-color STED imaging, separate channels are acquired sequentially through excitations of 447 nm and 491 nm lasers and a common depletion of 592 nm lasers. As illustrated in Fig. 6, the 447-nm light excites BD V500 as well as Oregon Green 488, thus leading to minor signal cross talk between the two dyes. To address this issue, emission signals from two dyes should be well balanced. A ciliary protein with high brightness should be assigned to the 447-nm channel to minimize signal cross talk. The 491-nm laser only excites Oregon Green 488, thus enabling intrinsically spectrally separable STED imaging without the need of unmixing post-processing.

3.4.1 Image Processing for STED

Different from dSTORM, the mechanical drift issue of STED imaging of fixed samples is generally negligible because an effective superresolution image can be obtained with a single-frame acquisition. The major problem of STED imaging is the poor signal-to-noise ratio of depleted signals resulting from the STED mechanism. In addition to the optimization of imaging parameters such as the laser power, the sampling step size, and the

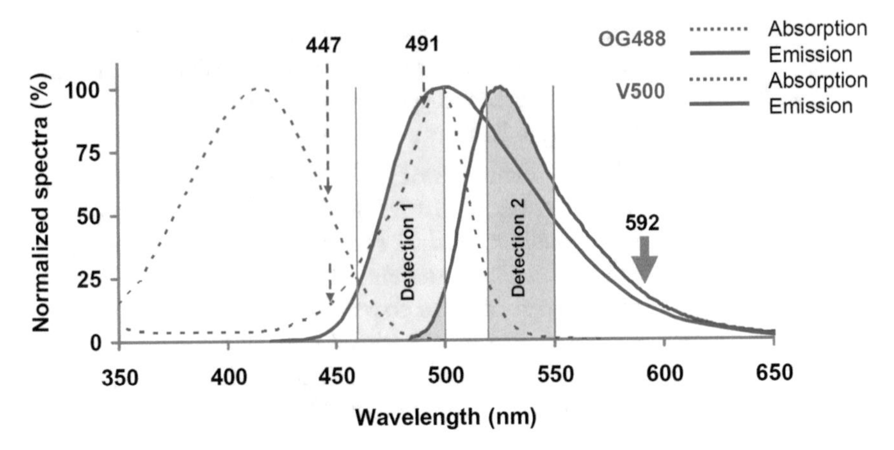

Fig. 6 The spectral concept behind the two-color STED system using two excitation lasers (447 and 491 nm) and one depletion laser (592 nm). In addition to exciting V500, the 447-nm laser can also excite Oregon Green 488 slightly, causing minor cross talk between two channels. Therefore, suitable detection windows to separate the signal of V500 from that of Oregon Green 488 are necessary. The cross talk can be further reduced by assigning the protein with brighter signals to the V500 channel

scanning rate, the resulting raw image can be processed to enhance its signal contrast by methods such as deconvolution. To adequately restore a noisy image with deconvolution, the PSF of the system should be first characterized and utilized in the computation algorithm. An inappropriate PSF could induce incorrect information of images. If the used PSF is too narrow (small full width at half maximum), it may amplify the small and noisy feature, causing image artifacts. In contrast, a wide PSF can remove subtle but real features due to over-filtering. In principle, a qualified deconvolved image should retain most of the original features captured in the raw data while improving their contrast.

4 Notes

1. Poly-L-lysine coating increases the adherence of hTERT RPE-1 cells to coverslips. Some cell types may require this coating while others may not, depending on the adherence of each cell type.
2. The fixation condition may vary for different kinds of proteins. For example, image quality using 4% PFA fixation is better for membrane-bound proteins as compared to methanol fixation; while methanol fixation is especially good to preserve actin and cytoskeletal proteins. It is advised to optimize the specific fixation condition for different proteins of interest.
3. Alexa Fluor 647 dye exhibits a photoswitching property better than any other existing dyes for dSTORM superresolution imaging. It is recommended to use Alexa Fluor 647 to mark the protein of higher interest between the two proteins when performing dual-color dSTORM imaging.

4. Microspheres serve as fiducial markers to calibrate the mechanical drift during imaging. Nano-gold particles can also be used. Because of the stringent requirement of resolution (~20 nm), fiducial markers are essential for a good performance of dSTORM superresolution imaging.
5. MEA concentration is critical for the control of the photoswitching mechanism and should be adjusted depending on samples. For example, if dyes do not reversibly switch, MEA concentration should be increased.
6. The stock of MEA, glucose oxidase or catalase can be prepared and kept in a freezer. The stocks such as 1 M MEA, 50 mg/ml glucose oxidase and 40 mg/ml catalase are recommended. These stocks usually can be stored for at least several weeks in a freezer. The fresh MEA within 1 week is recommended after it is thawed.
7. To maintain the imaging quality, the imaging buffer should be replaced with a fresh one every 1–2 h.
8. The subtle movement of immersion oil may occur right after the sample is placed, so it is recommended to let samples stand for at least 15 min before imaging.
9. The power of the light source should be kept at a low level to prevent photobleaching.
10. Adjustment of the exposure time is always necessary. This parameter can be first set to the upper bound of the range in order to gain the maximal signal to noise ratio. If too many emitters stay in their ON state, exposure time should be reduced.
11. In addition to the camera integration time, the illumination power can significantly affect whether single emitter events are achieved or not. We usually use the full power ~4 kW/cm^2 at which we successfully achieve single molecule detection. During imaging, the power is kept at the maximum.
12. If scattered localizations appear randomly all over the image, it indicates that the threshold is too low. A high contrast and continuous image is usually rendered through the optimization of the threshold level.
13. If one imaged the Cy3B dye first, the Alexa Fluor 647 will be photobleached by the strong power of laser when illuminating Cy3B.
14. All proteins of interest should be separately studied first with single-color dSTORM imaging (possibly with a ciliary marker using conventional microscopy to locate the protein of interest) to characterize their pattern and spatial distribution prior to two-color imaging.
15. Some unstable markers may move or dither. They should be excluded from the drift correction procedure.

16. The fine adjustment at a finer sampling size ~25–50 nm may be needed for some cases. Use a minimal excitation power that is barely enough to differentiate subtle pattern changes in order to avoid photobleaching.
17. The same field of view is not allowed to be imaged for too many times with an intense STED laser power. Successful STED images are mostly generated from the first three imaging scans.

Acknowledgments

This work was supported by the Ministry of Science and Technology, Taiwan (Grant No. 103-2112-M-001-039-MY3), Academia Sinica Career Development Award, and Academia Sinica Nano Program.

References

1. Williams CL, Li C, Kida K, Inglis PN, Mohan S et al (2011) MKS and NPHP modules cooperate to establish basal body/transition zone membrane associations and ciliary gate function during ciliogenesis. J Cell Biol 192:1023–1041
2. Sang L, Miller JJ, Corbit KC, Giles RH, Brauer MJ et al (2011) Mapping the NPHP-JBTS-MKS protein network reveals ciliopathy disease genes and pathways. Cell 145:513–528
3. Garcia-Gonzalo FR, Corbit KC, Sirerol-Piquer MS, Ramaswami G, Otto EA et al (2011) A transition zone complex regulates mammalian ciliogenesis and ciliary membrane composition. Nat Genet 43:776–784
4. Chih B, Liu P, Chinn Y, Chalouni C, Komuves LG et al (2011) A ciliopathy complex at the transition zone protects the cilia as a privileged membrane domain. Nat Cell Biol 14:61–72
5. Craige B, Tsao C-C, Diener DR, Hou Y, Lechtreck K-F et al (2010) CEP290 tethers flagellar transition zone microtubules to the membrane and regulates flagellar protein content. J Cell Biol 190:927–940
6. Tanos BE, Yang H-J, Soni R, Wang W-J, Macaluso FP et al (2013) Centriole distal appendages promote membrane docking, leading to cilia initiation. Genes Dev 27:163–168
7. Ye X, Zeng H, Ning G, Reiter JF, Liu A (2014) C2cd3 is critical for centriolar distal appendage assembly and ciliary vesicle docking in mammals. Proc Natl Acad Sci U S A 111:2164–2169
8. Burke MC, Li F-Q, Cyge B, Arashiro T, Brechbuhl HM et al (2014) Chibby promotes ciliary vesicle formation and basal body docking during airway cell differentiation. J Cell Biol 207:123–137
9. Yang TT, Hampilos PJ, Nathwani B, Miller CH, Sutaria ND et al (2013) Superresolution STED microscopy reveals differential localization in primary cilia. Cytoskeleton (Hoboken, NJ) 70:54–65
10. Nathwani B, Yang TT, Liao J-C (2013) Towards a subdiffraction view of motor-mediated transport in primary cilia. Cell Mol Bioeng 6:82–97
11. Dorn KV, Hughes CE, Rohatgi R (2012) A smoothened-Evc2 complex transduces the hedgehog signal at primary cilia. Dev Cell 23:823–835
12. Lau L, Lee YL, Sahl SJ, Stearns T, Moerner WE (2012) STED microscopy with optimized labeling density reveals 9-fold arrangement of a centriole protein. Biophys J 102:2926–2935
13. Hu Q, Milenkovic L, Jin H, Scott MP, Nachury MV et al (2010) A septin diffusion barrier at the base of the primary cilium maintains ciliary membrane protein distribution. Science 329:436–439
14. Sillibourne JE, Specht CG, Izeddin I, Hurbain I, Tran P et al (2011) Assessing the localization of centrosomal proteins by PALM/STORM nanoscopy. Cytoskeleton 68:619–627
15. Kobayashi T, Kim S, Lin Y-C, Inoue T, Dynlacht BD (2014) The CP110-interacting proteins Talpid3 and Cep290 play overlapping and distinct roles in cilia assembly. J Cell Biol 204:215–229

16. Lu Q, Insinna C, Ott C, Stauffer J, Pintado PA et al (2015) Early steps in primary cilium assembly require EHD1/EHD3-dependent ciliary vesicle formation. Nat Cell Biol 17:228–240
17. Hell S, Wichmann J (1994) Breaking the diffraction resolution limit by stimulated emission: stimulated-emission-depletion fluorescence microscopy. Opt Lett 19:780–782
18. Klar TA, Jakobs S, Dyba M, Egner A, Hell SW (2000) Fluorescence microscopy with diffraction resolution barrier broken by stimulated emission. Proc Natl Acad Sci U S A 97:8206–8210
19. Gustafsson MGL (2000) Surpassing the lateral resolution limit by a factor of two using structured illumination microscopy. J Microsc 198:82–87
20. Hell SW (2003) Toward fluorescence nanoscopy. Nat Biotechnol 21:1347–1355
21. Gustafsson MGL (2005) Nonlinear structured-illumination microscopy: wide-field fluorescence imaging with theoretically unlimited resolution. Proc Natl Acad Sci U S A 102:13081–13086
22. Betzig E, Patterson GH, Sougrat R, Lindwasser OW, Olenych S et al (2006) Imaging intracellular fluorescent proteins at nanometer resolution. Science 313:1642–1645
23. Rust MJ, Bates M, Zhuang XW (2006) Sub-diffraction-limit imaging by stochastic optical reconstruction microscopy (STORM). Nat Methods 3:793–795
24. Hess ST, Girirajan TPK, Mason MD (2006) Ultra-high resolution imaging by fluorescence photoactivation localization microscopy. Biophys J 91:4258–4272
25. Sharonov A, Hochstrasser RM (2006) Wide-field subdiffraction imaging by accumulated binding of diffusing probes. Proc Natl Acad Sci 103:18911–18916
26. Bretschneider S, Eggeling C, Hell SW (2007) Breaking the diffraction barrier in fluorescence microscopy by optical shelving. Phys Rev Lett 98:218103
27. Egner A, Geisler C, von Middendorff C, Bock H, Wenzel D et al (2007) Fluorescence nanoscopy in whole cells by asynchronous localization of photoswitching emitters. Biophys J 93:3285–3290
28. Heilemann M, van de Linde S, Schüttpelz M, Kasper R, Seefeldt B et al (2008) Subdiffraction-resolution fluorescence imaging with conventional fluorescent probes. Angew Chem Int Ed 47:6172–6176
29. Folling J, Bossi M, Bock H, Medda R, Wurm CA et al (2008) Fluorescence nanoscopy by ground-state depletion and single-molecule return. Nat Methods 5:943–945
30. Dertinger T, Colyer R, Iyer G, Weiss S, Enderlein J (2009) Fast, background-free, 3D super-resolution optical fluctuation imaging (SOFI). Proc Natl Acad Sci 106:22287–22292
31. Vicidomini G, Moneron G, Han KY, Westphal V, Ta H et al (2011) Sharper low-power STED nanoscopy by time gating. Nat Methods 8:571–573
32. Burnette DT, Sengupta P, Dai Y, Lippincott-Schwartz J, Kachar B (2011) Bleaching/blinking assisted localization microscopy for superresolution imaging using standard fluorescent molecules. Proc Natl Acad Sci 108:21081–21086
33. Quan T, Zhu H, Liu X, Liu Y, Ding J et al (2011) High-density localization of active molecules using Structured Sparse Model and Bayesian Information Criterion. Opt Express 19:16963–16974
34. Cox S, Rosten E, Monypenny J, Jovanovic-Talisman T, Burnette DT et al (2012) Bayesian localization microscopy reveals nanoscale podosome dynamics. Nat Methods 9:195–200
35. Rittweger E, Han K, Irvine S, Eggeling C, Hell S (2009) STED microscopy reveals crystal colour centres with nanometric resolution. Nat Photon 3:144–147
36. Willig K, Harke B, Medda R, Hell S (2007) STED microscopy with continuous wave beams. Nat Methods 4:915–918
37. Meyer L, Wildanger D, Medda R, Punge A, Rizzoli SO et al (2008) Dual-color STED microscopy at 30-nm focal-plane resolution. Small 4:1095–1100
38. Wildanger D, Medda R, Kastrup L, Hell SW (2009) A compact STED microscope providing 3D nanoscale resolution. J Microsc 236:35–43
39. Moneron G, Hell SW (2009) Two-photon excitation STED microscopy. Opt Express 17:14567–14573
40. Westphal V, Rizzoli S, Lauterbach M, Kamin D, Jahn R et al (2008) Video-rate far-field optical nanoscopy dissects synaptic vesicle movement. Science 320:246–249
41. Pellett PA, Sun X, Gould TJ, Rothman JE, Xu M-Q et al (2011) Two-color STED microscopy in living cells. Biomed Opt Express 2:2364–2371
42. Keppler A, Gendreizig S, Gronemeyer T, Pick H, Vogel H et al (2003) A general method for the covalent labeling of fusion proteins with small molecules in vivo. Nat Biotechnol 21:86–89

43. Los GV, Encell LP, McDougall MG, Hartzell DD, Karassina N et al (2008) HaloTag: a novel protein labeling technology for cell imaging and protein analysis. ACS Chem Biol 3:373–382
44. Zhang M, Chang H, Zhang Y, Yu J, Wu L et al (2012) Rational design of true monomeric and bright photoactivatable fluorescent proteins. Nat Methods 9:727–729
45. Wang S, Moffitt JR, Dempsey GT, Xie XS, Zhuang X (2014) Characterization and development of photoactivatable fluorescent proteins for single-molecule–based superresolution imaging. Proc Natl Acad Sci 111:8452–8457
46. Subach FV, Patterson GH, Manley S, Gillette JM, Lippincott-Schwartz J et al (2009) Photoactivatable mCherry for high-resolution two-color fluorescence microscopy. Nat Methods 6:153–159
47. Shroff H, Galbraith CG, Galbraith JA, White H, Gillette J et al (2007) Dual-color superresolution imaging of genetically expressed probes within individual adhesion complexes. Proc Natl Acad Sci 104:20308–20313
48. Dempsey GT, Vaughan JC, Chen KH, Bates M, Zhuang X (2011) Evaluation of fluorophores for optimal performance in localization-based super-resolution imaging. Nat Methods 8:1027–1036
49. Huang B, Jones SA, Brandenburg B, Zhuang X (2008) Whole-cell 3D STORM reveals interactions between cellular structures with nanometer-scale resolution. Nat Methods 5:1047–1052
50. Juette MF, Gould TJ, Lessard MD, Mlodzianoski MJ, Nagpure BS et al (2008) Three-dimensional sub-100 nm resolution fluorescence microscopy of thick samples. Nat Methods 5:527–529
51. Pavani SRP, Thompson MA, Biteen JS, Lord SJ, Liu N et al (2009) Three-dimensional, single-molecule fluorescence imaging beyond the diffraction limit by using a double-helix point spread function. Proc Natl Acad Sci 106:2995–2999
52. Shtengel G, Galbraith JA, Galbraith CG, Lippincott-Schwartz J, Gillette JM et al (2009) Interferometric fluorescent super-resolution microscopy resolves 3D cellular ultrastructure. Proc Natl Acad Sci 106:3125–3130
53. Henriques R, Lelek M, Fornasiero EF, Valtorta F, Zimmer C et al (2010) QuickPALM: 3D real-time photoactivation nanoscopy image processing in ImageJ. Nat Methods 7:339–340
54. Izeddin I, Boulanger J, Racine V, Specht CG, Kechkar A et al (2012) Wavelet analysis for single molecule localization microscopy. Opt Express 20:2081–2095

Chapter 12

CLEM Methods for Studying Primary Cilia

Frank P. Macaluso, Geoffrey S. Perumal, Johan Kolstrup, and Peter Satir

Abstract

CLEM (correlated light and electron microscope) imaging is a highly useful technique for examining primary cilia. With CLEM, it is possible to determine the distribution of tagged proteins along the ciliary membrane and axoneme with high precision. Scanning electron microscopy (SEM) permits measurement of ciliary length and orientation in relation to nearby cellular structures in a 3D image; in optimal cases, this can be combined with superresolution microscopy of selected ciliary components as they enter or leave the cilium. This chapter discusses CLEM methods. In the method described in detail, samples are completely processed for sequential fluorescence and SEM observation. This method is ideal for robust antibody localization and minimizes image manipulation in correlating the fluorescent and SEM images. Alternative methods prepare samples for fluorescence imaging followed by processing for SEM then observation in the SEM. This method is ideal for optimal fluorescence imaging, particularly live cell imaging.

Key words CLEM, SEM, Primary cilia, Axoneme, MEF, Superresolution

1 Introduction

CLEM (correlated light and electron microscope) imaging is a highly useful technique for examining primary cilia. Primary cilia, present on many types of mammalian cells, are usually labeled along their length with antibodies to acetylated α tubulin [1], AC3 (adenylyl cyclase 3) [2], or Arl13B [3]. In confocal microscopy, it is often difficult to determine ciliary orientation. Colocalization of components moving along the cilium is important for studies of ciliary signaling [4] and sometimes requires superresolution and fast timing [5, 6]. Since primary cilia normally extend several micrometers above the remaining cell surface, combining localization studies with scanning electron microscopy (SEM) can be quite revealing. The 3D SEM image clearly defines the extent and orientation of the cilium and makes the localization studies of fixed material more definitive. In this chapter, we discuss methods of CLEM preparation and imaging for primary cilia grown on

Peter Satir and Søren Tvorup Christensen (eds.), *Cilia: Methods and Protocols*, Methods in Molecular Biology, vol. 1454, DOI 10.1007/978-1-4939-3789-9_12, © Springer Science+Business Media New York 2016

cultured mouse embryo fibroblasts (MEFs) or kidney cells that should prove valuable for studies of molecular movement within the axoneme or the ciliary membrane.

Correlative light and electron microscopy (CLEM) bridges the gap between light and electron microscopy. Protein identification obtained with fluorescence probes in live and fixed samples is combined with nanometer scale structural information provided by electron microscopy of exactly the same region [7]. We have developed a sample preparation technique to maintain fluorescence signal while preserving morphology for scanning electron microscopy. CLEM experiments are facilitated by sequential imaging of the same field of view such that CLEM requires a matched set of light and EM instrumentation. We illustrate the method for one specific set of instrumentation (*see* **Note 1**).

The methods presented here enable considerable experimental flexibility. In the method described in detail, cells are processed completely. Cells are fixed and labeled with fluorescence secondary antibodies, then dehydrated and dried for SEM processing and sequential fluorescence and SEM imaging. This results in unambiguous alignment in the correlated images. The sample may be imaged repeatedly in fluorescence and SEM if additional data are required. The disadvantage is that modifications to an already established staining protocol may be required for the fluorochrome to remain fluorescent through SEM processing. Alternatively, as discussed in **Note 6**, cells may be fluorescently stained then imaged in a hydrated environment. Subsequently, the cells are fixed again, dehydrated, critical point dried, and metal coated for imaging in the SEM. The advantages are: established staining protocols may be utilized; dynamic live cell fluorescence experiments are readily employed; and cell morphology is optimized for both fluorescence and SEM imaging. The disadvantages are that repeated imaging of fluorescence is not possible after processing for SEM and the inevitable cell shrinkage during dehydration (however minimal) may add difficulty in correlating the images.

2 Materials

2.1 Cells and Culture Media

1. Mouse embryonic fibroblasts (MEFs) from the Tg737 mouse or its WT control, obtained from Gregory Pazour, University of Massachusetts Medical School (Worchester, MA).
2. Growth medium: Dulbecco's Modified Eagle's Medium DMEM, 45% F12 Nutrient Mixture (Thermofischer Scientific), 10% fetal bovine serum (FBS).
3. Starvation medium: Growth medium without FBS.
4. 10% Dimethyl sulfoxide (DMSO).

2.2 Chemicals, Buffers, and Solutions

1. Double-distilled water (ddH_2O).
2. Hank's Balanced Salt Solution (HBSS) mixture: 17% (50 ml) HBSS (Thermofischer Scientific), 6% (1.2 ml) $NaHCO_3$, and 450 ml ddH_2O.
3. Trypsin/Ethylene diamine tetraacetic acid (EDTA): 0.05%.
4. Phosphate buffered saline (PBS).
5. Bovine serum albumin (BSA): 0.25 g in 25 ml dH_2O.
6. Permeabilization solution (25 ml: 0.25 g BSA, 50 μl Triton X-100).
7. Paraformaldehyde (4% solution).
8. Glutaraldehyde (2% solution).
9. Graded series of ethanol: 30–95%.
10. SEM Buffer: 0.1 M sodium cacodylate, 0.2 M sucrose, 5 mM $MgCl_2$ (pH 7.4).
11. 0.1% sodium borohydrate.
12. HMDS: Reagent grade (ʾ99%) Hexamethyldisilazane.

2.3 Equipment

1. Culture petri dishes.
2. Water bath.
3. 15 and 45 ml capacity Falcon tubes.
4. Cryo tubes.
5. 6-well slides or trays.
6. 22 × 22 mm (No. 1.5) coverslips.
7. Coverslips with fiducial markings (*see* **Note 2**).

2.4 Antibodies and Staining Reagents

1. Primary antibodies (Table 1).
2. Secondary antibodies (Table 2).
3. DAPI (4′,6-Diamidino-2-Phenylindole, Dihydrochloride) (Thermofischer Scientific).

2.5 Microscopy and Image Acquisition

1. DIC and fluorescent high-end light microscope.
2. Zeiss AxioObserver Z1 microscope with AxioVision 4.8.2, CLEM holder accessories and Shuttle & Find.
3. Zeiss Supra 40 Field Emission Scanning Electron Microscope with Shuttle & Find.
4. Critical point drier.
5. Sputter coater.

Table 1
Some useful primary antibodies

Primary antibody (1°Ab)	IgG	Company	Catalog #	IF dilution
Acetylated a-tubulin	Mouse	Sigma	T7451	1:1000
Parafusin	Rabbit	Matthisen et al. (2001)[a]	Custom made	1:1000
Pericentrin	Goat	Santa Cruz	Sc-28145	1:250
Clathrin	Rabbit	Abcam		1:500
Ki67	Rabbit	Abcam	Ab15580	1:600
Autophagic vacuoles	Rabbit	Cell Signaling	2775	1:500
Lysosomal-associated membrane protein 1	Rat	Developmental Studies Hybridoma Bank	1D4B	1:1000
Detyrosinated α-tubulin	Rabbit	Abcam	Ab48389	11:500
RF12	Rabbit	Meier and Blobel, 1992[b]	Custom made	1:100
CALM (C-18)	Goat	Santa Cruz	Sc-6433	1:200
IFT88	Goat	Novus	NB100-2475	1:100
β-Actin	Mouse	Sigma	A5441	1:15000
PDGFR-α (C-20)	Rabbit	Santa Cruz	Sc-338	1:200

[a]Matthissen SH et al. (2001) Europ. J. Cell Biol. 80:775–83
[b]Meier UT and Blobel G (1992) Cell 70:127–38

Table 2
Secondary antibodies for IFM

Secondary antibody (2°Ab)	Abbreviation	Company	Catalog #	IF dilutions
Alexa Fluor 488 donkey anti-mouse	DAM488	Invitrogen	A21202	0.4583
Alexa Fluor 568 donkey anti-mouse	DAM568	Invitrogen	A10037	0.4583
Alexa Fluor 568 donkey anti-goat	DAG568	Invitrogen	A11057	0.4583
Alexa Fluor 488 donkey anti-rabbit	DAR488	Invitrogen	A21206	0.4583

3 Methods

All experimental procedures are carried out at room temperature unless otherwise stated.

3.1 Sub-culturing and Maintenance of Cells

1. Cell lines are obtained frozen in 10% dimethyl sulfoxide (DMSO), and 90% fetal bovine serum (FBS) in Cryo tubes labeled with cell type, passage, and date. After thawing 2–3 min in a 37 °C water bath, the cells are transferred from the cryo tube to a petri dish containing 10 ml pre-heated medium

(see below), and left at 37 °C to settle. After 3–4 h when the cells have attached to the petri dish floor, the medium is exchanged for fresh, pre-heated culture medium to optimize the growth environment.

2. At 90–100% confluency, cells are sub-cultured with trypsin/EDTA. Before treating the cells with trypsin, the growth medium is aspirated from the culture petri dish, and the cells are washed once with pre-heated HBSS mixture to remove components, such as protease inhibitors, calcium and magnesium ions, which could inhibit the enzymatic activity of trypsin. EDTA enhances the effect of trypsin by intercalating free calcium ions.
3. The suspension of detached cells is diluted with 10 ml pre-heated culture medium as described above. The cells are further suspended by gentle pipetting up and down, with the petri dish tilted. The suspension is transferred to either a 15 or 45 ml Falcon tube from which a desired amount is separated out into petri dishes containing coverslips with fiduciary markings or six-well trays for starvation experiments (*see* **Note 3**).

3.2 Culture Conditions for Formation of Primary Cilia

1. Cells destined for CLEM of primary cilia are grown on coverslips to a confluency of 50–100%. Some coverslips are reserved as a control.
2. Remove growth medium and wash cells with HBSS.
3. Add starvation medium and incubate cells for the desired time (normally up to 48 h). Cells are routinely checked for percentage ciliation after immunofluorescence labeling.

3.3 Preparation of Cells for Immunofluorescence Microscopy

1. Aspirate the starvation medium and wash cells gently in ice-cold PBS. Ice-cold 4% paraformaldehyde is then added and left for 15 min at 20 °C.
2. Wash cells twice in PBS and incubate for 12 min in Permeabilization solution.
3. Remove Permeabilization solution by suction and add Blocking buffer for 30 min or overnight at 4 °C. After incubation, transfer the coverslips with cells to humidity chambers.
4. Suitable primary antibodies are chosen. Table 1 includes a selection of primary antibodies suitable for CLEM of primary cilia and surrounding organelles. Dilute the selected primary antibodies appropriately according to suppliers' instructions (Table 1), and add the antibody solution to the coverslips followed by incubation for 1½ h.
5. Matching fluorescent secondary antibodies are chosen. Usually the secondary antibodies will recognize the species IgG of the primary antibodies. A selection of secondary antibodies useful for CLEM is shown in Table 2. Wash coverslips three times

with Blocking buffer and incubate with the secondary antibodies for 45–60 min followed by three PBS washes.

6. For staining of the nucleus, a 5 min BSA wash is applied. DAPI is added and left to incubate for 1 min.
7. Wash in PBS three times.

3.4 SEM Preparation

1. Fix cells with 2 % glutaraldehyde in SEM buffer for 30 min.
2. Wash 3 × 5 min with SEM buffer.
3. Wash with sodium borohydrate 3 × 5 min (*see* **Note 4**).
4. Wash coverslip well with ddH_2O 10 × 1 min.
5. Dehydrate through a graded series of ethanol, 10 min for each alcohol. Stop at 95 % ethanol (*see* **Note 5**).
6. Dry coverslips using HMDS 3 × 5 min. Allow the third wash to fully evaporate under a fume hood (*see* **Note 6**).

3.5 Fluorescence Microscopy

1. Mount the coverslip in the AxioObserver Shuttle & Find 22 × 22 coverslip holder. Fasten securely.
2. Insert the coverslip holder onto the microscope stage.
3. Insert the holder ensuring the Position 1 "L" is in the upper left corner.
4. Open Zeiss AxioVision 4.8.2 featuring Shuttle & Find. In transmitted light, locate the first fiducial "L" which is located in the upper left part of the coverslip holder (9 o'clock).
5. Select the calibrate button in the software. A calibration dialog box will open.
6. Align the crosshairs in the calibration function of the software with the first "L." Click the automatic button. The software will mark this location and automatically move to Position 2. Repeat this step for the second and third fiducial "L" marks (6 and 3 o'clock).
7. Following the third fiducial calibration, click the OK button. The software and microscope is now calibrated and ready to image. Use the 63× oil objective to image the cells.
8. Image primary cilia, clathrin, and nuclei using DsRed, Alexa 488, and DAPI channels, respectively. Refer to Tables 1 and 2 to use alternate antibodies for other proteins, organelles; this might affect which channels you use when imaging the aforementioned organelles.
9. Image at least 20 fields. Save all images as .ZVI files (*see* **Note 7**).

3.6 Characterization of Primary Cilia

1. Cilia are identified after immunofluorescent labeling using a cilia-specific primary antibody (*see* **Note 8**).
2. To characterize the preparation, measure the primary cilia frequency. Count randomly selected areas on each coverslip.

Primary cilia frequency is determined as the percentage of total number of primary cilia counted divided by total number of nuclei or cells.

3. Measure average ciliary length using AxioVision software.

3.7 Sputter Coating

1. Cover the entire CLEM holder, except the coverslip, with aluminum foil (*see* **Note 9**).
2. Sputter coat the coverslip with chromium (*see* **Note 10**).

3.8 SEM Imaging

1. Open Zeiss SmartSEM software and vent the chamber.
2. Insert CLEM slide holder into Zeiss SEM CLEM slide holder stage.
3. Load stage into chamber and pump.
4. If you have not done so yet, save all acquired fluorescent images to a folder on the SEM computer.
5. Open Zeiss AxioVision 4.8.2 featuring Shuttle & Find.
6. When the SEM is pumped and ready, set working distance to between 4 and 8 mm, high voltage to 2–3 kV and choose a detector (*see* **Note 11**).
7. Locate a cell on the first coverslip and focus and correct astigmatism. This is done to make locating the fiducial "L" marks easier.
8. Locate the first positioned "L" fiducial mark on the holder.
9. Select the calibrate button in AxioVision. A calibration dialog box will open.
10. Align the crosshairs in the calibration function of the software with the first "L." Click the automatic button on the Position 1 portion of the dialog box. The software will mark this location and automatically move to Position 2. Repeat this step for the second and third fiducial "L" marks.
11. Following the third fiducial calibration, click the OK button. The software and microscope is now calibrated and ready to image.
12. Click on the mark/recover button in AxioVision. A dialog box will appear and you will be asked to locate the folder (on the SEM computer) with all of your fluorescent images. In the image Source section of this dialog box, select the path with your image folder. A list of all of the fluorescent images will appear on the left.
13. Choose an image of interest. The image will appear in AxioVision.
14. Double click a point of interest on the fluorescent image (i.e. a primary cilium). This double clicking action will drive the now calibrated SEM to that point.

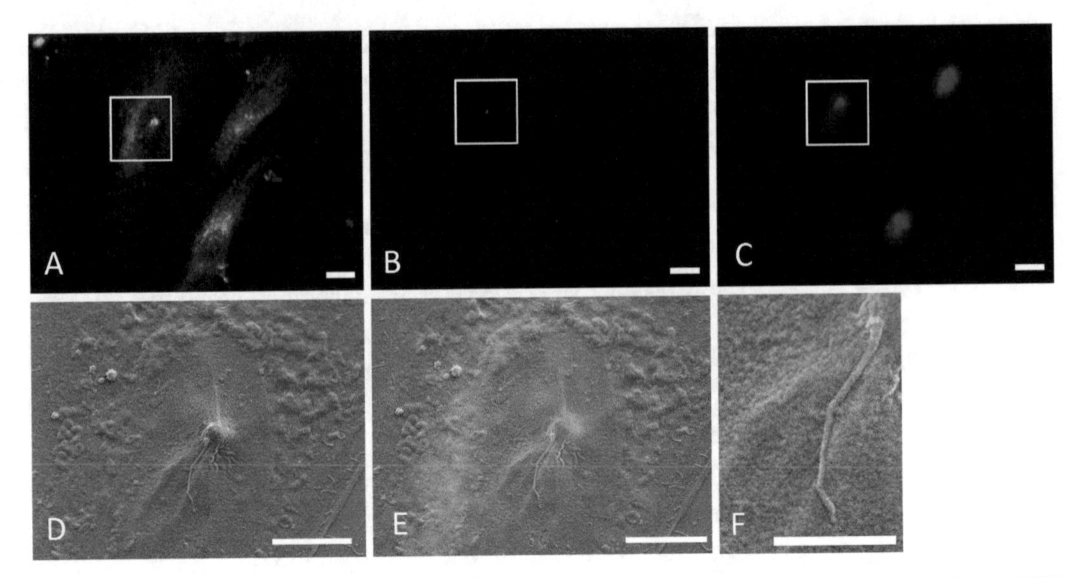

Fig. 1 Sequence of CLEM images showing individual acquisition (**a–d**) and correlated images (**e**, **f**). (**a**, **c**) IFM wide field, cell of interest (*boxed*); (**a**) clathrin (*green* channel), labels MEF clathrin-coated pits and vesicles (CCP/CCVs); (**b**) acetylated α-tubulin (*red* channel), labels the primary cilium; (**c**) DAPI (*blue* channel), labels the cell nucleus; (**d**) SEM of cell of interest; (**e**) correlated image, three fluorescent channels plus SEM image; (**f**) higher magnification of correlated image. Scale bar = 5 μm

15. Acquire all SEM images as .tif. Be somewhat consistent with selected magnifications (*see* **Note 12**).
16. Repeat **steps 13–15** for all desired fluorescent images. You will now have a library of fluorescent images and corresponding SEM images (*see* Fig 1a, b).

3.9 Merging of Fluorescent and SEM images

1. Select the Overlay tab in AxioVision. The Overlay dialog box will appear.
2. In the image source part of this dialog box, choose the folder containing your images (fluorescent and SEM).
3. The box has a dual selectable section which means that you can load a fluorescent and SEM image here. Being sure that the images are coming from the same field, select an SEM image in the top part and a corresponding fluorescent image in the bottom. Each half has a moveable red line flanked with *X*'s and the numbers 1 and 2. An anchor, located on either the top or bottom half, which is movable will let you know which half of the box is "active."
4. Locate an identifiable point in both images (*see* **Note 13**).
5. Drag the red *X* (the one numbered 1) to this point on the SEM image. Be as precise as possible. Do this for the fluorescent image as well.

6. Locate as second identifiable mark on both images.
7. Drag the red *X* (the one numbered 2) to this point on the SEM image. Again, be as precise as possible. Do this for the fluorescent image as well.
8. Access the accuracy of your *X* positionings by selecting the overlay tab and in the dialog box and choosing an appropriate pseudo-color. If the overlay looks off, toggle back and re-position the *X*'s. When satisfied, select execute. A composite CLEM image (Fig. 1c) is now produced.
9. Repeat **steps 4–8** for all desired images (*see* **Note 14**).

4 Notes

1. We have successfully used the matched set of Zeiss equipment: Zeiss AxioObserver Z1 and Zeiss Supra 40 FESEM (Carl Zeiss Microscopy, Thornwood, NY) both equipped with Shuttle & Find. These instruments and programs are named and discussed here, but alternate equipment is available from FEI and Leica Microsystems.
2. Coverslips and Mattek dishes are hand scored with an "L" in the 3, 6, and 9 o'clock positions. The three L's serve as fiducial markers which precisely triangulate the cell(s) of interest. Cell *X*/*Y* positions are embedded in each fluorescent image and essential for recall in the SEM.
3. Do not use a cell line exceeding passage 17. Instead thaw a new sample.
4. Make sure it is active before applying. Active is defined as bubbling comparable to a can of soda.
5. Critical: DO NOT dehydrate to 100% ethanol. This will quench ALL fluorescent signal.
6. Alternatively, critical point dry the coverslips using liquid carbon dioxide. SEM preparation can also be done after fluorescent imaging. In this case, the procedure may be modified for fluorescently staining under variable experimental conditions in a hydrated environment. Subsequently, the cells are prepared for imaging in the SEM by further fixation in glutaraldehyde, followed by dehydration in ethanol, critical point drying, and metal coating as described in Subheading 3.2.
7. This is a critical step. Shuttle & Find will not work if files are saved with any other extension.
8. Anti-acetylated α tubulin is commonly used as a primary antibody, although anti-Arl13b or AC3 antibodies are often used.
9. This is done to prevent the fiducial "L" marks from becoming snowed in with chromium.

10. Critical: DO NOT remove the coverslip from the holder until AFTER they have been imaged on the SEM. Doing so before then will make calibration on the SEM impossible.
11. SE2 is a typical choice.
12. This consistency is not critical, but it will make side-by-side comparison easier.
13. Greatly (and temporarily) increasing the fluorescent image contrast (located at the bottom left of the dialog box) will help with this.
14. The created composite image is limited to ONE channel. Multi-channel composite images must be created with a different imaging program, such as Adobe Photoshop.

References

1. Piperno G, Fuller MT (1985) Monoclonal antibodies specific for an acetylated form of alpha-tubulin recognize the antigen in cilia and flagella from a variety of organisms. J Cell Biol 101:2085–2094
2. Bishop GA, Berbari NF, Lewis J, Mykytyn K (2007) Type III adenylyl cyclase localizes to primary cilia throughout the adult mouse brain. J Comp Neurol 505:562–571
3. Caspary T, Larkins CE, Anderson KV (2007) The graded response to sonic Hedgehog depends on cilia architecture. Dev Cell 12:767–778
4. Satir P, Pedersen LP, Christensen ST (2010) The primary cilium at a glance. J Cell Sci 123:499–503
5. Lechtreck KF, Gould TJ, Witman GB (2013) Flagellar central pair assembly in *Chlamydomonas reinhardtii*. Cilia 2:15
6. Yang TT, Su J et al (2015) Superresolution pattern recognition reveals the architectural map of the ciliary transition zone. Sci Rep 5:Article 14096
7. de Boer P, Hoogenboom JP, Giepmans BNG (2015) Correlated light and electron microscopy: ultrastructure lights up! Nat Methods 12:503–513

Chapter 13

Methods for Visualization of Neuronal Cilia

Tamara Caspary, Daniela Marazziti, and Nicolas F. Berbari

Abstract

Neuroscientists have been captivated by cilia ever since these slender, microtubule-based projections on the cell body were found to play critical roles in neuronal specification, maintenance, and function. In mammals, the most common cilia marker, acetylated α-tubulin, is extremely difficult to detect in neuronal cilia. Here, we describe methods to detect neuronal cilia in culture, in fixed sections, and in vivo, taking advantage of transgenic mice carrying fluorescently tagged cilia proteins.

Key words Acetylated α-tubulin, Arl13b, ACIII, Cilia-GFP

1 Introduction

Although primary cilia have been observed in the central nervous system for half a century, their functions on developing and terminally differentiated neurons remained largely unknown. The original early neuro-anatomical electron microscopy studies describing the presence of primary cilia in the brain were extremely labor-intensive and had limited applications for functional studies. The study of cilia in other tissues, such as kidney, liver, and cartilage, was greatly facilitated by the identification of markers that preferentially localize to the cilia. Until recently, neuronal cilia markers were extremely limited. In fact, the classic cilia marker used by most cilia biologists, acetylated α-tubulin, is virtually undetectable in mammalian neuronal cilia. It is likely present since it can be detected in other vertebrates, but the tubulin acetylation in the mammalian brain may be harder for the antibody to access or may be modified in a slightly distinct manner that is not recognized by current antibodies. New antibodies for several posttranslational modifications of tubulin are under development, and some may prove useful for detecting neuronal cilia.

In the meantime, several developments have made the study of neuronal cilia feasible. First, primary neuronal cultures form cilia, enabling the use of additional approaches, such as live-cell imaging

Peter Satir and Søren Tvorup Christensen (eds.), *Cilia: Methods and Protocols*, Methods in Molecular Biology, vol. 1454,
DOI 10.1007/978-1-4939-3789-9_13, © Springer Science+Business Media New York 2016

of neuronal cilia. Furthermore, specific proteins and receptors are known to selectively localize to the membrane of neuronal cilia and can serve as markers for immunofluorescence approaches both in vitro and in vivo. Recently, the development of mouse alleles that express fluorescent cilia markers has proved to be very useful for the study of cilia within the developing and adult brain. Here, we provide an overview of these strategies and present current staining techniques for visualizing neuronal cilia in vitro and in vivo, and we touch on some of the genetic tools developed to study cilia in the brain.

2 Labeling and Visualization of Neuronal Cilia In Vitro

2.1 Rationale

The use of neuronal cultures to study cilia has several advantages, including the direct imaging of cilia on living neurons by expressing fluorophore-tagged, cilia-specific receptors. This approach allows us to visualize cilia and their receptors under different genetic and pharmacological manipulations. This approach can also be used to assess neurons derived from specific brain regions (e.g., melanin-concentrating hormone receptor 1 in neuropeptide Y-expressing neurons of the hypothalamus; Fig. 1a). Furthermore, it can be used to directly assess changes in the electrophysiological properties of neurons with perturbed cilia functions. This approach also has the advantages of economy and speed for answering questions about neuronal cilia prior to lengthy and costly in vivo investigations.

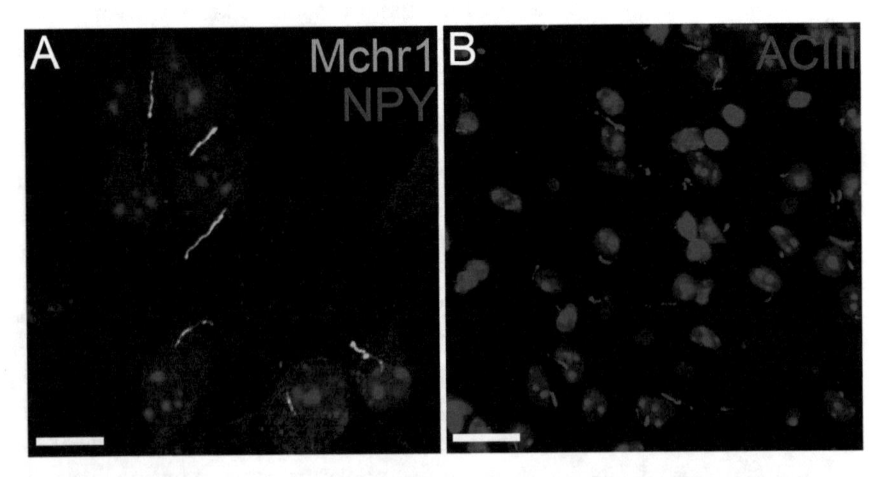

Fig. 1 Immunofluorescence of neuronal cilia in vitro and in vivo. (**a**) Day 10 primary hypothalamic neuronal cultures showing neuropeptide Y-positive stained neurons (NPY, *red*) with melanin-concentrating hormone receptor 1-positive cilia (Mchr1, *green*). Scale bar 11 μm. (**b**) Section of mouse hypothalamus showing adenylyl cyclase III-positive cilia (ACIII, *red*). Scale bar 21 μm. Hoechst-stained nuclei are *blue*

2.2 Materials

Dissecting media

Leibovitz's L-15 medium (Fisher 11415064).

0.2 mg/ml bovine serum albumin (Sigma-Aldrich A9647).

Dissociation media

Dissecting medium with 0.375 mg/ml papain (Sigma-Aldrich P4762).

To insure papain goes into solution, incubate for at least 20 min at 37 °C with occasional vortexing.

Trituration M5-5 media

90 ml Earle's minimal essential medium without L-glutamine (Fisher 11090081).

5 ml fetal bovine serum (heat-inactivated) (Fisher 16140063).

5 ml horse serum (heat-inactivated) (Fisher 26050070).

0.2 ml glutaMAX supplement (Fisher 35050061)

1 ml penicillin/streptomycin (Fisher 15140122).

1 ml 30% (w/v) glucose (Sigma-Aldrich G5767)

0.25 ml insulin/selenite/transferrin (Sigma-Aldrich I1884).

Plating Neurobasal A media

97.4 ml Neurobasal A media (Fisher 50842920).

2 ml B27 supplement (Fisher 501215421).

0.25 ml glutaMAX supplement (Fisher 35050061).

0.25 ml insulin/selenite/transferrin (Sigma-Aldrich I1884).

0.1 ml gentamycin (Fisher 15750060).

All media should be sterile-filtered and can be stored at 4 °C for up to 1 week.

2.3 Method

The following method for neuronal cultures was developed and adapted from several published reports, and its specific use for the study cilia on neurons and glia was developed in the lab of Kirk Mykytyn. It has been used successfully to visualize cilia on neurons and glia derived from the hippocampus and several other brain regions [1–5].

On their day of birth (P0), pups are decapitated and brains are isolated and placed in sterile tissue culture dishes containing sterile dissection media. The brain region of interest is then isolated under a dissecting microscope and torn into smaller pieces and moved to a tissue culture hood. Using a sterile technique, the tissue pieces are transferred into a 15-ml conical tube containing 1 ml of dissociation media and incubated for 15 min at 37 °C with 95% O_2/5% CO_2 blowing gently over the surface of the solution. It is important when making the dissociation solution that it be incubated for at least 20 min at 37 °C to ensure the papain is fully dissolved before sterile filtration. After incubation, the tissue is washed three times with 1–2 ml warm M5-5 trituration media, and then pipetted up and down three times with a Pasteur pipette. The pieces are allowed to settle, and the supernatant is discarded. This removes less-viable cells and undesired debris. The remaining pieces are then triturated with a full-bore but flame-polished Pasteur pipette, followed by pipettes of approximately 2/3 bore diameter, and finally with a pipette of approximately 1/3 bore diameter size. The reduced diameter and polished pipettes are generated by briefly heating over a Bunsen burner. Each trituration step consists of pipetting the pieces up and down approximately ten times with 1–2 ml trituration M5-5 media. After each trituration the pieces are allowed to settle, and the supernatant is transferred to a fresh conical tube. The combined supernatants are then slowly spun at $80 \times g$ for 5 min. The supernatant is removed, and the remaining cells are suspended in Neurobasal A plating media, and then plated onto poly-D-lysine-coated coverslips in a 24-well dish. The number of wells the neurons are distributed into depends on the starting tissue amount, the region of the brain that was isolated, and the desired density. In general, we plate the neurons from a hippocampus or hypothalamus from one mouse onto 2–3 coverslips in 1 ml of medium/well. To ensure the cultures are not dominated by dividing glial cells, the DNA synthesis inhibitor cytosine arabinofuranoside (ARA-C, Sigma-Aldrich C1768) is added to a final concentration of 10 μM after 2–3 days. After 3–5 days in culture, the neurons can be transfected using different techniques, such as Lipofectamine 2000, lentiviral transduction, or calcium phosphate transfection, to express cilia-specific receptors. Twenty-four to 48 h after transfection, the cilia markers can be readily observed.

After 1 week in culture, neurons are well ciliated; however, cultures can be maintained for up to 3 weeks, if desired. It has been our experience that, during this time, cilia length will also increase. They can be fixed for immunofluorescence using the protocol that follows.

Fix with a solution of 4% paraformaldehyde and 10% sucrose in PBS for 10 min at room temperature, followed by a 5-min PBS wash. Neurons on slips can then be further post-fixed with cold MeOH at 20 °C for 15 min and permeabilized with 0.1% Triton X-100 in PBS for 7 min. After permeabilization, the cells should be put in a blocking solution of PBS with 2% serum, 0.02% sodium azide, and 10 mg/ml BSA for 1 h at room temperature. Slips can be stored in this solution at 4 °C for several weeks prior to subsequent immunofluorescence analysis.

All antibody incubations and washes should be carried out in PBS with 2% normal serum from the host species of the secondary antibody, 0.02% sodium azide, and 10 mg/ml BSA. All primary antibody incubations work best if carried out for 16–24 h at 4 °C, with subsequent secondary incubations using standard techniques.

2.4 Summary

This culture system has several advantages for the study of neuronal cilia. The cultures are primary, and after a week to 10 days in vitro, begin to form synapses and are electrophysiologically active. Second, the cultures allow for analysis of individual neurons in the context of cell biology questions surrounding the cilium. Finally, the cultures can be derived from either congenital or conditionally mutant mice and can also be the subject of pharmacological studies as they pertain to primary cilia. With all these features, this system makes for a more physiologically relevant approach over certain other studies in cell lines with greater economy and speed than studies in animal models.

3 Labeling and Visualization of Neuronal Cilia in Sections

3.1 Rationale

The study of primary cilia in both the developing and adult nervous system has suffered from a paucity of tools for their visualization. In recent years, we have begun to resolve that problem with the elucidation of several ciliary-specific proteins and receptors that can be visualized in cilia throughout the brain. Some of the proteins enriched in neuronal cilia include Arl13b, ACIII, Mchr1, Kiss1r, Drd1, 5HT6, and Sstr3 [4, 6–11] (ACIII in hypothalamic cilia, Fig. 1b and Table 1). This growing understanding of which proteins specifically localize to cilia in the brain has allowed investigations using different genetic and pharmacological manipulations in vivo aimed at understanding the function of cilia on these cells. This approach has the advantage of looking at the cilia from animals displaying altered behavior and phenotypes thought to be due to cilia perturbation.

3.2 Materials

Block, permeabilization, and hybridization solution

Phosphate-buffered saline (PBS; pH 7.6).

0.3% Triton X-100.

2–5% normal serum from secondary host.

0.02% sodium azide.

10 mg/ml BSA.

3.3 Cryosection Method

In our experience, perfusion of the animal with PBS and fixative is necessary for studies that require visualization of cilia in the CNS. Animals are anesthetized by a 0.1 ml/10 g intraperitoneal injection of 2.5% tribromoethanol, and then sacrificed by cardiac puncture and perfused with phosphate-buffered saline (PBS; pH 7.6), followed by 4% paraformaldehyde (PFA) in PBS. Once brains are isolated, they are then further fixed in 4% PFA for 16–24 h at 4 °C. Cryoprotection is then necessary to protect the sample during sectioning and is achieved by incubation in 30%

Table 1
Neuronal cilia antibodies

Antibody	Company	Catalog number	Dilution	Species	Neurons in vivo	Reference
Arl13b	Neuromab	Clone N295B/66	1:500	Mouse monoclonal	Several regions	[23–25]
Arl13b	Proteintech	17711-1-AP	1:1000	Rabbit polyclonal	Several regions	[26–28]
ACIII	Santa Cruz Biotechnology	SC-588	1:500	Rabbit polyclonal	Several regions	[7]
SSTR3	Santa Cruz Biotechnology	SC-11617	1:500	Rabbit polyclonal	Neocortex, hippocampus, olfactory bulb	[29]
MCHR1	Santa Cruz Biotechnology	SC-5534	1:500	Goat polyclonal	Hypothalamus, nucleus accumbens	[29]
KISS1R	Strategic Diagnostics		1:5000	Goat polyclonal	GnRH neurons	[11]

sucrose in PBS at 4 °C. Samples are ready to process for sectioning once they have sunk in the sucrose solution: generally 16–24 h at 4 °C for mouse brains, depending on the size of the sample.

Once cryoprotected, the sample should be embedded and frozen in optimal cutting temperature (OCT) compound in the desired orientation in a Cryomold. The frozen block is then removed from the mold and mounted on a sample stub and sectioned in a cryostat at a thickness of 10–30 μm, for immunofluorescence studies, or thicker (>60 μm) for floating section applications (*see* [7, 12] for expansion on neuronal cilia immunohistochemical staining approaches). Sections for immunofluorescence are collected onto positively charged slides, briefly dried, and can be stored in a slide case at −20 °C for several months.

To label the sections, slides are rinsed in PBS to remove any remaining OCT, and then placed in permeabilization solution for approximately 1 h at room temperature. As with the protocol described for the in vitro neuron labeling, the serum needs to be from the same species as the secondary antibody host. This will reduce nonspecific binding of the secondary antibody and lower the background labeling. Primary antibodies are then diluted in hybridization solution and applied to the tissue sections and incubated for 16–24 h at 4 °C. We have found that an easy way to perform antibody incubations is to place the slides in a humidified chamber (i.e., a lidded plastic container lined with a moist paper towel), add just enough diluted antibody solution to cover the sections, and then place a piece of parafilm cut to the size of the slide directly on the solution. The parafilm ensures the solution completely coats the sections and prevents the solution from evaporating or leaking from the slide. Furthermore, a much smaller volume is required (100–200 μl per slide), and therefore less primary antibody is needed. After primary antibody incubation, the sections are washed three times with blocking solution, and then incubated with corresponding secondary antibodies diluted in block for 1 h at room temperature. If desired, a nuclear stain can be included in the diluted secondary antibody solution. The sections are then washed two times with block for 5 min at room temperature, and then two times with PBS, DABCO-mounting media (10 mg of DABCO (D2522; Sigma-Aldrich, St. Louis, MO) in 1 ml of PBS and 9 ml of glycerol). Slides should be sealed using nail polish and stored at −20 °C for future analysis. Slides are ready for analysis once the nail polish has dried.

3.4 Paraffin Materials

4 % paraformaldehyde (prepare fresh).

Safeclear.

100 % EtOH.

Blocking solution

3 % BSA, 0.05 % Tween-20 in 1×TBS.

Blocking solution without serum can be stored in aliquots at −20 °C.

Make with and without 5 % serum.

Tris–EDTA pH 9.0

10× solution: 12.1 g Tris; 3.7 g EDTA in 1000 ml distilled water.

3.5 Paraffin Method

We find that paraffin sectioning is often required to preserve morphology, especially in the adult brain. We perfuse and fix as above, but instead of sucrose sinking, put the sample through the standard dehydration and xylene washes to prepare the sample for paraffin sectioning [13]. After cutting 8-μm thick sections, we let the slides dry overnight since the sections would fall off if wet. The slides can either be stored in a parafilm-sealed slide box with a small amount of desiccant at −20 °C indefinitely or processed for staining at this point.

When ready for staining, we place the slides for 20 min at 37 °C on a slide warmer. We next take the slides through a series of washes to rehydrate the sample (Safeclear 10 min × 3; 100 % ethanol 2 min × 3, 90 % ethanol 1 min, 70 % ethanol 1 min, 30 % ethanol 1 min, 1×TBS 5 min). We have found that several antibodies, most notably Arl13b, require antigen retrieval. To do this we submerge the slides in Tris–EDTA pH 9.0. We microwave the slides in the solution at maximum power for 10 min, let them rest for 10 min, and then microwave another 10 min at maximum power. The slides then rest another 30 min at room temperature, before being permeabilized in 0.1 % Triton (1×TBS containing 0.1 % Triton-X). We block for 1 h at RT in 5 % normal serum, 3 % BSA, 0.05 % Tween-20 in 1×TBS. As above, we use serum from the same animal as the secondary antibody host to minimize background. The slides are then incubated in the dark overnight in the cold room, with the primary antibody diluted in the blocking solution.

The next day we wash the slides 3 × 5 min with 1×TBS. We then incubate the slides at RT for 2–3 h with the secondary antibody diluted in the blocking solution *lacking* 5 % normal serum. The slides are washed a final 3 × 5 min with 1×TBS, stained with Hoechst, and mounted using ProLong antifade (Life Technologies) before imaging.

3.6 Summary

This method for labeling neuronal cilia in vivo has proved useful for answering several questions about cilia in the central nervous system. One example is the confirmation of cilia loss and region-specific cilia loss in the brain using conditional mouse alleles of intraflagellar transport genes [14, 15]. Other examples include analysis of cilia length and composition in mouse models of the ciliopathies.

4 Expression of Fusion Proteins for Neuronal Cilia Visualization In Vivo

One emerging set of tools for the study of neuronal cilia has been the use of mouse alleles that express cilia-specific proteins fused to fluorescent proteins (Table 2). In this section, we briefly describe published alleles and discuss their advantages and potential drawbacks.

The Cilia^{GFP} mouse uses a Cre-inducible somatostatin receptor 3 protein fused to an enhanced green fluorescent protein (Sstr3-GFP) allele, which has been targeted to the ROSA26a locus [16]. The Cilia^{GFP} allele is inducible using Cre recombinase for tissue or cell type-specific analysis, or the label can be constitutively expressed. For example, after Cre-induced expression using a proopiomelanocortin (POMC) promoter to drive Cre expression, only the POMC neuronal cilia are visible (Fig. 2a, b). The Cilia^{GFP} allele also allows for direct visualization of mammalian cilia in vivo in live tissue, such as cilia in hippocampal slices and the ependymal cilia of the lateral ventricle (Fig. 2c, d).

There are three transgenic lines using fusions of fluorescent proteins with the small cilia membrane-associated protein Arl13b. David Clapham's and Nathalie Spassky's labs generated distinct Arl13b-GFP lines ([17] and Nathalie Spassky, personal communication), while Kathryn Anderson's lab made an Arl13b-mCherry transgenic mouse [18]. Overall, each allele appears to be ubiquitously expressed and can be used to study cilia in neural development, as well as in adult brains. However, not all tissues have been examined. Multiciliated ependymal cells beautifully express Arl13b-GFP and Sstr3-GFP, while other multi-ciliated cells, such as the trachea, express suboptimal levels of Sstr3-GFP, perhaps because the finite Sstr3-GFP is diluted when in many cilia.

While all of these alleles have proved useful for the study of cilia in the brain, their use requires caution because the effects of overexpressing these cilia markers are not entirely clear. It remains possible, even likely, that overexpression of these membrane-associated ciliary proteins may alter the function of the cilia or the complement of its signaling proteins. For example, overexpression of the ciliary G-protein-coupled receptor Sstr3-gfp in the testes of Cilia^{GFP} mice leads to infertility due to sperm immotility.

Table 2
Mouse alleles for neuronal cilia studies

Cilia marker	Allele	Availability	Reference
Arl13b-mCherry	Tg(Arl13b-mCherry)1Kv	Jackson Labs Stock 027967	[18]
Arl13-GFP	*Arl13b-EGFP*^{tg}	Via Clapham Lab	[17]
Sstr3-GFP	Gt(ROSA)26Sor$^{tm1.1(Sstr3/GFP)Bky}$	Jackson Labs Stock 024540	[16]
Arl13b-GFP	Arl13b-EGFP$^{Tg\text{-}NS}$	Via Spassky Lab	Nathalie Spassky

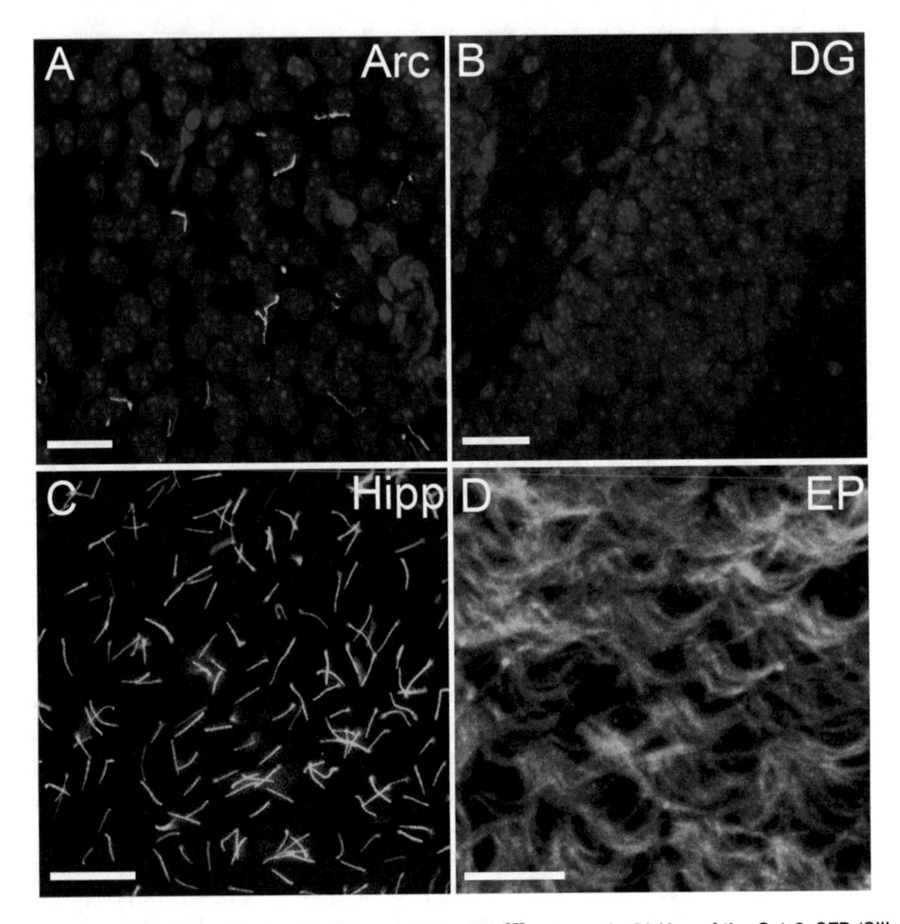

Fig. 2 Central nervous system cilia visualization using the Cilia^{GFP} mouse. (**a**, **b**) Use of the Sstr3-GFP (Cilia^{GFP}) Cre-inducible allele to visualize cilia in the brains of adult mice. (**a**) A subset of neurons within the arcuate nucleus of the hypothalamus (Arc) in an adult mouse display GFP-positive cilia, when a proopiomelanocortin promoter Cre line is used to induce Sstr3-GFP expression. (**b**) Shows a negative control region of the hippocampus, the dentate gyrus (DG), with no GFP-positive cilia. Hoechst-positive nuclei are *blue*. (**c**, **d**) Use of the congenitally "on" Cilia^{GFP} allele in live tissue shows extensive expression and cilia-specific GFP localization in the mouse brain. (**c**) Hippocampus (Hipp), (**d**) ependymal cell cilia (EP). All scale bars 21 μm. Images adapted from O'Connor et al. [16]

Together, these models allow for cilia analysis in living and fixed brain samples. They can enhance studies requiring live tissues, such as the measurement of ligand-induced translocation of proteins into and out of the cilium, in electrophysiology studies for patch-clamping techniques on ciliated neurons or even the cilium itself, or in pharmacological studies of factors regulating cilia length in the brain.

5 Discussion and Conclusions

It is an exciting time for studies aimed at understanding the functions of neuronal cilia. While their existence has been known for decades, it is only recently, with the identification of neuronal cilia-enriched receptors and signaling proteins, that functional studies could begin. Although their functions remain elusive, several basic observations made with current tools hint at their importance. Cilia in different regions of the adult mammalian brain display significantly different lengths. For example, primary cilia in the hypothalamus are significantly longer than those in the dentate gyrus of the hippocampus. The impact of these length variations is not understood; however, it is clear that cilia length is regulated [19–22]. Another example of an interesting yet mysterious phenomenon observed in neuronal cilia is the presence of specific G-protein-coupled receptors (GPCR) in certain regions of the brain. For example, the G-protein-coupled receptor serotonin receptor 6 (5HT6) is observed in cilia of the olfactory tubercle, while another cilia-enriched GPCR, melanin-concentrating hormone receptor 1 (Mchr1), is enriched in the cilia of the nucleus accumbens and the hypothalamus [6, 8]. It is enticing to speculate on the functional significance of these length and receptor variations in the developing and adult brain. Through the continued development of tools to visualize cilia, functional studies will begin to reveal the roles for these small but important cellular appendages in the central nervous system.

Acknowledgments

Special thanks to Cheryl Timms Strauss for editing.

References

1. Askwith CC, Wemmie JA, Price MP, Rokhlina T, Welsh MJ (2004) Acid-sensing ion channel 2 (ASIC2) modulates ASIC1 H+-activated currents in hippocampal neurons. J Biol Chem 279(18):18296–18305
2. Berbari NF, Bishop GA, Askwith CC, Lewis JS, Mykytyn K (2007) Hippocampal neurons possess primary cilia in culture. J Neurosci Res 85(5):1095–1100
3. Brewer GJ, Torricelli JR, Evege EK, Price PJ (1993) Optimized survival of hippocampal neurons in B27-supplemented neurobasal, a new serum-free medium combination. J Neurosci Res 35(5):567–576
4. Domire JS, Green JA, Lee KG, Johnson AD, Askwith CC, Mykytyn K (2011) Dopamine receptor 1 localizes to neuronal cilia in a dynamic process that requires the Bardet-Biedl syndrome proteins. Cell Mol Life Sci 68(17):2951–2960
5. Wemmie JA, Chen J, Askwith CC, Hruska-Hageman AM, Price MP, Nolan BC, Yoder PG, Lamani E, Hoshi T, Freeman JH Jr, Welsh MJ (2002) The acid-activated ion channel ASIC contributes to synaptic plasticity, learning, and memory. Neuron 34(3):463–477
6. Berbari NF, Johnson AD, Lewis JS, Askwith CC, Mykytyn K (2008) Identification of ciliary localization sequences within the third intracellular loop of G protein-coupled receptors. Mol Biol Cell 19(4):1540–1547
7. Bishop GA, Berbari NF, Lewis J, Mykytyn K (2007) Type III adenylyl cyclase localizes to primary cilia throughout the adult mouse brain. J Comp Neurol 505(5):562–571
8. Brailov I, Bancila M, Brisorgueil MJ, Miquel MC, Hamon M, Verge D (2000) Localization of 5-HT(6) receptors at the plasma membrane of neuronal cilia in the rat brain. Brain Res 872(1-2):271–275
9. Caspary T, Larkins CE, Anderson KV (2007) The graded response to Sonic Hedgehog depends on cilia architecture. Dev Cell 12(5):767–778

10. Handel M, Schulz S, Stanarius A, Schreff M, Erdtmann-Vourliotis M, Schmidt H, Wolf G, Hollt V (1999) Selective targeting of somatostatin receptor 3 to neuronal cilia. Neuroscience 89(3):909–926
11. Koemeter-Cox AI, Sherwood TW, Green JA, Steiner RA, Berbari NF, Yoder BK, Kauffman AS, Monsma PC, Brown A, Askwith CC, Mykytyn K (2014) Primary cilia enhance kisspeptin receptor signaling on gonadotropin-releasing hormone neurons. Proc Natl Acad Sci U S A 111(28):10335–10340
12. Domire JS, Mykytyn K (2009) Markers for neuronal cilia. Methods Cell Biol 91:111–121
13. Carson FL, Cappellano CH (2015) Histotechnology a self instructional text, 4th edn. ASCP, Chicago, IL, p 368
14. Berbari NF, Malarkey EB, Yazdi SM, McNair AD, Kippe JM, Croyle MJ, Kraft TW, Yoder BK (2014) Hippocampal and cortical primary cilia are required for aversive memory in mice. PLoS One 9(9), e106576
15. Berbari NF, Pasek RC, Malarkey EB, Yazdi SM, McNair AD, Lewis WR, Nagy TR, Kesterson RA, Yoder BK (2013) Leptin resistance is a secondary consequence of the obesity in ciliopathy mutant mice. Proc Natl Acad Sci U S A 110(19):7796–7801
16. O'Connor AK, Malarkey EB, Berbari NF, Croyle MJ, Haycraft CJ, Bell PD, Hohenstein P, Kesterson RA, Yoder BK (2013) An inducible CiliaGFP mouse model for in vivo visualization and analysis of cilia in live tissue. Cilia 2(1):8
17. Delling M, DeCaen PG, Doerner JF, Febvay S, Clapham DE (2013) Primary cilia are specialized calcium signalling organelles. Nature 504(7479):311–314
18. Bangs FK, Schrode N, Hadjantonakis AK, Anderson KV (2015) Lineage specificity of primary cilia in the mouse embryo. Nat Cell Biol 17(2):113–122
19. Berman SA, Wilson NF, Haas NA, Lefebvre PA (2003) A novel MAP kinase regulates flagellar length in Chlamydomonas. Curr Biol 13(13):1145–1149
20. Nguyen RL, Tam LW, Lefebvre PA (2005) The LF1 gene of Chlamydomonas reinhardtii encodes a novel protein required for flagellar length control. Genetics 169(3):1415–1424
21. Tam LW, Ranum PT, Lefebvre PA (2013) CDKL5 regulates flagellar length and localizes to the base of the flagella in Chlamydomonas. Mol Biol Cell 24(5):588–600
22. Tam LW, Wilson NF, Lefebvre PA (2007) A CDK-related kinase regulates the length and assembly of flagella in Chlamydomonas. J Cell Biol 176(6):819–829
23. Falcon-Urrutia P, Carrasco CM, Lois P, Palma V, Roth AD (2015) Shh signaling through the primary cilium modulates rat oligodendrocyte differentiation. PLoS One 10(7), e0133567
24. Menzl I, Lebeau L, Pandey R, Hassounah NB, Li FW, Nagle R, Weihs K, McDermott KM (2014) Loss of primary cilia occurs early in breast cancer development. Cilia 3:7
25. Piotrowska-Nitsche K, Caspary T (2012) Live imaging of individual cell divisions in mouse neuroepithelium shows asymmetry in cilium formation and sonic hedgehog response. Cilia 1:1
26. Mukhopadhyay S, Wen X, Ratti N, Loktev A, Rangell L, Scales SJ, Jackson PK (2013) The ciliary G-protein-coupled receptor Gpr161 negatively regulates the sonic hedgehog pathway via cAMP signaling. Cell 152(1-2):210–223
27. Seo S, Zhang Q, Bugge K, Breslow DK, Searby CC, Nachury MV, Sheffield VC (2011) A novel protein LZTFL1 regulates ciliary trafficking of the BBSome and smoothened. PLoS Genet 7(11), e1002358
28. Stratigopoulos G, Martin Carli JF, O'Day DR, Wang L, Leduc CA, Lanzano P, Chung WK, Rosenbaum M, Egli D, Doherty DA, Leibel RL (2014) Hypomorphism for RPGRIP1L, a ciliary gene vicinal to the FTO locus, causes increased adiposity in mice. Cell Metab 19(5):767–779
29. Berbari NF, Lewis JS, Bishop GA, Askwith CC, Mykytyn K (2008) Bardet-Biedl syndrome proteins are required for the localization of G protein-coupled receptors to primary cilia. Proc Natl Acad Sci U S A 105(11):4242–4246

Chapter 14

Methods to Study Centrosomes and Cilia in *Drosophila*

Swadhin Chandra Jana, Susana Mendonça, Sascha Werner, and Monica Bettencourt-Dias

Abstract

Centrioles and cilia are highly conserved eukaryotic organelles. *Drosophila melanogaster* is a powerful genetic and cell biology model organism, extensively used to discover underlying mechanisms of centrosome and cilia biogenesis and function. Defects in centrosomes and cilia reduce fertility and affect different sensory functions, such as proprioception, olfaction, and hearing. The fly possesses a large diversity of ciliary structures and assembly modes, such as motile, immotile, and intraflagellar transport (IFT)-independent or IFT-dependent assembly. Moreover, all the diverse ciliated cells harbor centrioles at the base of the cilia, called basal bodies, making the fly an attractive model to better understand the biology of this organelle. This chapter describes protocols to visualize centrosomes and cilia by fluorescence and electron microscopy.

Keywords *Drosophila*, **Centrosome,** Basal body, Cilia, Intraflagellar transport, Sensory neuron, Sperm

1 Introduction

Centrioles are microtubule (MT)-based structures that are required to form two highly conserved eukaryotic organelles: centrosomes and cilia. The centrosome acts as major microtubule-organizing center in cycling as well as several differentiated cells. The cilium is a MT-based protrusion that provides differentiated cells the ability to move, stir particles around and respond to various external signals. These organelles are altered in several human diseases, such as microcephaly, cancer, and ciliopathies (for review [1, 2]). Therefore, a better understanding of the structure and physiology of these organelles is needed. In the last decades, the biology of these two organelles was studied in many model organisms, including *Chlamydomonas*, *C. elegans*, *Drosophila*, zebrafish, and mouse [3–7].

The fruit fly, *Drosophila melanogaster*, is extensively used to investigate the fundamental mechanisms of many biological processes including the development of organ and tissues, cell division,

Peter Satir and Søren Tvorup Christensen (eds.), *Cilia: Methods and Protocols*, Methods in Molecular Biology, vol. 1454,
DOI 10.1007/978-1-4939-3789-9_14, © Springer Science+Business Media New York 2016

organelle biology and animal behavior [8–11]. The fruit fly is also an excellent genetic and cell biology model organism for studying the molecular mechanisms in centrosomes and cilia biology using fluorescent microscopy and electron microscopy for a variety of reasons highlighted below [12–15].

1. Most of the *Drosophila* proteins that are required for centrosomes and cilia biogenesis are conserved among eukaryotes, including humans (Table 1), making the studies in the fly highly relevant [16–18].
2. Unlike many other models (e.g., zebrafish, mice), the fruit fly mutants of centrosomal and ciliary proteins are not embryonic lethal [19]. Thus, it is possible to investigate the role of centrosomes and cilia in organ and tissue development in a whole animal by eliminating these organelles.
3. *Drosophila* possesses cells that harbor both immotile cilia (9 + 0: e.g., in external sensory neurons) and motile cilia (9 + 2 in sperm). Ciliary functions can be tested in the fly by measuring the response to sensory stimuli. This can be done directly in physiological assays at the level of sensory neurons (e.g., calcium imaging, electrophysiology) or in a more indirect manner by quantifying behavioral responses. An obvious readout for cilia function in sperm is fertility. In embryo, larvae and adult, the sensory reception is mediated by cilia grown on type-I sensory neurons of the peripheral nervous system. Type-I sensory neurons can grossly be divided into two categories: (a) external sensory neurons (that grow 9 + 0 type axonemes without dynein arms) and (b) chordotonal neurons (that grow 9 + 0 type axonemes with dynein arms). All different external sensory neurons grow structurally and functionally diverse cilia (Figs. 1, 2, 3, and 5) [15, 20]. Notably, all cilia on sensory neurons require intraflagellar transport (IFT) for their assembly [21]. Furthermore, *Drosophila* testes harbor sperm cells and their precursors that also grow cilia. While the cilia grown in sperm cells are motile (9 + 2), sperm precursor cells (also called spermatocytes) have immotile cilia (9 + 0/1, Figs. 1, 4, and 6) [22, 23]. The cilia in testes assemble in an IFT-independent manner [21]. Many structurally and functionally diverse cilia can be found in different unicellular and multicellular animals. This diversity can be represented by one single model organism, *Drosophila* [24]. Additionally, cilia assembly and maintenance in many organisms are regulated by complex combinations of IFT- and non-IFT-dependent transports which are hard to distinguish [25]. Therefore, the fly is particularly suitable to distinguish IFT-related functions from other cilia-associated functions of any centrosomal and ciliary molecule.

Table 1
Overview of some available tools to study centrosome/basal body and cilia biology in *Drosophila*

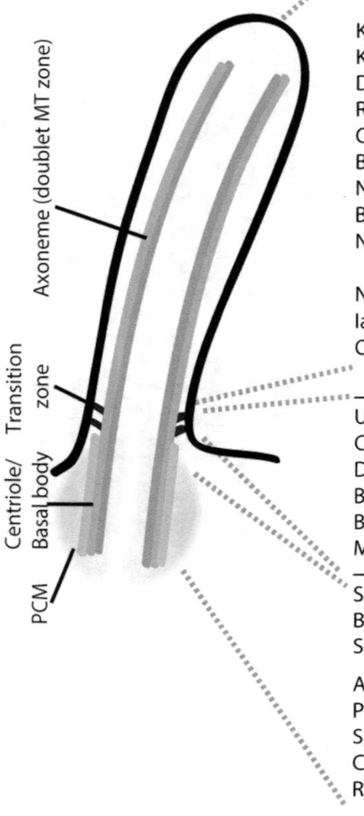

Axoneme (doublet MT zone)
Transition zone
Centriole/ Basal body
PCM

Name	Human Orthologue	Gene*	Mutant	RNAi	Reported in	Markers	References
α-tub84B	α-tubulin	CG1913	--	√	all cells	GFP-tag(UAS-)	18
DCX-EMAP	--	CG42247	√	√	chordotonal, companiform SNs	GFP-tag(UAS-)	35
KLP64D	KIF3A	CG10642	√	√	embryo, all Ns	GFP-tag(UAS-)	20
KLP68D	KIF3B	CG7293	√	√	embryo, all Ns	YFP-tag(UAS-)	20
DmKAP	KAP	CG11759	√	√	embryo, all Ns	--	20
RempA	IFT140	CG11838	√	√	all ciliated SNs	YFP-tag(Endo-)	36
Oseg1	IFT122	CG7161	√	√	all ciliated SNs	--	18
Btv	DYNC2H1	CG15148	√	√	all ciliated SNs	antibody	37
NompB	IFT88	CG12548	√	√	all ciliated cells	GFP-tag(Endo-)	21
Bug22	CFAP20	CG5343	√	√	chordotonal SNs	antibody	38
NompC	--	CG11020	√	√	chordotonal, companiform SNs	GFP-tag(Endo-), antibody	39, 40
Nan	--	CG5842	√	√	chordotonal SNs	antibody	41
Iav	--	CG4536	√	√	chordotonal SNs	GFP-tag(Endo-), antibody	42
Orco	--	CG10609	√	√	olfactory SNs	GFP-tag(UAS-), antibody	43
UNC	Ofd1	CG1501	√	√	all ciliated SNs, sperm	GFP-tag, Endo-	44
Chibby	Chibby	CG13415	√	√	chordotonal SNs, sperm	GFP/tdTom-tag, Endo-	15
Dilatory	CEP131/AZI1	CG1625	√	√	chordotonal SNs, sperm	FLAG-tag, Endo-	33
B9d1	B9d1	CG14870	--	√	sperm	GFP, MYC-tag, Endo-	45
B9d2	B9d2	CG42730	--	√	sperm	GFP-tag, Endo-	45
Mks1	Mks1	CG15730	--	√	sperm	GFP-tag, Endo-	45
SAS-4	SAS-4	CG10061	√	√	all cells	GFP-tag(Endo-, UAS-), antibody	19
BLD10	CEP135	CG17081	√	√	neuroblasts, sperm	GFP-tag(Endo-, UAS-), antibody	16, 46
SAS-6	SAS-6	CG15524	√	√	embryo, sperm	GFP-tag(Endo-, UAS-), antibody	47
Asl	CEP152	CG2919	√	√	all cells	antibody	48
PLP	Pericentrin	CG33957	√	√	all cells	GFP-tag(UAS-), antibody	49
Spd-2	CEP192	CG17286	√	√	all cells	GFP-tag(Ubq-), antibody	50
Cnn	Centrosomin	CG4832	√	√	all cells	GFP-tag(Ubq-), antibody	50
Rootletin	Rootletin	CG6129	√	√	all cilated SNs	GFP-tag(Endo-, UAS-), antibody	51

PCM- pericentriolar material * - Drosophila gene numbers √ present -- absent SNs- sensory neurons Ns- neurons GFP- Green fluorescent protein
tdTom- tandem dimer tomato UAS- upstream activating sequence promotor Endo- endogenous promotor Ubq- ubiquitous promoter

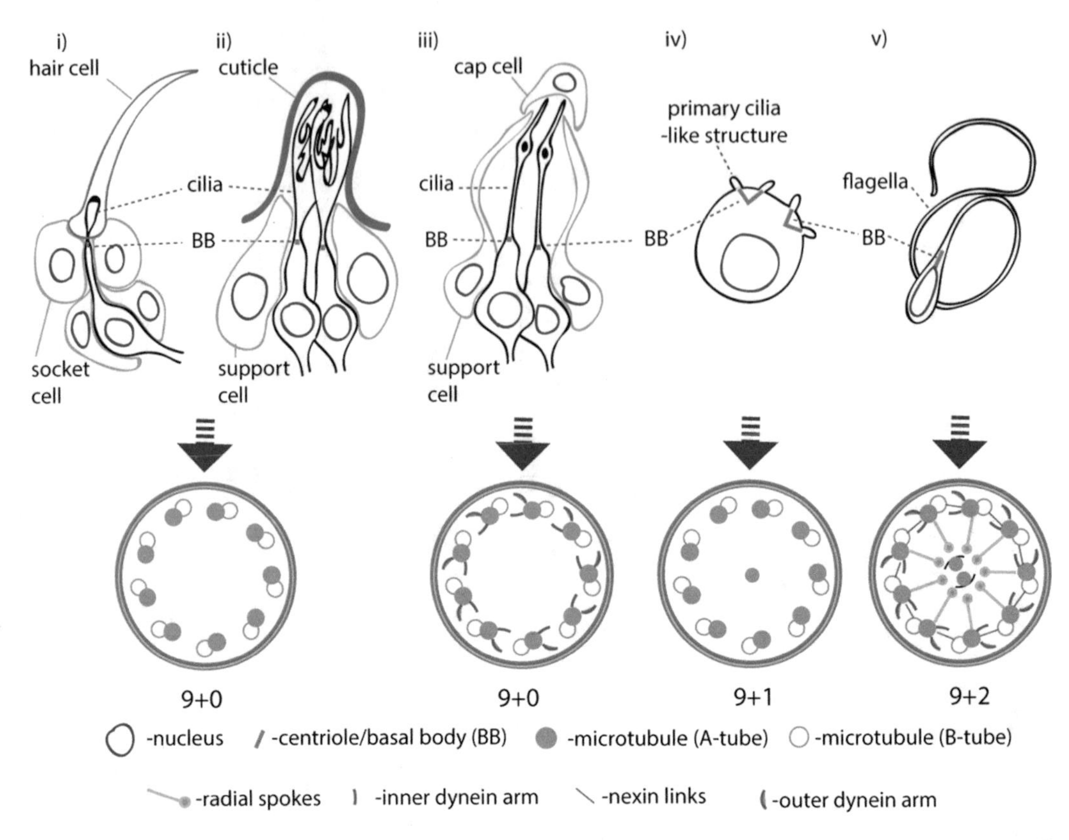

Fig. 1 Drosophila harbors many cells that grow different types of cilia. Schematic representation of some ciliated cells present in adult *Drosophila melanogaster*. Cilia grow from the distal tip of the dendrites of the Type-1 sensory neurons in the fruit fly. These are generally divided in two groups: neurons in external sensory (ES) organs (*i*, *ii*) and chordotonal organs (*iii*). Cilia in ES organs usually display a 9 + 0 microtubule organization (without dynein arms). The proximal segment of the chordotonal cilia shows 9 + 0 microtubule architecture with dynein arms. Thus, these cilia are believed to be motile and mutants without dynein arms suffer hearing loss. In neurons of the peripheral nervous system (PNS), cilia assembly relies on intraflagellar transport (IFT). Primary cilia-like structures and flagella are found in spermatocytes (*iv*) and in elongating spermatids (*v*). While primary cilia-like structures in spermatocytes (*iv*) display 9 + 0 or 9 + 1 MT organization, a 9 + 2 architecture with a central pair, radial spokes, nexin links, and dynein arms are observed in spermatid flagella. Sperm flagella assembly does not rely on IFT and is carried out inside the cytoplasm

4. During the development and differentiation of all ciliated cells, the centrosome assembles, centrioles migrate to the apical parts of the cell where one gets anchored and modified, and grows a cilium [2]. Additionally, in *Drosophila* mature sperm cells the centrosome loses most of its components (centriolar components as well as pericentriolar material (PCM)). This process is called centrosome reduction [26]. Thus, studies in the neurons and testes can allow one to address multiple aspects of "the life of a centrosome" such as centriole duplication, centriole stability, centriole elongation, centriole separation, PCM recruitment and removal, and ciliogenesis among others.

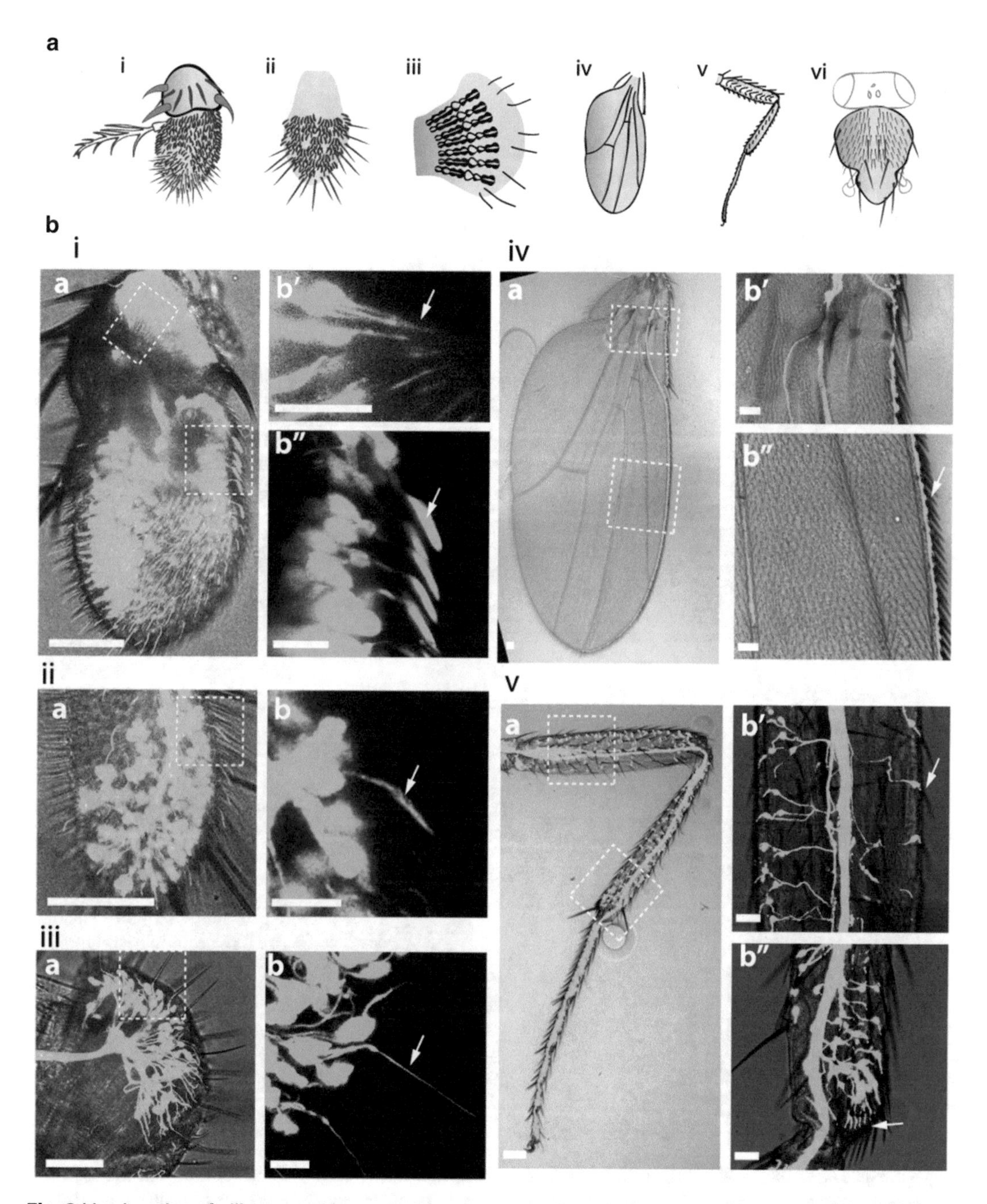

Fig. 2 Live imaging of cilia present in many sensory neurons in adult *Drosophila.* (**A**) Cartoons of an antenna (*i*), a maxillary palp (*ii*), a proboscis (*iii*), a wing (*iv*), a leg (*v*), and a thorax (*vi*) that harbor different types of external sensory neurons. Two antennae, two maxillary palps, and one proboscis are present in the fly head. The thorax harbors six legs and two wings. (**B**) Representative images of sensory neurons (marked by *UAS-GFP* using *Gal4Cha*) in different external sensory organs, such as second and third antennal segment (*i*), maxillary palp (*ii*), proboscis (*iii*), wing (*iv*) and leg (*v*). The bright-field image combined with the fluorescent image of neurons (*i–v*) helps understanding the organization of the tissue with respect to the cuticle in the given organ. In *i–v*, a magnified view of the marked region(s) in (*a*) is presented in (*b* or *b′*, *b″*) to highlight the cilia in the sensory neurons (*arrows*). In principle, fluorescently tagged centrosomal components can also be studied. The third antennal segment (*i*) and maxillary palp (*ii*) harbor different types of olfactory sensory neurons. Chordotonal neurons (*b′*) are present in the second antennal segment (*i*). The proboscis (*iii*) and wing (*iv*) harbor gustatory, chemosensory, and mechanosensory neurons, and the leg (*v*) also harbors mechanosensory neurons. Scale bars represent 50 μm in *i-a*, *ii-a*, *iii-a*, *iv*, and *v*, and 10 μm in *i-b′*-b′, *ii-b*, and *iii-b*

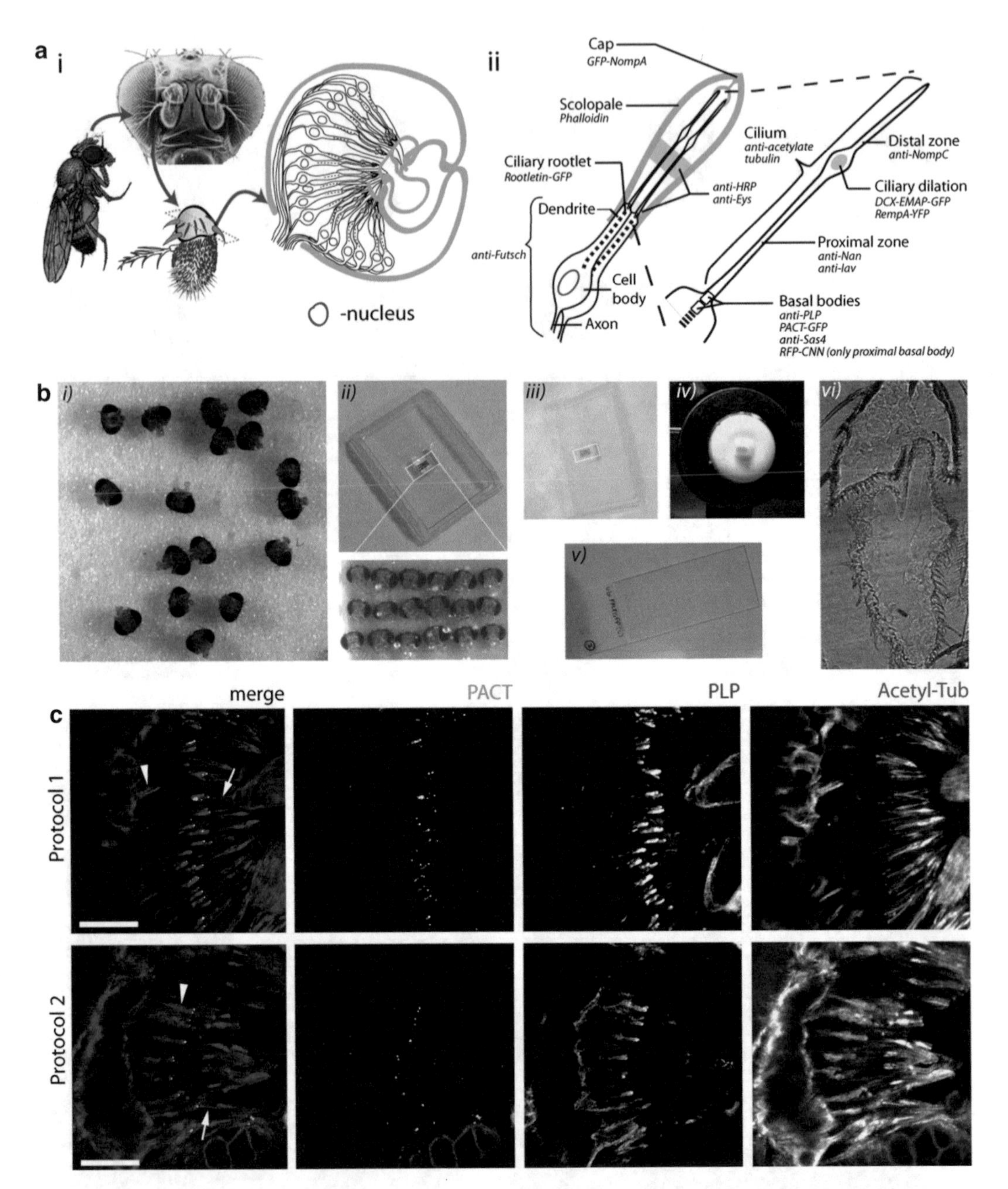

Fig. 3 Imaging basal body and cilium in auditory neurons in *Drosophila* antenna. (**a**) The scheme depicts the organization of (*i*) the Johnston's organs (JOs) in the second segment of the adult (or pupal) antenna present in the head of the fly, (*ii*) the chordotonal neurons in the JO of the antenna, and the cilium in those neurons. Some markers for different regions of the centrosomes and cilia are shown (for detailed list of the components of these organelles *see* Table 1). (**b**) Around 16–20 heads of pupae or 2–4 days old flies are cut (*i*) followed by aligning heads in a mold filled with OCT. (*ii*). Then the mold is frozen on dry ice. After the OCT has solidified, the block is mounted upside down (*iii*) on the cryostat holder (*iv*). 10–12 μm thin sections are collected on the coated side of the Poly-L-lysine slide (Sigma, USA). Then a boundary is made using a Dako pen. The boundary acts as a barrier preventing the aqueous solutions from running of the slide during fixing and immunostaining (*v*). Also *see* [29]. (*vi*) A representative bright-field picture of the second and third antennae in a section. (**c**) Representative pictures of a set of chordotonal neurons in JO of second antennal segment in which centrioles (PACT in *green*), pericentriolar material (PLP in *red*) and cilia (acetylated tubulin in *magenta*) are marked. Representative images obtained with protocol 1 and 2 are shown (*see* Subheading 3.1). PACT is the conserved centriole localizing C-terminal domain of pericentrin-like protein (PLP); thus PACT::GFP marks centrioles when expressed in *Drosophila*. *Arrowheads* and *arrows* mark the dendrite and cilia of the auditory neurons, respectively

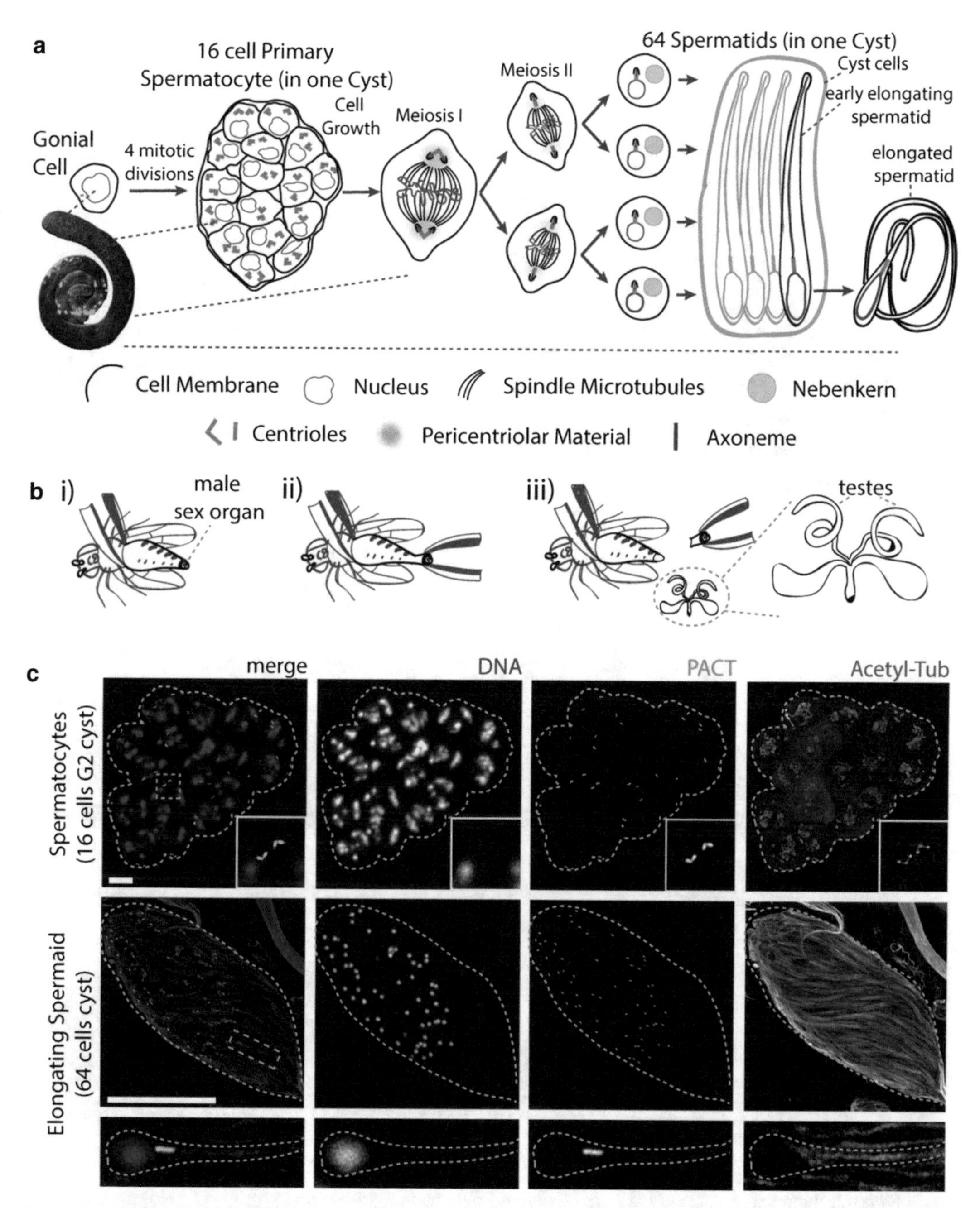

Fig. 4 Imaging centriole and cilia in testes in adult (or pupae) *Drosophila.* (**a**) Schematic representation of *Drosophila* spermatogenesis. A stem cell after division gives rise to a gonial cell that in turn undergoes four rounds of incomplete mitotic divisions to produce a 16-cell cyst of primary spermatocytes. Primary spermatocytes go through a long G2 phase when centrioles elongate and migrate to the cell membrane where they grow a cilium. Each spermatocyte then undergoes two consecutive meiotic divisions without either DNA replication or centriole duplication. As a result, each early spermatid forms with a single basal body nucleating the sperm flagellum. (**b**) Scheme shows different steps of testes dissection from an adult male fly. Dissection is done in a drop of testis buffer. Hold the fly with legs facing up (*i*) and pull out the male genital part (*ii*). Testes will come out with it (*iii*). If they do not, insert the forceps inside the fly's abdomen and pull them out. Testes can be immunostained either by keeping them intact or squashing them (for details *see* Subheading 3.2, [30]). (**c**) Representative images of spermatocytes and elongating spermatids from 16 cell G2 and 64 cells cyst, respectively, with markers for DNA (DAPI in *blue*), centrioles (PACT in *green*) and cilia (acetylated tubulin in *magenta*)

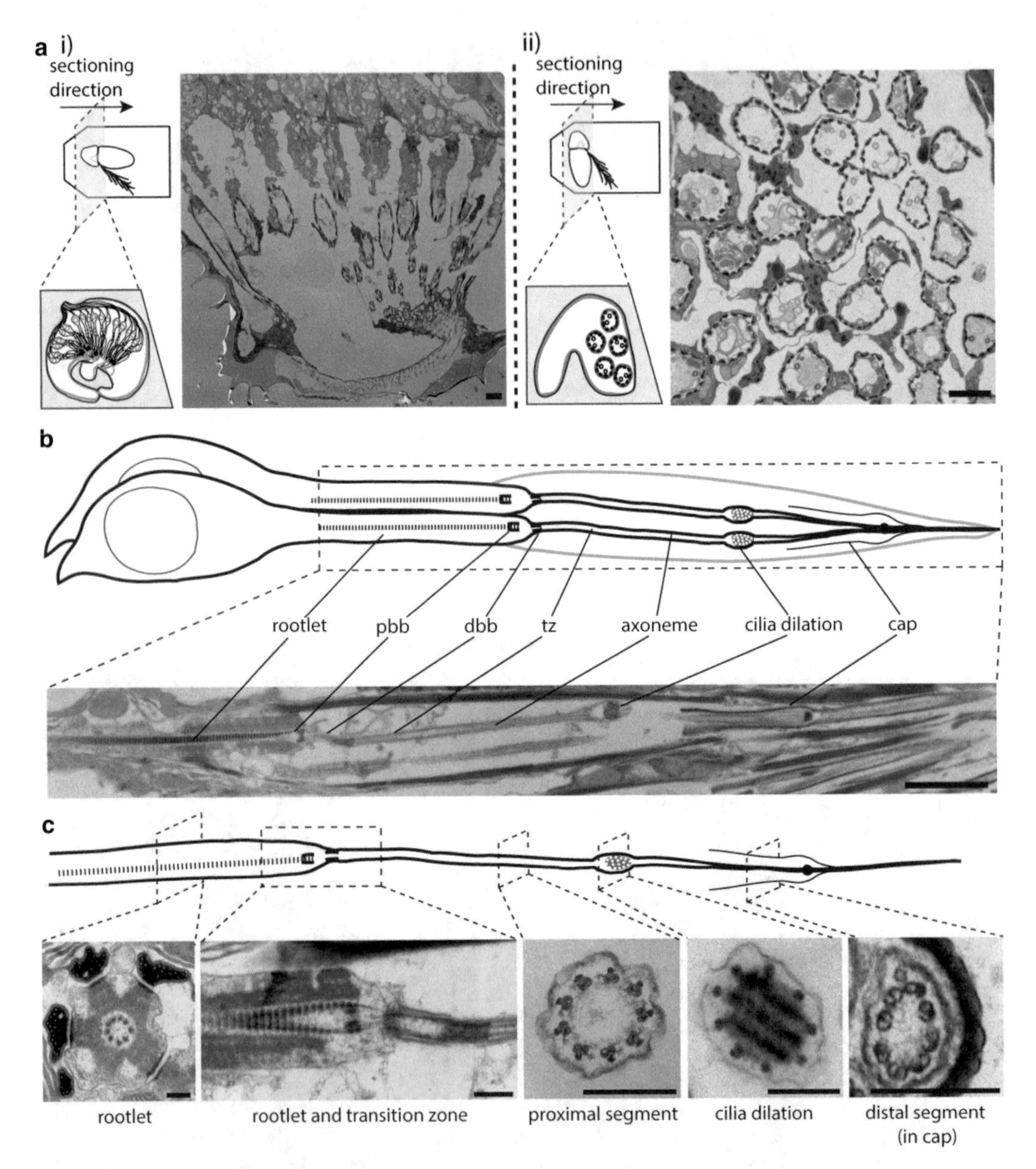

Fig. 5 Electron microscopy analysis of basal body and cilia from auditory neurons. (**a**) (*i*) Scheme depicts the orientation of the second segment in the antenna in the resin mold and the direction of sectioning to obtain longitudinal sections of chordotonal neurons (or auditory neurons). On the right side of the scheme, a representative low magnification electron micrograph displays the longitudinal organization of the Johnston's organs that harbor chordotonal neurons. (*ii*) Scheme represents the orientation of the antennal second segment in the resin mold and the direction of sectioning to achieve cross sections of auditory neurons. On the right side of the scheme, a low magnification electron micrograph represents cross organization of the Johnston's Organ. Scale bars represent 2 μm (**b**) Scheme of the two auditory neurons present in a scolopale and a representative electron micrograph of a longitudinal view of those neurons. Scale bar represents 2 μm. (**c**) Scheme of the auditory cilium and representative electron micrographs of cross-sectional and longitudinal view of the different regions of cilia, such as ciliary rootlet, transition zone, the proximal segment of the cilia, ciliary dilation and distal segment of the cilia. Scale bars represent 250 nm

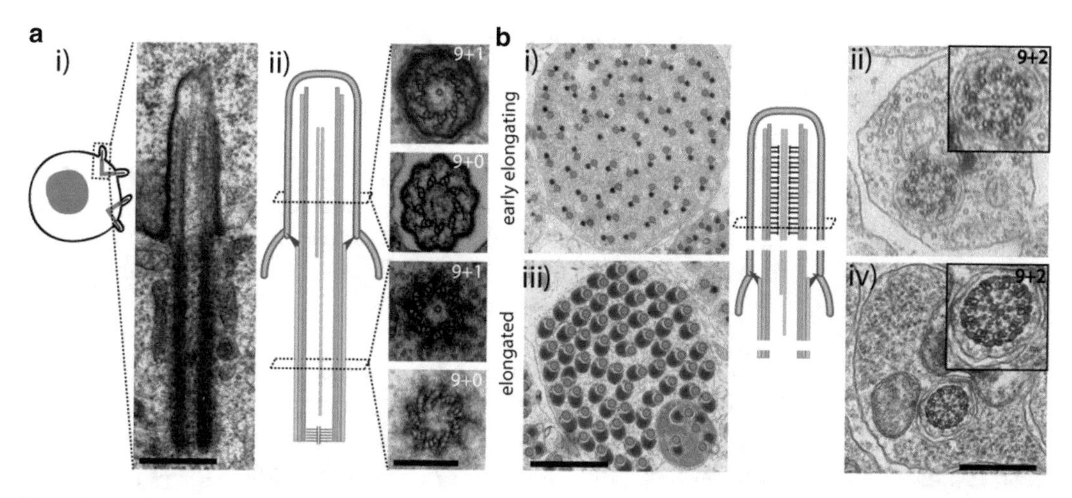

Fig. 6 Electron microscopy analysis of centrosomes and cilia from testes. (**a**) Ultrastructural organization of primary cilia-like structure and basal body in G2 stage spermatocyte. (*i*) Longitudinal section of a primary spermatocyte cilium and the basal body. (*ii*) This cilium contains nine radially arranged doublet microtubules either with a central singlet microtubule that extends from basal body to the axoneme (9+1) or without it (9+0). Representative cross-sectional images of axoneme and basal body in both types of cilia are shown. Interestingly, while the basal body is formed with triplet microtubules, the cilium is formed with doublet microtubules. (**b**) Ultrastructural organization of flagella in elongating (*i–ii*) and elongated (*iii–iv*) spermatids. Representative images of early elongating (*i*) and elongated (*iii*) cysts containing 64 spermatids. Representative cross-sectional images of cilia (9+2) in both types of spermatids are shown. Note changes in axoneme organization during transition from early elongating to elongated spermatid, including the acquisition of dynein arms

5. *Drosophila* harbors different types of centrosomes. They vary in their length and the types of microtubules they are made of. For example, centrioles in the embryo are short and made of nine doublet microtubules, whereas those in sperm cells are uniquely long and consist of nine triplet microtubules (Fig. 4 and 6) [27]. Long centrioles in sperm allow rapid and easy analysis of this organelle by imaging, providing an appropriate system to understand the mechanisms underlying centriole elongation [22, 28]. Studies in centriolar cells with or without cilia in the embryo are widely appreciated [28]. Moreover, dissection and imaging of the tissues that have ciliated cells in larvae, pupae, and adults are simple (Fig. 2) [20, 29].
6. *Drosophila* is a widely preferred model organism for biological studies due to its short generation time and ease to do genetics. Importantly, mutants and RNAi lines for most of the known centriolar and ciliary components are available (Table 1). Flies that express fluorescently tagged centrosomal and ciliary proteins are available to the community and several of those fluorescently tagged components are expressed under the control of endogenous promoters. In addition,

antibodies against many of those proteins are available (*see* Table 1).

In this chapter, we describe protocols to visualize centrosomes and cilia directly or by immunofluorescence either in Type-I sensory neurons or in testes in both pupae and adults. Moreover, we provide detailed protocols to perform electron microscopic analyses of centrosome and cilia ultrastructure in different ciliated cells in pupae or adult. Finally, we show and discuss results and limitations of using those protocols.

1.1 Visualization of Centrosomes and Cilia Using Fluorescence Microscope

Centrosomes and cilia can be studied at all different stages of *Drosophila* development: in embryos, larvae, pupae, or adults [28–30]. In pupae and adults, the cuticles that encapsulate the sensory neurons are opaque and display autofluorescence making antibody staining and fluorescence live imaging difficult. Recently, several methods for studying centrosome and cilia in sensory neurons in pupae/adults were reported [15, 20]. Here, we present protocols (*see* Subheading 3.1) to visualize cilia by live-imaging of pupal/adult tissues, such as different antennal segments, maxillary palp, proboscis, wing, and leg (Fig. 2). We also present protocols (*see* Subheadings 3.1 and 3.2) for immunostaining of sensory neurons (Fig. 3) and sperm cells (Fig. 4) in pupal/adult antenna and testes, respectively. Protocols to study these organelles in embryo and larvae, and for live imaging of the centrosome in testes can be found elsewhere [29, 31, 32].

The *Drosophila* second antennal segment harbors approximately 500 sensory neurons called chordotonal neurons, which grow cilia and are required for hearing and coordination in the adult flies [33]. Several markers can be used to study the different centrosome or ciliary compartments in sensory neurons (Table 1, Fig. 3a). For example, the pericentriolar material and ciliary axoneme can be specifically labeled using antibodies that recognize pericentrin-like protein (PLP) and acetylated tubulin, respectively (Fig. 3c). Here, we describe two methods of fixation of antenna for immunostaining of the chordotonal neurons. Although we do not find remarkable differences using both protocols in the localization of the studied components, we did not systematically compare large set of centrosome and cilium components. Many antibodies and flies expressing several fluorescent reporters can be used to label specific compartments of the centrosomes (e.g., centriole and pericentriolar material) and cilia (e.g., transition zone, proximal segment of the cilia, ciliary dilation, distal segment of the cilia, IFT-components etc.) (Fig. 3a; for detailed list *see* Table 1). Since, all the ciliated sensory neurons present in adult *Drosophila* develop during the pupal stage, and these neurons are assumed to be long-lived, the techniques discussed here (Subheading 3.1) can be used to study the assembly and maintenance of these organelles [15,

20]. However, a live-imaging protocol to track the same set of ciliated sensory neurons in pupal stage or adult for several days is still unavailable. Such tool would help to better understand the centrosomes and cilia biology.

In testes, a stem cell divides to form a cyst cell and a gonial cell. The gonial cell subsequently undergoes four rounds of incomplete mitotic divisions to produce a cyst of 16 primary spermatocytes (Fig. 4a). Primary spermatocytes go through a long G2 phase, during which two pairs of small centrioles (0.2 μm) elongate strongly until they reach about eight times their initial size (1.5 μm) and grow small protrusions that are structurally similar to primary cilia (Fig. 4a; Subheading 3.2). In spermatocytes and sperm cells, these organelles can be labeled using PACT::GFP (marks centrioles) and acetylated tubulin (marks centrioles and cilia) (Fig. 4c; for a detailed list of markers used in sperm cells *see* Table 1). The protocols mentioned here can be used to analyze centrioles and cilia in the mutants that affect the assembly, conversion, and stability of these organelles [22]. However, to better understand the morphogenesis and functional transformations of these organelles in sperm cells it would be desirable to develop protocols for live imaging of centrosomes and cilia during spermatogenesis.

1.2 Ultrastructure of Centrosomes and Cilia in *Drosophila*

Centrioles are small MT-based cylinders with ~250 nm in diameter and 300–500 nm in length, while cilia are MT-based long cylinders with ~250 nm in diameter and variable lengths [1, 2]. Thus, many of the centriole and cilium substructures are under the resolution limit of the light microscope, so a detailed ultrastructural analysis is critical to understand their mechanisms of assembly and maintenance. Although, the protocol to study centrosomes and cilia ultrastructure using transmission electron microscopy in embryo and larvae is not yet well established, these organelles have been investigated by several groups in the pupal/adult antenna and testes [15, 20, 22] (also *see* Subheading 3.3).

Due to the special 3D-orientation of the neurons in the second antenna segment (*see* Fig. 3a), both cross and longitudinal sections should be analyzed to study centrosomes and cilia. Notably, longitudinal and cross sections of antennae can be used to study the cross and longitudinal section of the centrosomes/cilia, respectively (Fig. 5a) [29]. Each chordotonal neuron grows a sensory cilium (Fig. 5b). Each cilium has different components, such as a long rootlet, two basal bodies, a transition zone, a proximal segment of the cilia, a ciliary dilation, and a distal segment of the cilia (Fig. 5c). The proximal part of the chordotonal cilium consists of radially arranged nine doublet microtubules with dynein arms which are required for the function of the cilium, but is devoid of central pair of singlet microtubules (9 + 0) (Fig. 5c). However, little is known about the ultrastructure of the basal bodies and cilia in different sensory neurons.

Centrosomes and cilia undergo multiple morphological and functional transformations during spermatogenesis (Fig. 4a). In spermatocytes, centrioles/basal bodies are formed of nine triplet microtubules, whereas primary cilia exhibit either nine doublet microtubules (9 + 0) or nine doublet microtubules with a central singlet microtubule (9 + 1) (Fig. 6; for protocol *see* Subheading 3.3). These small primary-like cilia are devoid of dynein arms and hence considered immotile. Each spermatocyte then undergoes two consecutive meiotic divisions without DNA replication and centriole duplication. As a result, in each early spermatid a single basal body (or centriole) nucleates the sperm flagellum. Interestingly, the elongating sperm flagellum that is motile consists of nine radially arranged doublet microtubules and a central pair of singlet microtubules (9 + 2) [22, 23]. However, how immotile cilia and motile flagella are assembled and how different types of cilia grow from same basal body is still unclear.

Using the protocols (*see* Subheading 3) mentioned in this chapter, it is possible to execute both live imaging of centrosomes and cilia for few minutes, and to immunodetect and perform ultrastructure analysis of these organelles. However, these organelles are small in size (especially in diameter), their assembly is slow and they are long-lived. Therefore, in the future it will be critical to develop protocols to both image these organelles using super-resolution microscopy and perform live imaging in the ciliated cells for several days. Moreover, novel ultrastructural analysis of these organelles in different tissues are desired, where samples are prepared using high-pressure freezing techniques or studied by cryo-electron microscope.

2 Materials

We prepare all solutions using deionized water (d-water, prepared using Millipore Elix 10, USA) or Milli-Q water (prepared by Millipore Synthesis 10, UAS). All reagents should be prepared and stored at room temperature (RT) (unless indicated otherwise). All waste disposal regulations must be followed when disposing of biohazardous waste materials.

2.1 *Drosophila* Stocks and Husbandry

The flies of known genotype can be obtained from the Bloomington stock center, USA, Vienna Drosophila Rnai Center, Austria and National Institute of Genetics-FLY, Japan. Grow *Drosophila* in vials or bottles containing standard culture media that are used in the above mentioned centers. Finally, collect the specific stages of *Drosophila* required for the studies.

1. Culture medium for *Drosophila*: To make one liter of medium, weigh 80 g molasses, 22 g Bett syrup, 80 g corn meal, 18 g yeast granules, 10 g soy flour, and 8 g Agar-Agar in a plastic beaker. Mix all ingredients in a beaker before adding the boiling d-water.
2. Gradually add 1050 ml boiling d-water and transfer the mixed medium to a 2 l Schott flask.
3. Autoclave the medium at 121 °C for 30 min.
4. Finally, add 12 ml 15% niapagin (made in ethanol) and 8 ml propionic acid, when the medium temperature reaches 45–50 °C. This recipe is used in Vienna Drosophila RNAi Center, Austria.

2.2 Components for Direct Visualization and Immunostaining of Sensory Neurons

1. Use halo carbon oil (Sigma, USA) to mount the tissues for live imaging.
2. Phosphate buffered saline (PBS): Dissolve phosphate buffer tablets (Sigma, USA) in d-water. According to company datasheet, when one tablet is dissolved in 200 ml of d-water yields 10 mM phosphate buffer, 2.7 mM potassium chloride, and 137 mM sodium chloride, pH 7.4, at 25 °C.
3. PBST-X buffer: 0.1 or 0.3% PBST-X is made by dissolving required amount of Triton X-100 in PBS solution. 0.1% PBST-X generally should be used in antenna and testes immunostaining protocols (*see* Subheadings 3.1 and 3.2 and **Note 1**).
4. Fixative solutions (for immunostaining): Mix one volume of 16% formaldehyde (methanol-free, ultrapure EM grade, Polyscience Inc, PA, USA) three volumes of PBS to make 4% formaldehyde fixative.
5. Stefanini's fixative: For 1 ml of fixative, combine 250 μl of 16% PFA with 150 μl of saturated picric acid, 150 μl 0.5 M PIPES buffer (pH 7.6), and 450 μl d-water (supplemented with 0.1% Triton-X—*see* **Note 2**).
6. Use optimal cutting temperature (OCT) compound to embed the heads.

2.3 Components for Immunostaining of Testes

1. PBS: *See* Subheading 2.2, **item 2**.
2. 1× Testes buffer: This buffer contains 183 mM KCl, 47 mM NaCl, and 10 mM Tris–HCl, pH 6.8.
3. Fixative solutions: *See* Subheading 2.2, **item 4**.

4. Siliconized coverslips: Wash new coverslips in ethanol for 10 min in a shaker. Air-dry and incubate in Sigmacoat solution (Sigma, USA) for 1 min. Air-dry coverslips.

2.4 Components for TEM Sample Preparation

1. Sodium bi-phosphate buffer: Make stock solution A and B first:

   Stock solution A: 0.2 M monobasic sodium phosphate, monohydrate (dissolve 27.6 g NaH_2PO_4 in 1 l Milli-Q water).

   Stock solution B: 0.2 M dibasic sodium phosphate (dissolve 28.4 g Na_2HPO4 in 1 l Milli-Q water).

   Mix 28 ml of stock solution A with 72 ml of stock solution B to obtain 100 ml of 0.2 M of phosphate buffer pH 7.2.

2. PBS: *See* Subheading 2.2, **item 2**.
3. Fixative solutions: 2% formaldehyde, 2.5% glutaraldehyde in 0.1 M bi-phosphate buffer is generally used for fixing the samples. To make 100 ml of fixative solution, mix 12.5 ml of 16% formaldehyde (EM grade), 10 ml of 25% glutaraldehyde (EM grade), 27.5 ml of d-water, and 50 ml of 0.2 M sodium bi-phosphate buffer.
4. Uranyl acetate (UA) solution: Use 2% UA in d-water (20 g/l) and 2% UA in 70% methanol (20 g/l) for sample fixation/processing and for the staining of ultra-thin sections, respectively.
5. Use propylene oxide (SPI supplies) for TEM sample preparation.
6. EPON™ Epoxy resin dilutions: Make stock solutions first:

   Stock EPON I: Mix 62 ml of EMbed 812 and 100 ml of dodecenyl succinic anhydride (DDSA).

   Stock EPON II: Mix 100 ml of EMbed 812 and 90 ml of Nadic methyl anhydride (NMA).

   Make the EPON working solutions according to Luft (1961): Mix 30 ml of EPON I with 70 ml of EPON II and add 1.5 ml of tris-(dimethylaminomethyl) phenol (DMP-30) (accelerator) in the mixture (*see* **Note 3**).

7. Reynolds lead citrate solution: Make 30 ml de-carbonated water by boiling the d-water in a microwave and cooling it in a freshly washed sealed container. Add 30 ml of decarbonated water to 1.33 g of lead nitrate and 1.76 g of sodium citrate in a volumetric flask and shake vigorously for 1 min followed by sonication for 30 min with shaking the container every 5 min. Add 8.0 ml of 1 N sodium hydroxide (NaOH) to the opaque mixture. The mixture should then turn clear. Make up the volume to 50 ml with the de-carbonated water. Store the solution in a tightly sealed volumetric flask (*see* **Note 4**).

2.5 Instruments

1. Use any surgical blade (Swann Morton Ltd, UK) to cut heads from the flies.
2. Use plastic molds (Simport, Canada) or similar to make blocks with the samples for cryosectioning.
3. Use a Leica CM 3050S Cryostate (Leica BioSystems, Germany) or similar to cut the cryo-frozen tissues of *Drosophila* head.
4. Use a Dako Pen (Dako Agilent Technologies, USA) or similar to make hydrophobic boundaries on the glass slides for immunostaining.
5. Use a Leica TCS Sp5 II confocal microscope, Leica BioSystems, Germany or similar to image all the fluorescent samples.
6. Use a GenLab oven (60 °C incubator) or similar to polymerize the resins.
7. Use a Hitachi Transmission Electron Microscope with AMT 2kX2k digital camera or similar to collect EM images.

3 Methods

All our procedures can be carried out at room temperature (RT) unless otherwise specified.

3.1 Direct Visualization or Immunostaining of Centrosomes and Cilia in Sensory Neurons in the Pupae and Adult

1. For live imaging: To study the centriole and ciliary components in live tissues, dissect required tissues, such as antenna, proboscis, wing, and leg (Fig. 2). Mount the samples in halocarbon oil (Sigma, USA) (*see* **Note 5**).
2. Sample preparation before sectioning:

   Protocol 1: Cut the heads from adult flies using a surgical blade (Swann Morton Ltd, UK) or dissect out heads from 30 to 100 h old pupae. Align ~16–20 heads in a plastic mold containing OCT (Fig. 3b). Arrange heads in an appropriate orientation (either antenna facing up or one eye facing up) to obtain the desired sections of antenna. Then freeze the mold in dry ice (*see* **Note 6**). This protocol is adapted from [20].

   Alternatively, fix whole flies (without the wings) in 4% formaldehyde in PBST-X buffer for 1 h at 4 °C and mount heads in a mold as described above (*see* **Note 7**).

   Protocol 2: Fix ~16–20 fly heads in 2 ml Stefanini's fixative for 40 min on ice (*see* **Note 8**). Subsequent steps should be done at RT. Wash the heads for 3×20 min in 0.1% PBST-X to remove the picric acid that might interfere with immunostaining. Incubate the fixed heads in 10% sucrose in PBST-X on a rotator for 1 h. Then incubate heads in 25% sucrose in PBST-X and place them on a rotator for overnight at 4 °C. After removing the residual sucrose solution, embed and orient the heads in OCT as described before (*see* **Notes 7** and **9**). This protocol is adapted from [34].

3. Collect sections (around 10–12 μm thick) on poly-L-lysine coated glass slides (Sigma-Aldrich, USA). Multiple sections can be collected on a slide and samples can be collected in two/three rows (Fig. 3b). Also, optionally collect consecutive sections in two separate slides to immunostain the same sample with separate sets of antibodies.
4. Air-dry the slide(s) with sections for 1 min and make a border around the sections using a Dako pen (*see* Fig. 3b). This pen mark creates a hydrophobic boundary that prevents aqueous solutions from running off the slide.
5. Keep the slide(s) in a humid chamber pre-wetted with PBS for further use. Then incubate sections in the fixative solution (for IF) for 1 h at 4 °C (*see* **Note 10**).
6. Wash sections with PBTX buffer for 4 × 10 min at RT.
7. Then block sections using 0.1% PBST-X + 5% bovine serum albumin (BSA) for 1 h at RT (*see* **Note 1**).
8. Incubate sections with primary antibody solutions (with required dilutions) made in 0.1% PBST-X + 5% BSA for overnight at 4 °C (*see* **Note 1**).
9. Wash slides with 0.1% PBST-X + 5% BSA for 3 × 10 min at RT (*see* **Note 1**).
10. Stain the sections with required secondary antibody solutions (with required dilutions) made using 0.1% PBST-X + 5% BSA for 2 h at RT (**Notes 1** and **11**).
11. Wash samples with 0.1% PBST-X at RT for 3 × 10 min.
12. For DNA staining, incubate sections in a DAPI (1 μl DAPI/ml PBS) or Hoechst solution for 15 min at RT.
13. Wash again for 10 min with PBS buffer at RT.
14. Mount sections using either 70% glycerol (Sigma-Aldrich, USA) or Vectashield (Vector Laboratories, USA), cover and seal the coverslips with nail polish. Alternatively, instead of staining the sample(s) using DAPI solutions (mentioned in **step 12**), mount samples using Vectashield with DAPI (Vector Laboratories, USA). Samples are ready to be imaged using any fluorescence microscope.

3.2 Visualizing Directly or Immunostaining of Centrosomes and Cilia in Testes

For dissection:

1. Dissect testes from larvae, pupae and adult males in a drop 1× testis buffer (*see* Subheading 2.2, **item 4**) containing EDTA (2 μl 0.5 mM EDTA in 1 ml testes buffer) for immunostaining. For pupae or adult, hold the fly with the legs facing up and

pull out the male genitalia (dark brown spot present below the abdomen). The testes will come out with it. If they do not, insert the forceps inside abdomen of the fly and pull out the testes. For immunostaining intact testes, ignore **steps 2–5** and fix the samples in 4% formaldehyde as described in **step 6**. The detailed protocols for imaging live sperm cells [32] and dissection of testes of different stages [28, 30, 32] can be found elsewhere.

For immunostaining:

2. Transfer testes to a 4 μl drop of 1× testis buffer placed on a poly-L-lysine slide. Use up to three pairs of testes per slide and two slides per antibody.
3. Randomly squash testes with forceps and cover them with a 18 × 18 mm siliconized coverslip (*see* **Note 12**).
4. Put slides in liquid nitrogen (*see* **Note 13**).
5. Remove coverslips by flipping them off with a scalpel.
6. For fixation: Acetone-methanol fixation: Place slides in dry ice-cold methanol in an upright slide box for 8 min. Then, transfer slides to dry ice-cold acetone in an upright slide box and incubate for 10 min.
7. Formaldehyde fixation: Place slides in 4% formaldehyde in PBS (*see* Subheading 2.3, **item 3**) for 15 min at RT.
8. Make a border around the sections using Dako pen (*see* Subheading 3.1, **step 4**).
9. Wash samples for 3 × 5 min in PBS at RT.
10. Block samples in PBS + 1% BSA for 1 h at RT.
11. Keep slides in a humid chamber pre-wetted with PBS and incubate slides with primary antibody in PBS + 1% BSA overnight at 4 °C (*see* **Notes 1** and **14**).
12. Wash slides for 4 × 10 min each with PBS + 1% BSA.
13. Incubate samples with secondary antibody (with required dilution) in PBS + 1% BSA for 2 h at RT (*see* **Notes 11** and **14**).
14. Wash slides for 2 × 15 min with PBS + 1% BSA and then 1 × 15 min with PBS.
15. For DNA staining, incubate samples in a DAPI (1 μl DAPI/ml PBS) or Hoechst solution for 15 min at RT.
16. Wash samples for 1 × 5 min with PBS. Rinse the slides in water and mount slides in Vectashield. Cover samples with a clean coverslip (non-siliconized) and seal them. Samples are ready to be imaged in any fluorescence microscope.

3.3 Ultrastructure Analysis of Centrosomes and Cilia in Antenna and Testes

1. Dissection: Remove heads and dissect testes from a pupae or adult in PBS as mentioned in Subheading 3.1, **step 2** and Subheading 3.2 **step 1**, respectively.
2. For antenna: First, immerse the heads in fixative (2% formaldehyde, 2.5% glutaraldehyde in 0.1 M sodium bi-phosphate buffer pH 7.2–7.4) for 30 min. Pierce the third antennal segments using a thin tungsten needle (*see* **Note 15**). Finally, fix heads in 2% formaldehyde, 2.5% glutaraldehyde in 0.1 M sodium bi-phosphate buffer (pH 7.2–7.4) overnight at 4 °C with rotation.

   For testes: Fix whole testes in 2.5% glutaraldehyde in PBS at pH 7.2–7.4 for 2 h at 4 °C with rotation.
3. Wash samples for 5 × 5 min in sodium bi-phosphate buffer (pH 7.2–7.4).
4. Post-fix heads and testes in 1% osmium tetroxide (OsO_4), for 1 h 30 min and 1 h on ice, respectively.
5. Wash samples with d-water for 5 × 5 min.
6. Incubate samples in 2% uranyl acetate for 20 min in RT with rotation.
7. Wash samples in d-water for 3 × 10 min.
8. Dehydrate samples in a graded series of alcohol (50, 70, 90, and 100%) for 10 min in each solution. Repeat the dehydration in 100% alcohol for two more times to obtain complete dehydration.
9. Treat samples in propylene oxide for 2 × 15 min.
10. Incubate heads in 1:1 propylene oxide: EPON resins for 3 h with rotation. For testes, incubate samples for 1 h.
11. Incubate heads in EPON resin for overnight at 4 °C and testes for 1 h at RT.
12. Next day, incubate heads in fresh resin (same batch as used in **step 11**) for 2 h at RT with rotation. This step is not required for testes samples.
13. Align samples in the molds with resin and polymerize them overnight in 70 °C oven.
14. Sectioning: Cut semithin sections (500 nm) using a diamond knife on an ultramicrotome and collect sections on a glass slide. Stain sections with TBO to check the gross morphology of the tissues under a compound microscope. After reaching the area of interest, cut serial ultrathin sections (~70 nm) and collect them on a formvar-coated copper grid.
15. Uranyl acetate staining: Stain grids (with sections) by putting them (upside down) on a drop of 2% uranyl acetate in 70% methanol for 3 min at RT (*see* **Note 16**).

16. Rinse sections by passing the grid through two drops of 70% methanol.
17. Wash grids in four drops of d-water and air-dry the grids.
18. Lead citrate staining: In a closed chamber with NaOH pastilles (used to absorb humidity), put the grids (upside down) on a drop of lead citrate for 1 min (*see* **Note 17**).
19. Wash grids in five drops of d-water and air-dry the grids. The sections are ready to be imaged in a transmission electron microscope (TEM).

4 Notes

1. For immunostaining of some epitopes (proteins) in the antenna, a higher amount of the detergent (0.3% Triton X-100) can be required. Although, for immunostaining of most of the proteins in the testes, the detergent is not required, for some epitopes detergent (0.1% Triton X-100) might be required.
2. The Triton-X in the buffer makes the fixative access the cells faster through the cuticle.
3. These two stock solutions should be stored at 4 °C separately. Since both EPON I and EPON II are strongly hydroscopic, mix the stock solutions just before the mixture is used.
4. Since lead citrate reacts with carbon dioxide (CO_2) to form lead carbonate precipitate, the lead citrate solution should be stored in a sealed container.
5. To avoid scattering effects, carefully remove all the bubbles that may adhere to the tissue of interest.
6. Carefully remove all the bubbles adhered to the tissue of interest before freezing the sample.
7. It might be preferable to cut both the eyes as they possess pigmented cells that show strong autofluorescence and they do not harbor any ciliated cells.
8. Heads should sink during the fixation. Occasionally shake the samples with care to help fixing the heads.
9. The incubation in sucrose helps preserving the tissue architecture. Removing the residual sucrose solution from heads helps to embed samples more evenly in the OCT, which should improve sectioning.
10. Avoid any type of shaking during the entire process of immunostaining.
11. As fluorophores tagged to secondary antibodies are light-sensitive, keep the chamber in the dark from now on.

12. Pre-siliconized coverslips should be washed in ethanol and air-dry before using (*see* Subheading 2.3, **item 4**).
13. To avoid tissue damage before fixation, do not keep slides in liquid nitrogen for more than 1 h.
14. A volume of 50–75 μl of antibody solution per slide should be enough.
15. Since olfactory neurons are encapsulated by hard cuticles in the third antennal segment, appropriate piercing of the third segment is very critical. This helps the fixative to properly infiltrate the tissue.
16. Protect the grids from the light to avoid precipitate formation.
17. Use a closed chamber with NaOH to protect the sections from CO_2, avoiding precipitate formation.

Acknowledgement

We apologize to colleagues whose work was not discussed or cited due to space constraints. We thank the IGC imaging unit (Light microscopy and Electron microscopy) for the help with image acquisition, and the IGC fly facility, and MBD Lab for discussions. S.C.J. and S.W. are supported by the FCT (Fundação Portuguesa para a Ciência e Tecnologia) Fellowships SFRH/BPD/87479/2012 and SFRH/BD/52176/2013, respectively. The laboratory and MBD are supported by an EMBO installation grant and an ERC starting grant.

References

1. Bettencourt-Dias M, Hildebrandt F, Pellman D, Woods G, Godinho SA (2011) Centrosomes and cilia in human disease. Trends Genet 27:307–315
2. Jana SC, Marteil G, Bettencourt-Dias M (2014) Mapping molecules to structure: unveiling secrets of centriole and cilia assembly with near-atomic resolution. Curr Opin Cell Biol 26:96–106
3. Kozminski KG, Beech PL, Rosenbaum JL (1995) The Chlamydomonas kinesin-like protein FLA10 is involved in motility associated with the flagellar membrane. J Cell Biol 131:1517–1527
4. Signor D, Wedaman KP, Rose LS, Scholey JM (1999) Two heteromeric kinesin complexes in chemosensory neurons and sensory cilia of Caenorhabditis elegans. Mol Biol Cell 10: 345–360
5. Dubruille R, Laurencon A, Vandaele C, Shishido E, Coulon-Bublex M, Swoboda P, Couble P, Kernan M, Durand B (2002) Drosophila regulatory factor X is necessary for ciliated sensory neuron differentiation. Development 129:5487–5498
6. Amack JD, Yost HJ (2004) The T box transcription factor no tail in ciliated cells controls zebrafish left-right asymmetry. Curr Biol 14:685–690
7. Tonna EA, Lampen NM (1972) Electron microscopy of aging skeletal cells. I. Centrioles and solitary cilia. J Gerontol 27:316–324
8. Postlethwait JH, Schneiderman HA (1969) A clonal analysis of determination in Antennapedia a homoeotic mutant of Drosophila melanogaster. Proc Natl Acad Sci U S A 64:176–183

9. Ursprung H, Conscience-Egli M, Fox DJ, Wallimann T (1972) Origin of leg musculature during Drosophila metamorphosis. Proc Natl Acad Sci U S A 69:2812–2813
10. Horikawa M, Fox AS (1964) Culture of embryonic cells of Drosophila melanogaster in vitro. Science 145:1437–1439
11. Hotta Y, Benzer S (1972) Mapping of behaviour in Drosophila mosaics. Nature 240:527–535
12. Li K, Xu EY, Cecil JK, Turner FR, Megraw TL, Kaufman TC (1998) Drosophila centrosomin protein is required for male meiosis and assembly of the flagellar axoneme. J Cell Biol 141:455–467
13. Callaini G, Riparbelli MG (1990) Centriole and centrosome cycle in the early Drosophila embryo. J Cell Sci 97(Pt 3):539–543
14. Mahowald AP, Strassheim JM (1970) Intercellular migration of centrioles in the germarium of Drosophila melanogaster. An electron microscopic study. J Cell Biol 45:306–320
15. Enjolras C, Thomas J, Chhin B, Cortier E, Duteyrat JL, Soulavie F, Kernan MJ, Laurencon A, Durand B (2012) Drosophila chibby is required for basal body formation and ciliogenesis but not for Wg signaling. J Cell Biol 197:313–325
16. Carvalho-Santos Z, Machado P, Branco P, Tavares-Cadete F, Rodrigues-Martins A, Pereira-Leal JB, Bettencourt-Dias M (2010) Stepwise evolution of the centriole-assembly pathway. J Cell Sci 123:1414–1426
17. Carvalho-Santos Z, Azimzadeh J, Pereira-Leal JB, Bettencourt-Dias M (2011) Evolution: tracing the origins of centrioles, cilia, and flagella. J Cell Biol 194:165–175
18. Avidor-Reiss T, Maer AM, Koundakjian E, Polyanovsky A, Keil T, Subramaniam S, Zuker CS (2004) Decoding cilia function: defining specialized genes required for compartmentalized cilia biogenesis. Cell 117:527–539
19. Basto R, Lau J, Vinogradova T, Gardiol A, Woods CG, Khodjakov A, Raff JW (2006) Flies without centrioles. Cell 125:1375–1386
20. Jana SC, Girotra M, Ray K (2011) Heterotrimeric kinesin-II is necessary and sufficient to promote different stepwise assembly of morphologically distinct bipartite cilia in Drosophila antenna. Mol Biol Cell 22:769–781
21. Han YG, Kwok BH, Kernan MJ (2003) Intraflagellar transport is required in Drosophila to differentiate sensory cilia but not sperm. Curr Biol 13:1679–1686
22. Carvalho-Santos Z, Machado P, Alvarez-Martins I, Gouveia SM, Jana SC, Duarte P, Amado T, Branco P, Freitas MC, Silva ST et al (2012) BLD10/CEP135 is a microtubule-associated protein that controls the formation of the flagellum central microtubule pair. Dev Cell 23:412–424
23. Riparbelli MG, Callaini G, Megraw TL (2012) Assembly and persistence of primary cilia in dividing Drosophila spermatocytes. Dev Cell 23:425–432
24. Briggs LJ, Davidge JA, Wickstead B, Ginger ML, Gull K (2004) More than one way to build a flagellum: comparative genomics of parasitic protozoa. Curr Biol 14:R611–R612
25. Broekhuis JR, Verhey KJ, Jansen G (2014) Regulation of cilium length and intraflagellar transport by the RCK-kinases ICK and MOK in renal epithelial cells. PLoS One 9, e108470
26. Blachon S, Cai X, Roberts KA, Yang K, Polyanovsky A, Church A, Avidor-Reiss T (2009) A proximal centriole-like structure is present in Drosophila spermatids and can serve as a model to study centriole duplication. Genetics 182:133–144
27. Gottardo M, Callaini G, Riparbelli MG (2015) The Drosophila centriole - conversion of doublets into triplets within the stem cell niche. J Cell Sci 128:2437–2442
28. Martins AR, Machado P, Callaini G, Bettencourt-Dias M (2010) Microscopy methods for the study of centriole biogenesis and function in Drosophila. Methods Cell Biol 97:223–242
29. Vieillard J, Duteyrat JL, Cortier E, Durand B (2015) Imaging cilia in Drosophila melanogaster. Methods Cell Biol 127:279–302
30. Basiri ML, Blachon S, Chim YC, Avidor-Reiss T (2013) Imaging centrosomes in fly testes. J Vis Exp 2013:e50938
31. Ma L, Jarman AP (2011) Dilatory is a Drosophila protein related to AZI1 (CEP131) that is located at the ciliary base and required for cilium formation. J Cell Sci 124:2622–2630
32. Noguchi T, Koizumi M, Hayashi S (2011) Sustained elongation of sperm tail promoted by local remodeling of giant mitochondria in Drosophila. Curr Biol 21:805–814
33. Kavlie RG, Albert JT (2013) Chordotonal organs. Curr Biol 23:R334–R335
34. Mishra M (2015) A quick method to investigate the Drosophila Johnston's organ by confocal microscopy. Journal of Microscopy and Ultrastructure 3:1–7
35. Bechstedt S, Albert JT, Kreil DP, Muller-Reichert T, Gopfert MC, Howard J (2010) A doublecortin containing microtubule-associated protein is implicated in mechano-

transduction in Drosophila sensory cilia. Nat Commun 1:11

36. Lee E, Sivan-Loukianova E, Eberl DF, Kernan MJ (2008) An IFT-A protein is required to delimit functionally distinct zones in mechanosensory cilia. Curr Biol 18:1899–1906
37. Eberl DF, Hardy RW, Kernan MJ (2000) Genetically similar transduction mechanisms for touch and hearing in Drosophila. J Neurosci 20:5981–5988
38. Mendes Maia T, Gogendeau D, Pennetier C, Janke C, Basto R (2014) Bug22 influences cilium morphology and the post-translational modification of ciliary microtubules. Biol Open 3:138–151
39. Walker RG, Willingham AT, Zuker CS (2000) A Drosophila mechanosensory transduction channel. Science 287:2229–2234
40. Liang X, Madrid J, Saleh HS, Howard J (2011) NOMPC, a member of the TRP channel family, localizes to the tubular body and distal cilium of Drosophila campaniform and chordotonal receptor cells. Cytoskeleton (Hoboken) 68:1–7
41. Newton FG, zur Lage PI, Karak S, Moore DJ, Gopfert MC, Jarman AP (2012) Forkhead transcription factor Fd3F cooperates with Rfx to regulate a gene expression program for mechanosensory cilia specialization. Dev Cell 22:1221–1233
42. Gong Z, Son W, Chung YD, Kim J, Shin DW, McClung CA, Lee Y, Lee HW, Chang DJ, Kaang BK et al (2004) Two interdependent TRPV channel subunits, inactive and Nanchung, mediate hearing in Drosophila. J Neurosci 24:9059–9066
43. Benton R, Sachse S, Michnick SW, Vosshall LB (2006) Atypical membrane topology and heteromeric function of Drosophila odorant receptors in vivo. PLoS Biol 4, e20
44. Baker JD, Adhikarakunnathu S, Kernan MJ (2004) Mechanosensory-defective, male-sterile unc mutants identify a novel basal body protein required for ciliogenesis in Drosophila. Development 131:3411–3422
45. Basiri ML, Ha A, Chadha A, Clark NM, Polyanovsky A, Cook B, Avidor-Reiss T (2014) A migrating ciliary gate compartmentalizes the site of axoneme assembly in Drosophila spermatids. Curr Biol 24:2622–2631
46. Mottier-Pavie V, Megraw TL (2009) Drosophila bld10 is a centriolar protein that regulates centriole, basal body, and motile cilium assembly. Mol Biol Cell 20:2605–2614
47. Rodrigues-Martins A, Bettencourt-Dias M, Riparbelli M, Ferreira C, Ferreira I, Callaini G, Glover DM (2007) DSAS-6 organizes a tube-like centriole precursor, and its absence suggests modularity in centriole assembly. Curr Biol 17:1465–1472
48. Dzhindzhev NS, Yu QD, Weiskopf K, Tzolovsky G, Cunha-Ferreira I, Riparbelli M, Rodrigues-Martins A, Bettencourt-Dias M, Callaini G, Glover DM (2010) Asterless is a scaffold for the onset of centriole assembly. Nature 467:714–718
49. Galletta BJ, Guillen RX, Fagerstrom CJ, Brownlee CW, Lerit DA, Megraw TL, Rogers GC, Rusan NM (2014) Drosophila pericentrin requires interaction with calmodulin for its function at centrosomes and neuronal basal bodies but not at sperm basal bodies. Mol Biol Cell 25:2682–2694
50. Conduit PT, Raff JW (2015) Different Drosophila cell types exhibit differences in mitotic centrosome assembly dynamics. Curr Biol 25:R650–R651
51. Laurencon A, Dubruille R, Efimenko E, Grenier G, Bissett R, Cortier E, Rolland V, Swoboda P, Durand B (2007) Identification of novel regulatory factor X (RFX) target genes by comparative genomics in Drosophila species. Genome Biol 8:R195

Chapter 15

Analysis of Axonemal Assembly During Ciliary Regeneration in *Chlamydomonas*

Emily L. Hunter, Winfield S. Sale, and Lea M. Alford

Abstract

Chlamydomonas reinhardtii is an outstanding model genetic organism for study of assembly of cilia. Here, methods are described for synchronization of ciliary regeneration in *Chlamydomonas* to analyze the sequence in which ciliary proteins assemble. In addition, the methods described allow analysis of the mechanisms involved in regulation of ciliary length, the proteins required for ciliary assembly, and the temporal expression of genes encoding ciliary proteins. Ultimately, these methods can contribute to discovery of conserved genes that when defective lead to abnormal ciliary assembly and human disease.

Key words *Chlamydomonas*, Ciliary regeneration, pH shock, Deciliation, Axonemal assembly

1 Introduction

Cilia are found on nearly every differentiated cell where they play essential motile and signaling roles [1, 2]. Studies over the last 15 years have linked defects in genes required for assembly of the ciliary axoneme or ciliary membranes with a wide range of human developmental disorders and diseases in adults [3–5]. However, we are just beginning to understand the genes and the mechanisms responsible for ciliary assembly [6–9].

Analysis of mutations in *Chlamydomonas* has revealed many conserved genes required for cilia formation and control of cilia length [10–13]. For example, the genes that encode the intraflagellar transport (IFT) proteins [14], ciliary transition zone proteins [15–17], and proteins required for assembly of ciliary dynein motors [18–22] were discovered in *Chlamydomonas*. In addition, ciliary regeneration following ciliary excision in *Chlamydomonas* provides a powerful experimental model for defining mechanisms of ciliary assembly [10, 23, 24] and control of ciliary length [8, 25–30]. Ciliary assembly during regeneration occurs distal to the transition zone, the site of ciliary excision [31–34]. Thus, these methods preclude assembly analysis of structures associated with

Peter Satir and Søren Tvorup Christensen (eds.), *Cilia: Methods and Protocols*, Methods in Molecular Biology, vol. 1454,
DOI 10.1007/978-1-4939-3789-9_15, © Springer Science+Business Media New York 2016

the basal body and transition zone (*see* [35]). Rather, ciliary regeneration in *Chlamydomonas* provides an experimental paradigm for analysis of ciliary gene regulation and protein synthesis [36], localization of protein assembly in the cilium [37, 38], regulation of IFT [38–40], the regulation of ciliary length [30], and definition of proteins located at the ciliary tip [41]. In addition, regeneration can provide understanding of the sequence of axonemal protein assembly [23, 42].

The focus of this chapter is on the temporal analysis of ciliary protein assembly utilizing *Chlamydomonas.* The methods developed and described here allow for biochemical analysis of the cilium at precise lengths during regeneration. Ciliary detachment is induced by pH shock [33, 43] and the deciliated cells are kept on ice to control the start time of regeneration. Synchronous regeneration is stimulated by bringing the culture to room temperature for collection of similar length cilia at particular time points. Cilia of designated lengths can then be isolated and fractionated into axoneme and membrane-matrix fractions.

2 Materials

1. *Chlamydomonas* strains obtained from the *Chlamydomonas* Resource Center (http://chlamycollection.org, University of Minnesota). Wild-type strains CC-124 and CC-125 are described here for study of ciliary regeneration, however numerous *Chlamydomonas* mutants are available through the *Chlamydomonas* Resource Center. For example, mutants defective in assembly of the cilium can be analyzed using the following regeneration protocol.
2. Cell culture grown in 300 mL L-media ([44] plus 12.2 mM sodium acetate, *see* **Notes 1** and **2**). Liquid cultures grown at room temperature (20–25 °C) and constantly aerated on a light:dark cycle (14:10 h.). On day 3, the culture is medium green, inoculate 3 L L-media with the 300 mL culture.
3. 1 L cold (4 °C) deciliation buffer: 10 mM Tris–HCl, pH 7.5, 5% sucrose, 1 mM $CaCl_2$.
4. 1 L cold (4 °C) L-media.
5. Sorvall RC6+ floor model high-speed centrifuge with SLA-1500 and SS-34 rotors and corresponding tubes.
6. Chilled centrifuge bottles (*see* **Note 3**).
7. 1 L room temperature (20–25 °C) L-media.
8. 10 mL 0.5 M Glacial acetic acid.
9. 10 mL 0.5 M KOH.
10. Thermometer.

11. Aeration mechanism or stir bar.
12. Light microscope equipped with phase contrast and 100× total magnification.

3 Methods

Prior to concentrating and harvesting cells, check the culture for cell health and remove a 300 mL aliquot as the "pre-deciliation" sample for comparison to later time points (*see* **Note 2**).

3.1 Initial Deciliation

1. Concentrate and harvest cells gently at 2500 rpm for 5 min at 4 °C.
2. Resuspend cell pellets in a total of 200 mL cold deciliation buffer.
3. Transfer cells to a beaker on ice containing a stir bar (*see* **Note 3**).
4. While gently stirring cells, record starting pH.
5. Add 0.5 M acetic acid in 100 μl increments until the pH drops to 4.5.
6. Check cells for deciliation.
7. Neutralize the cells with an equal volume of 0.5 M KOH.
8. Harvest cells in pre-chilled bottles at 3000 rpm for 5 min at 4 °C (*see* **Note 3**).

3.2 Synchronous Regeneration

1. On ice, resuspend cells with ~150 mL cold L-media and harvest at 3000 rpm for 5 min at 4 °C to wash free cilia from the cell pellet (*see* **Note 4**).
2. On ice, resuspend cell pellets in 100 mL of cold L-media and transfer to a room temperature beaker (*see* **Note 5**).
3. Move cells in beaker to constant light, aerate or stir, and monitor temperature; the temperature will be about 10 °C at this point.
4. While aerating or stirring, add ~1 L room temperature L-media; the temperature will progressively increase to 18–20 °C and regeneration of cilia will begin.

3.3 Collection of Time Points

1. At desired time point(s) (Fig. 1), harvest ~300 mL cells in pre-chilled centrifuge bottles at 3000 rpm for 5 min at 4 °C (*see* **Note 6**).
2. Resuspend cell pellets in cold deciliation buffer and deciliate as in Subheading 3.1 (*see* **Note 7**).
3. Harvest cells in pre-chilled bottles at 5000 rpm for 5 min at 4 °C.
4. Transfer supernatant, containing regenerated cilia, to clean chilled tubes for further purification of axonemal and membrane plus matrix fractions (*see* **Note 8** [45]).

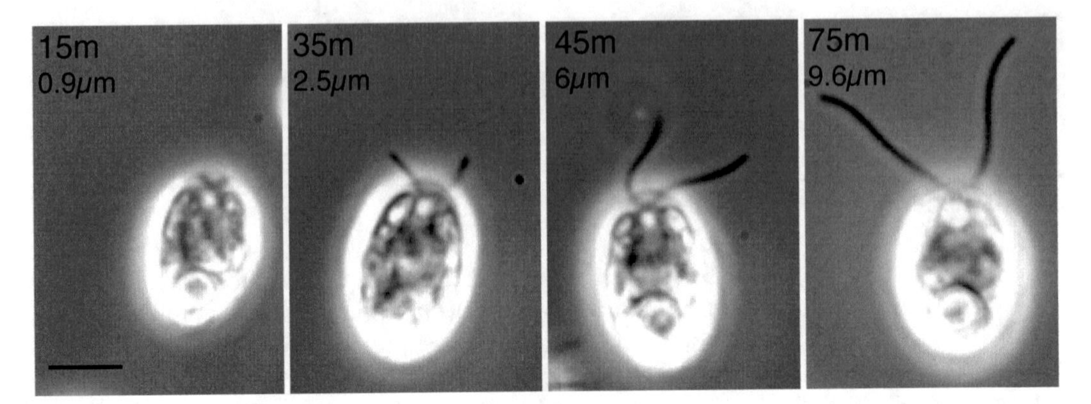

Fig. 1 Ciliary regeneration in *Chlamydomonas*. Phase-contrast images of *Chlamydomonas* cells approximately 15, 35, 45, and 75 min post-deciliation by pH shock. The average lengths between the two cilia are 0.9, 2.5, 6, and 9.6 μm, respectively. Scale bar = 5 μm. Cells shown are representative of a synchronously regenerating culture

4 Notes

1. The choice of L media is based on ease of deciliation when cells are cultured in this medium. Tris-acetate-phosphate (TAP) media is Tris-buffered and therefore will not allow for effective deciliation [46]. The pH of cells in deciliation buffer cultured in TAP media will drop upon addition of acetic acid (Subheading 3.1, **step 5**), but the cells will not shed their cilia. Sager and Granick's Medium I [47] or Minimal (M) media [43] are likely compatible with this protocol, though not tested here.
2. An early- to mid-log phase (light-medium green in color) culture is ideal for this protocol. We inoculate 2 × 300 mL cultures from <1-week-old cells on plates, dilute into 2 × 1.5 L. Healthy non-contaminated cells deciliate very efficiently by pH shock and remain live and intact upon neutralization. Thus, healthy cells will regenerate their cilia nicely and ciliary fractions collected will contain minimal cell debris.
3. The key to synchronous regeneration of *Chlamydomonas* is to keep cells ice cold (4 °C) during and after initial deciliation (Subheadings 3.1 and 3.2). Fill several buckets with ice, pre-chill centrifuge bottles on ice, keep deciliation buffer and bottles of L-media at 4 °C. For deciliation, we place the beaker of cells in cold deciliation buffer in a shallow Tupperware container filled with ice (Subheading 3.1, **step 3**). It is not necessary to work in a cold room if materials are pre-chilled.
4. After initial deciliation (Subheading 3.1), cell pellets contain some amount of free cilia. By washing the cell pellets with cold L-media, the cilia are further removed thus ensuring that subsequent samples of regenerating cilia collected do not contain full-length steady-state cilia.

5. If the deciliated cells resuspended in cold L-media are transferred to a cold 1 L beaker, addition of ~1 L room temperature media will not bring the culture temperature up quickly enough to 18–20 °C. Cells will thus not regenerate their cilia.
6. T_0 is the time at which the cold deciliated cells are brought to 18–20 °C. Within 10 min, short (~0.5 μm) cilia will be noticeable. Cilia grow to full-length within 90 min.
7. If sample purity (free of cell wall material and debris) is not a concern, deciliation by the dibucaine method [43] is an option.
8. Typical yield from a 3 L culture is 100–200 μg axonemes per time point.

Acknowledgements

This work was supported by grants from the NIH (GM051173, WSS; Training Grant K12 GM000680, LMA; Training Grant 5T32 GM00836725, ELH) and the American Heart Association (14PRE19510013, ELH).

References

1. Drummond IA (2012) Cilia functions in development. Curr Opin Cell Biol 24(1):24–30. doi:10.1016/j.ceb.2011.12.007, S0955-0674(11)00169-4 [pii]
2. Satir P, Heuser T, Sale WS (2014) A structural basis for how motile cilia beat. Bioscience 64(12):1073–1083. doi:10.1093/Biosci/Biu180
3. Brown JM, Witman GB (2014) Cilia and diseases. Bioscience 64(12):1126–1137. doi:10.1093/Biosci/Biu174
4. Oh EC, Katsanis N (2012) Cilia in vertebrate development and disease. Development 139(3):443–448. doi:10.1242/dev.050054
5. Hildebrandt F, Benzing T, Katsanis N (2011) Ciliopathies. N Engl J Med 364(16):1533–1543. doi:10.1056/NEJMra1010172
6. Garcia-Gonzalo FR, Reiter JF (2012) Scoring a backstage pass: mechanisms of ciliogenesis and ciliary access. J Cell Biol 197(6):697–709. doi:10.1083/Jcb.201111146
7. Kim S, Dynlacht BD (2013) Assembling a primary cilium. Curr Opin Cell Biol 25(4):506–511. doi:10.1016/J.Ceb.2013.04.011
8. Avasthi P, Marshall WF (2012) Stages of ciliogenesis and regulation of ciliary length. Differentiation 83(2):30–42. doi:10.1016/J.Diff.2011.11.015
9. Ishikawa H, Marshall WF (2011) Ciliogenesis: building the cell's antenna. Nat Rev Mol Cell Biol 12(4):222–234. doi:10.1038/nrm3085
10. Silflow CD, Lefebvre PA (2001) Assembly and motility of eukaryotic cilia and flagella. Lessons from *Chlamydomonas reinhardtii*. Plant Physiol 127(4):1500–1507
11. Dutcher SK (2014) The awesome power of dikaryons for studying flagella and basal bodies in Chlamydomonas reinhardtii. Cytoskeleton (Hoboken) 71(2):79–94. doi:10.1002/cm.21157
12. Lin H, Dutcher SK (2015) Genetic and genomic approaches to identify genes involved in flagellar assembly in Chlamydomonas reinhardtii. Methods Cell Biol 127:349–386. doi:10.1016/bs.mcb.2014.12.001, S0091-679X(14)00046-6 [pii]
13. Stolc V, Samanta MP, Tongprasit W, Marshall WF (2005) Genome-wide transcriptional analysis of flagellar regeneration in Chlamydomonas reinhardtii identifies orthologs of ciliary disease genes. Proc Natl Acad Sci U S A 102(10):3703–3707

14. Pedersen LB, Rosenbaum JL (2008) Intraflagellar transport (IFT) role in ciliary assembly, resorption and signalling. Curr Top Dev Biol 85:23–61. doi:10.1016/S0070-2153(08)00802-8, S0070-2153(08)00802-8 [pii]
15. Awata J, Takada S, Standley C, Lechtreck KF, Bellve KD, Pazour GJ, Fogarty KE, Witman GB (2014) NPHP4 controls ciliary trafficking of membrane proteins and large soluble proteins at the transition zone. J Cell Sci 127(21):4714–4727. doi:10.1242/Jcs.155275
16. Reiter JF, Blacque OE, Leroux MR (2012) The base of the cilium: roles for transition fibres and the transition zone in ciliary formation, maintenance and compartmentalization. EMBO Rep 13(7):608–618. doi:10.1038/embor.2012.73, embor201273 [pii]
17. Craige B, Tsao CC, Diener DR, Hou Y, Lechtreck KF, Rosenbaum JL, Witman GB (2010) CEP290 tethers flagellar transition zone microtubules to the membrane and regulates flagellar protein content. J Cell Biol 190(5):927–940. doi:10.1083/jcb.201006105, jcb.201006105 [pii]
18. Kamiya R, Yagi T (2014) Functional diversity of axonemal dyneins as assessed by in vitro and in vivo motility assays of Chlamydomonas mutants. Zool Sci 31(10):633–644. doi:10.2108/Zs140066
19. Kobayashi D, Takeda H (2012) Ciliary motility: the components and cytoplasmic preassembly mechanisms of the axonemal dyneins. Differentiation 83(2):S23–S29. doi:10.1016/j.diff.2011.11.009, S0301-4681(11)00199-X [pii]
20. Viswanadha R, Hunter EL, Yamamoto R, Wirschell M, Alford LM, Dutcher SK, Sale WS (2014) The ciliary inner dynein arm, I1 Dynein, is assembled in the cytoplasm and transported by IFT before axonemal docking. Cytoskeleton 71(10):573–586. doi:10.1002/Cm.21192
21. Ahmed NT, Gao C, Lucker BF, Cole DG, Mitchell DR (2008) ODA16 aids axonemal outer row dynein assembly through an interaction with the intraflagellar transport machinery. J Cell Biol 183(2):313–322
22. Desai PB, Freshour JR, Mitchell DR (2015) Chlamydomonas axonemal dynein assembly locus ODA8 encodes a conserved flagellar protein needed for cytoplasmic maturation of outer dynein arm complexes. Cytoskeleton 72(1):16–28. doi:10.1002/Cm.21206
23. Remillard SP, Witman GB (1982) Synthesis, transport, and utilization of specific flagellar proteins during flagellar regeneration in Chlamydomonas. J Cell Biol 93(3):615–631
24. Lefebvre PA (1995) Flagellar amputation and regeneration in Chlamydomonas. Methods Cell Biol 47:3–7
25. Engel BD, Ludington WB, Marshall WF (2009) Intraflagellar transport particle size scales inversely with flagellar length: revisiting the balance-point length control model. J Cell Biol 187(1):81–89. doi:10.1083/jcb.200812084, jcb.200812084 [pii]
26. Engel BD, Ishikawa H, Wemmer KA, Geimer S, Wakabayashi K, Hirono M, Craige B, Pazour GJ, Witman GB, Kamiya R, Marshall WF (2012) The role of retrograde intraflagellar transport in flagellar assembly, maintenance, and function. J Cell Biol 199(1):151–167. doi:10.1083/jcb.201206068, jcb.201206068 [pii]
27. Dentler W (2013) A role for the membrane in regulating Chlamydomonas flagellar length. PLoS One 8(1), e53366. doi:10.1371/journal.pone.0053366, PONE-D-12-27278 [pii]
28. Tam LW, Wilson NF, Lefebvre PA (2007) A CDK-related kinase regulates the length and assembly of flagella in Chlamydomonas. J Cell Biol 176(6):819–829. doi:10.1083/jcb.200610022, jcb.200610022 [pii]
29. Hilton LK, Gunawardane K, Kim JW, Schwarz MC, Quarmby LM (2013) The kinases LF4 and CNK2 control ciliary length by feedback regulation of assembly and disassembly rates. Curr Biol 23(22):2208–2214. doi:10.1016/J.Cub.2013.09.038
30. Lefebvre PA (2009) Flagellar Length Control. In: Witman GB (ed) The Chlamydomonas sourcebook, vol 3, 2nd edn. Academic, Amsterdam, pp 115–129. doi:10.1016/B978-0-12-370873-1.00042-3
31. Quarmby LM (2004) Cellular deflagellation. Int Rev Cytol 233:47–91. doi:10.1016/S0074-7696(04)33002-0, S0074769604330020 [pii]
32. Parker JD, Hilton LK, Diener DR, Rasi MQ, Mahjoub MR, Rosenbaum JL, Quarmby LM (2010) Centrioles are freed from cilia by severing prior to mitosis. Cytoskeleton (Hoboken) 67(7):425–430. doi:10.1002/cm.20454
33. Quarmby L (2009) Deflagellation. In: Witman GB (ed) The Chlamydomonas sourcebook, vol 3. Academic, Amsterdam, pp 43–69. doi:10.1016/B978-0-12-370873-1.00040-X
34. Satir B, Sale WS, Satir P (1976) Membrane renewal after dibucaine deciliation of Tetrahymena. Freeze-fracture technique, cilia, membrane structure. Exp Cell Res 97:83–91
35. Diener DR, Lupetti P, Rosenbaum JL (2015) Proteomic analysis of isolated ciliary transition zones reveals the presence of ESCRT proteins. Curr Biol 25(3):379–384. doi:10.1016/J.Cub.2014.11.066

36. Silflow CD, Rosenbaum JL (1981) Multiple alpha- and beta-tubulin genes in *Chlamydomonas* and regulation of tubulin mRNA levels after deflagellation. Cell 24(1):81–88
37. Johnson KA, Rosenbaum JL (1992) Polarity of flagellar assembly in Chlamydomonas. J Cell Biol 119(6):1605–1611
38. Craft JM, Harris JA, Hyman S, Kner P, Lechtreck KF (2015) Tubulin transport by IFT is upregulated during ciliary growth by a cilium-autonomous mechanism. J Cell Biol 208(2):223–237. doi:10.1083/Jcb.201409036
39. Liang YW, Pang YN, Wu Q, Hu ZF, Han X, Xu YS, Deng HT, Pan JM (2014) FLA8/KIF3B phosphorylation regulates kinesin-II interaction with IFT-B to control IFT entry and turnaround. Dev Cell 30(5):585–597. doi:10.1016/J.Devcel.2014.07.019
40. Avasthi P, Onishi M, Karpiak J, Yamamoto R, Mackinder L, Jonikas MC, Sale WS, Shoichet B, Pringle JR, Marshall WF (2014) Actin is required for IFT regulation in Chlamydomonas reinhardtii. Curr Biol 24(17):2025–2032. doi:10.1016/J.Cub.2014.07.038
41. Tammana TVS, Tammana D, Diener DR, Rosenbaum J (2013) Centrosomal protein CEP104 (Chlamydomonas FAP256) moves to the ciliary tip during ciliary assembly. J Cell Sci 126(21):5018–5029. doi:10.1242/Jcs.133439
42. Alford LM, Mattheyses AL, Hunter EL, Lin H, Dutcher SK, Sale WS (2013) The Chlamydomonas mutant pf27 reveals novel features of ciliary radial spoke assembly. Cytoskeleton (Hoboken) 70(12):804–818. doi:10.1002/cm.21144
43. Witman GB (1986) Isolation of Chlamydomonas flagella and flagellar axonemes. Methods Enzymol 134:280–290
44. Huang B, Rifkin MR, Luck DJ (1977) Temperature-sensitive mutations affecting flagellar assembly and function in Chlamydomonas reinhardtii. J Cell Biol 72(1):67–85
45. Pazour GJ, Agrin N, Leszyk J, Witman GB (2005) Proteomic analysis of a eukaryotic cilium. J Cell Biol 170(1):103–113
46. Gorman DS, Levine RP (1965) Cytochrome f and plastocyanin: their sequence in the photosynthetic electron transport chain of *Chlamydomonas reinhardi*. Proc Natl Acad Sci U S A 54(6):1665–1669
47. Sager R, Granick S (1953) Nutritional studies with *Chlamydomonas reinhardtii*. Ann N Y Acad Sci 466:18–30

Chapter 16

Planaria as a Model System for the Analysis of Ciliary Assembly and Motility

Stephen M. King and Ramila S. Patel-King

Abstract

Planarian flatworms are carnivorous invertebrates with astounding regenerative properties. They have a ventral surface on which thousands of motile cilia are exposed to the extracellular environment. These beat in a synchronized manner against secreted mucus thereby propelling the animal forward. Similar to the nematode *Caenorhabditis elegans*, the planarian *Schmidtea mediterranea* is easy to maintain in the laboratory and is highly amenable to simple RNAi approaches through feeding with dsRNA. The methods are simple and robust, and the level of gene expression reduction that can be obtained is, in many cases, almost total. Moreover, cilia assembly and function is not essential for viability in this organism, as animals readily survive for weeks even with the apparent total absence of this organelle. Both genome and expressed sequence tag databases are available and allow design of vectors to target any desired gene of choice. Combined, these feature make planaria a useful model system in which to examine ciliary assembly and motility, especially in the context of a ciliated epithelium where many organelles beat in a hydrodynamically coupled synchronized manner. In addition, as planaria secrete mucus against which the cilia beat to generate propulsive force, this system may also prove useful for analysis of mucociliary interactions. In this chapter, we provide simple methods to maintain a planarian colony, knockdown gene expression by RNAi, and analyze the resulting animals for whole organism motility as well as ciliary architecture and function.

Key words Axoneme, Cilia, Dynein, Microtubule, Planaria

1 Introduction

As the cilium (flagellum) has been extremely highly conserved throughout evolution [1], model organisms have played an immensely important role in the many studies that have lead to our current understanding of the mechanisms by which assembly and motility of this organelle is achieved. Each different model system has certain advantages that allow one to employ specific experimental approaches [2]. For example, the haploid unicellular

Peter Satir and Søren Tvorup Christensen (eds.), *Cilia: Methods and Protocols*, Methods in Molecular Biology, vol. 1454,
DOI 10.1007/978-1-4939-3789-9_16, © Springer Science+Business Media New York 2016

green alga *Chlamydomonas reinhardtii*, which has two flagella, has been widely used as this organism is amenable to both classical and molecular genetic approaches and because it can be grown cheaply in large quantity for biochemical studies [3]. However, RNAi-based and gene targeting methods in this organism have proven more difficult and as a biflagellate it cannot be used to study synchronous beating (so-called metachronal synchrony) when hundreds or thousands of cilia lining a surface are active simultaneously. This is a key feature of ciliary motility that is of the upmost importance in situations where a fluid needs to be moved across an epithelium, as for example occurs with mucus transport in the airways [4] and the movement of cerebrospinal fluid through the brain ventricles [5].

Planarian flatworms are members of the Turbellaria, one of four classes within the Platyhelminthes. They are free-living carnivorous scavengers and dwell in aquatic environments. Various planarian species have been studied for many years due to their extraordinary regenerative properties; remarkably, when cut into multiple pieces, each section will undergo complete reorganization of the original body plan resulting in multiple fully formed animals. As approximately 40% of cells in a planarian are neoblasts (i.e. stem cells), these animals provide an exciting model to understand stem cell biology and differentiation [6]. The normal mode of planarian locomotion is a smooth gliding driven by the action of cilia on the ventral surface beating against secreted mucus. Planaria also are able to employ their body musculature to essentially squeeze themselves across a surface using peristaltic waves of contraction.

In addition to their importance in stem cell research, in recent years, it has become apparent that planarian flatworms (*Schmitdea mediterranea* is the species most commonly used [7]) also provide a useful experimental model for analysis of ciliary biology [8]. For example, they have been used to help understand centriole architecture and evolution [9], to assess how cilia are involved in hedgehog [10] and Wnt signaling [11], to examine how ciliary metachronal synchrony is achieved [12], and have clear potential for the investigation of ciliary–mucus interactions. Advantages of this organism include a completely exposed ciliated epithelium on the ventral surface, robust and simple methods for RNAi-mediated knockdown of genes of interest [13], the availability of both EST and genome databases, and the ease of maintaining and propagating the animal in the laboratory. Disruption of planarian ciliary assembly or motility routinely results in clear robust phenotypes with mRNA levels in many cases reduced to almost undetectable levels. Importantly, we have found that experimentally induced knockdown of genes essential for ciliogenesis or ciliary motility does not compromise the overall viability of this

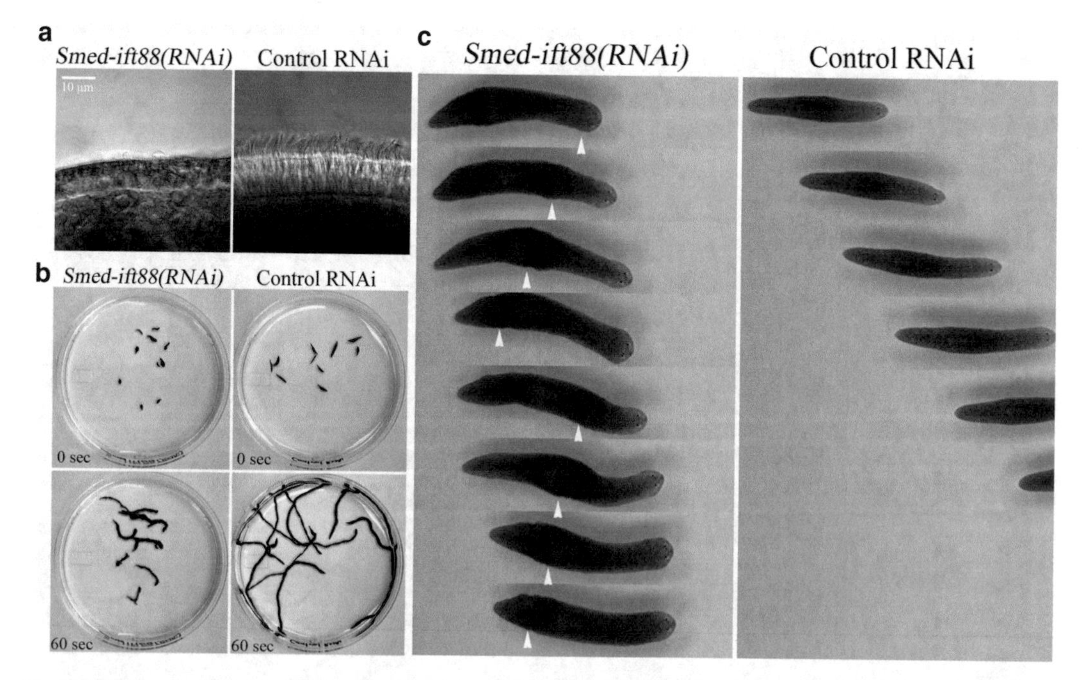

Fig. 1 Motility of planaria with normal and defective cilia. (**a**) Differential interference contrast images of the dorsal/ventral margin of control planaria and those defective in intraflagellar transport due to the knockdown of IFT88. The RNAi animals have essentially no cilia. (**b**) The *upper panels* show the first frame from videos of control and *Smed-ift88(RNAi)* animals. The *lower panels* show the time progression and were prepared by overlaying frames from the first 60 s to illustrate the paths taken by each animal. Note that the RNAi animals move much more slowly (0.46 mm/s versus 1.47 mm/s). (**c**) Video frames showing the movement of control and *Smed-ift88(RNAi)* planaria. The controls move with a rapid gliding motion, whereas the experimental animals use waves of muscle contraction (*white arrowheads*) to squeeze themselves across the surface. This figure was adapted and modified from [12]

organism (*see* Fig. 1 where *Smed-ift88(RNAi)* animals that have compromised intraflagellar transport are shown to be completely aciliate; *see* **Note 1**).

In our laboratory, we routinely use RNAi methods in *S. mediterranea* to rapidly test (within 1 month or less) whether a novel protein of interest is important for the assembly and/or motility of cilia. Based on the results, one can then decide whether detailed waveform analysis in the knockdown animals or biochemical analysis of the orthologous protein in another model system (such as *C. reinhardtii*) is warranted. In this article, we detail methods that allow for the propagation of planaria, and RNAi-mediated knockdown of genes of interest that when combined with light (Fig. 2) and electron microscopic (Fig. 3) imaging approaches enable one to characterize the resulting phenotypes.

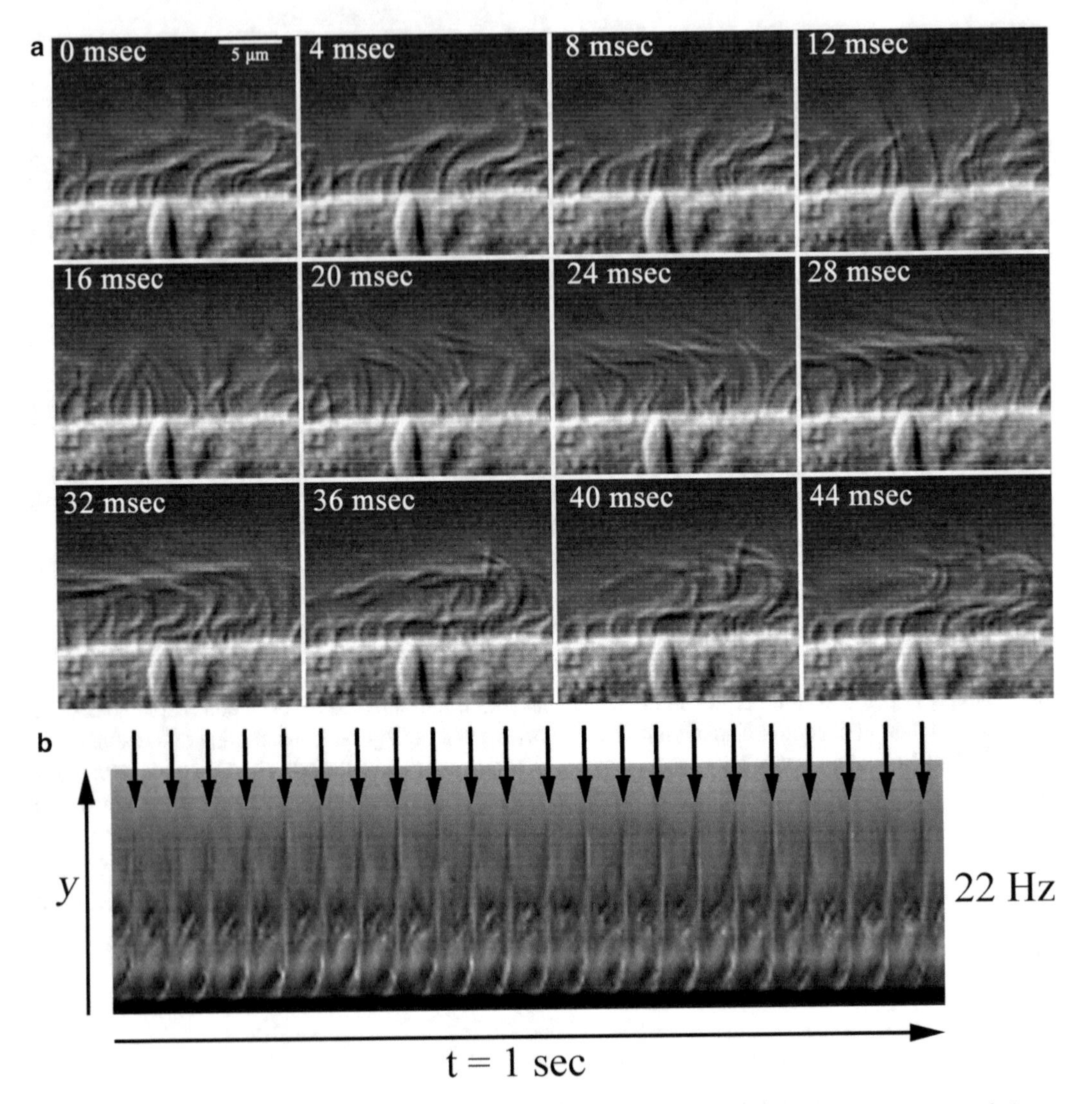

Fig. 2 Analysis of ciliary waveform and beat frequency. (**a**) Differential interference contrast images of planarian cilia during a single beat cycle taken from a video shot at 250 frames/s. The *panels* shown are 4 ms apart. (**b**) Kymograph generated from 1 s of decompiled video. This method allows the beat frequency (in this case 24 Hz) to be measured directly. This figure was adapted and modified from [12]

2 Materials

2.1 Propagating Planaria

1. There is currently no commercial source of *S. mediterranea*. A founding colony of animals can be obtained from one of the laboratories currently using this organism. There are two strains of this planarian: one is hermaphroditic and lays eggs that take approximately 3 weeks to hatch; the second is asexual, due to a chromosomal translocation, and naturally reproduces by fission. In our laboratory, we routinely use the sexual strain.

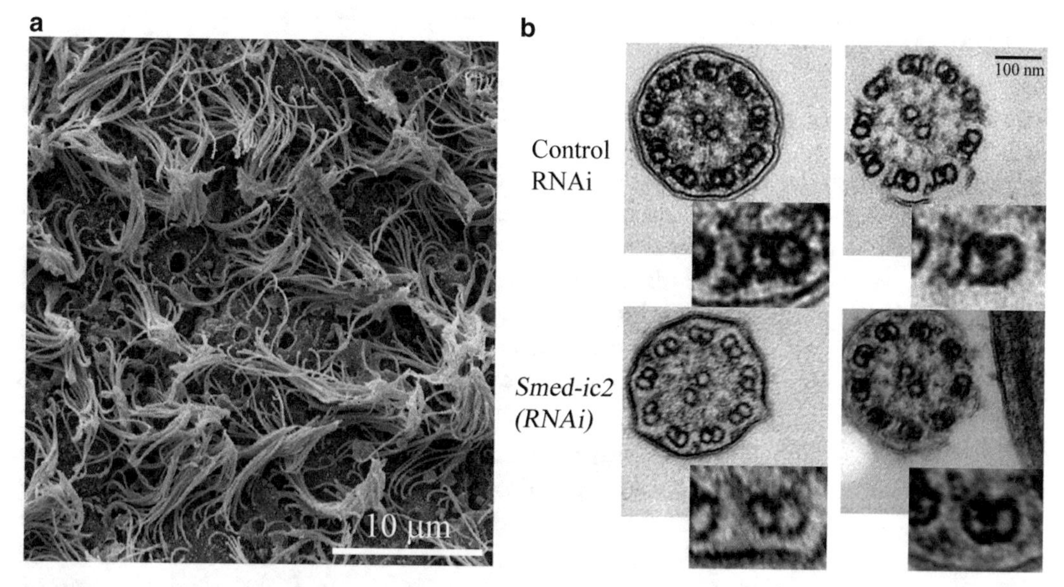

Fig. 3 Electron microscopic analysis of planarian cilia. (**a**) Scanning electron micrograph of the ventral surface of a control planarian illustrating the many closely apposed cilia that are approximately 10 μm long. The pores visible on the surface are used to secrete mucus. The bar = 10 μm. (**b**) Thin section transmission electron micrographs of control cilia and those from *Smed-ic2(RNAi)* animals that are unable to assemble outer dynein arms. The bar = 100 nm. This figure was adapted and modified from [12, 15]

2. 1× Montjuïch Salt solution: 1.6 mM NaCl, 1 mM $CaCl_2$, 1 mM $MgSO_4$, 0.1 mM $MgCl_2$, 0.1 mM KCl, 1.2 mM $NaHCO_3$. This solution can be made as a 5× concentrate and stored at 4 °C for up to a year or more. Occasionally a precipitate will form but this does not appear to alter its suitability for planarian maintenance.
3. High-quality calves liver, minced following removal of the connective tissue. This can be stored in aliquots at −80 °C for over a year.
4. Standard plastic food containers.
5. Dark incubator or other isolated location with constant temperature of ~22 °C.

2.2 RNA$_i$ Knockdowns

1. A ~300 bp segment of DNA that is specific to the cDNA of interest.
2. Plasmid L4440 which contains opposing T7 promoters.
3. Escherichia coli strain HT115(DE3).

2.3 Electron Microscopy

1. Relaxant Solution: 1% HNO_3, 0.85% formaldehyde, 50 mM $MgSO_4$ [14].
2. Phosphate-buffered saline pH 7.2.
3. Glutaraldehyde, osmium tetroxide, sodium cacodylate, ethanol, Epon resin, uranyl acetate.

3 Methods

3.1 Maintaining Planaria in the Laboratory

1. Keep planaria under water in 1× Montjuïch salt solution in the dark. For routine feeding approximately once per week, add a bolus of minced liver and leave for 1–2 h to allow the animals to feed freely. If the liver starts to float away, remove some of the culture liquid. At the end of the feeding session, remove and discard the remaining liver. Then carefully remove the culture solution and replace with fresh culture solution after wiping the container clean with a tissue. This is easiest done if all the planaria are collected in one corner of the dish by gently squirting them with fluid from a transfer pipette.
2. If you wish to monitor whether all the animals actually feed, then add a few drops of red food color to the minced liver. Planaria that have ingested the dye will exhibit a distinct reddish tinge.
3. To increase the number of animals in the colony, simply cut large planaria into several pieces using a sharp razor blade or scalpel. Leave the pieces in a separate container for 2–3 weeks. During this time, the flatworms will reorganize their body plan and each piece will regenerate to yield a complete animal. Do not feed during regeneration, and change the medium more often to remove any debris that may have resulted.

3.2 Generating RNAi Knockdowns

1. Obtain a region of ~300 bp or more that is specific to the cDNA of interest by RT-PCR from isolated planarian mRNA (*see* **Note 2**). Include XbaI and XhoI sites inside a GCGC clamp on the forward and reverse primers, respectively.
2. Subclone into the plasmid L4440 which contains a multiple cloning site flanked by opposing T7 promoters. When transformed into *E. coli* strain HT115(DE3) that lacks RNase III and induced with IPTG, this plasmid allows for the generation of large quantities of dsRNA.
3. Grow the *E. coli* HT115(DE3) strain transformed with either control or experimental vectors in 2× YT medium. When the culture reaches an $OD_{600} = 0.4–0.6$ add 1 mM isopropylthiogalactoside (IPTG) and induce for 2–3 h. Pellet the cells by centrifugation and mix with liver homogenate that has been diluted 1:1 with water. A pellet from 2 ml of bacterial culture is sufficient for 50 μl of diluted liver and can feed approximately ten animals. It is highly recommended to add food color at this step as this allows one to ensure that all animals in the control and experimental groups have ingested the dsRNA.
4. Repeat the feeding every 3–4 days for at least four feedings and monitor the effects on a variety of mRNAs including the target by RT-PCR. In our laboratory, we always employ three

groups of flatworms kept in separate containers: (a) untreated animals, (b) animals fed with an empty vector control, and (c) the experimental group. In some cases where protein turnover is slow, additional feeding for up to 3–4 weeks is often required.

5. The final degree of mRNA knockdown compared to control mRNAs can be readily evaluated by standard quantitative RT-PCR methods.

3.3 Evaluating Organismal Motility

1. Examination of the rate at which planaria move around in a petri dish and the mode of motility provide a rapid and simple screen for defects in ciliary motility.
2. Place a standard 9 cm-diameter petri dish containing about 8–10 planaria under a low magnification video camera (*see* **Note 3**) so that the entire petri dish fills the frame. Place a ruler to one side of the dish for scale.
3. Use a transfer pipette to gently squirt media at the animals to move them from the edge of the petri dish, where they tend to accumulate, toward the center.
4. Once the planaria start to move, record 30–60 s of movement at 30 frames/s.
5. Rates of movement for individual animals can readily be determined using tracking software (such as Metamorph or ImageJ), or even by using transparent acetate sheets to manually mark the movement over a given time period (*see* **Note 4**).
6. To prepare still images for publication that illustrate the movement (Fig. 1b), videos should be decompiled and then every 100th frame overlaid to form a composite image using the image mask feature of Adobe Photoshop. This provides a complete record of the track taken by each planarian in the dish over a given time period.

3.4 Light Microscopic Analysis of Ciliary Assembly and Activity

1. To examine planaria for the presence of cilia (Fig. 1a), place an animal in a microfuge tube and fill with relaxant solution (*see* **Note 5**).
2. Use differential interference contrast optics and a 60× objective lens to observe the head region where the dorsal and ventral surfaces come together. The cilia here are the easiest to image.
3. For video microscopy of ciliary activity, we usually confine the animals on a microscope slide with a coverslip sealed around with paraffin wax Alternatively, one can use a hole punch to make a small round hole in a piece of parafilm and place the animal in the center with a small drop of culture fluid prior to adding the coverslip (*see* **Note 6**).
4. Use a high-speed video camera to record ciliary motility at the dorsal/ventral junction near the head (*see* **Note 7**). We have found that 250 frames/s provides good quality video and

enables both waveform and beat frequency to be accurately determined (*see* Fig. 2).

5. Use standard kymograph analysis in ImageJ or other software to determine the beat frequency; for untreated animals this is usually 22–24 Hz.

3.5 Electron Microscopy of Cilia Assembly and Architecture

1. For scanning electron microscopic (SEM) analysis of the cilia on the ventral surface, first fix several planaria in relaxant solution. Replace the solution once and leave overnight at room temperature.
2. Fix with 2.5% glutaraldehyde in 0.1 M Na cacodylate buffer pH 7.4 for 1 day with several changes of the fixative. Wash several times with cacodylate buffer and postfix with 1% osmium tetroxide. After additional buffer washes, dehydrate through an ethanol series (10, 25, 50, 50, 85, 95, and 100% ethanol) and dry using a critical point dryer (Autosamdri-815 or equivalent). The planaria are now ready for mounting on the sample stub, sputter coating and imaging in the SEM (*see* Fig. 3).
3. For transmission electron microscopy of ciliary substructure, dissect the anterior region of a small animal, and fix in 1% glutaraldehyde in phosphate-buffered saline for 15 min. Replace with 1% glutaraldehyde in 50 mM Na cacodylate for 1–2 h, changing the solution approximately every 20 min (*see* **Note 8**) (*see* Fig. 3).
4. Postfix with 1% OsO_4 for 1 h and stain en bloc with 1% methanolic uranyl acetate. Following dehydration in ethanol, embed in Epon resin. Thin sections are poststained with uranyl acetate and lead citrate prior to imaging.

4 Notes

1. It is important to maintain a consistent naming scheme for RNAi-mediated gene knockdowns in *S. mediterranea*. The standard accepted nomenclature currently used is all italics and starts with *Smed* followed by a hyphen, the gene name in lower case letters and RNAi in parentheses, e.g. *Smed-genex(RNAi)*.
2. Expressed sequence tag and genomic databases for *S. mediterranea* are available at http://smedgd.neuro.utah.edu/ and a 454 sequence dataset can be obtained from https://planarian.bio.ub.edu/datasets/454/Planaria_FTP/README.html.
3. We currently use a USB 2.0 1024×768 pixel color camera (DFK31BU03.H; The Imaging Source, Charlotte, NC) with a progressive scan CCD image sensor and IC Capture 2.2 software (provided with the camera).

4. Wild-type planaria typically glide at ~1.4 mm/s whereas those without cilia use peristaltic muscle contractions, and traverse the dish much more slowly 0.5 mm/s or less.
5. When placed in standard aldehyde-based fixatives, planaria curl up and release large quantities of mucus. To obviate this problem, for both light microscopy and scanning electron microscopy, the animals are initially placed in relaxant solution. Although they will initially curl, after some hours they will straighten out and can then be processed for imaging.
6. Obtaining high-quality images of ciliary beating is challenging and time-consuming as the animals are strong and move vigorously around when illuminated. Currently there is not an easy way to fully immobilize the planaria for live imaging and so patience is required.
7. We use an X-PRI high speed CMOS camera (AOS Technologies AG, Switzerland) operating at 250 frames/s (*see* Fig. 2).
8. Standard glutaraldehyde/osmium tetroxide fixation protocols can be used provided that multiple glutaraldehyde changes are made to counteract fixative depletion by the secreted mucus. However, these methods sometimes do not result in high-quality preservation of the entire tissue, and so if available high-pressure freezing and freeze substitution are preferable.

Acknowledgement

Our laboratory is supported by grant GM051293 (to S.M.K.) from the National Institutes of Health.

References

1. Mitchell DR (2007) The evolution of eukaryotic cilia and flagella as motile and sensory organelles. Adv Exp Med Biol 607:130–140
2. King SM, Pazour GJ (2009) Cilia: model organisms and intraflagellar transport. Methods in cell biology, vol 93. Elsevier, Burlington, MA
3. Witman GB (1986) Isolation of Chlamydomonas flagella and flagellar axonemes. Methods Enzymol 134:280–290
4. Sears PR, Davis CW, Chua M, Sheehan JK (2011) Mucociliary interactions and mucus dynamics in ciliated human bronchial epithelial cell cultures. Am J Physiol 301:L181–L186
5. Sawamoto K, Wichterle H, Gonzalez-Perez O, Cholfin JA, Yamada M, Spassky N, Murcia NS, Garcia-Verdugo JM, Marin O, Rubenstein JLR, Tessier-Lavigne M, Okano H, Alvarez-Buylla A (2006) New neurons follow the flow of cerebrospinal fluid in the adult brain. Science 311:629–632
6. Rink JC (2013) Stem cell systems and regeneration in planaria. Dev Gene Evol 223: 67–84
7. Newmark P, Sanchez-Alvarado A (2002) Not your father's planarian: a classic model enters the era of functional genomics. Nat Rev Genet 3:210–219
8. Rompolas P, King SM (2009) *Schmidtea mediterranea*: a model system for analysis of motile cilia. Methods Cell Biol 93:81–98
9. Azimzadeh J, Wong ML, Downhour DM, Sanchez-Alvarado A, Marshall WF (2012) Centrosome loss in the evolution of planarians. Science 335:461–463

10. Rink J, Gurley K, Eliot S, Sanchez-Alvarado A (2008) Planarian Hh signaling regulates regeneration polarity and links Hh pathway evolution to cilia. Science 326:1406–1410
11. Almuedo-Castillo M, Salo E, Adell T (2011) Dishevelled is essential for neural connectivity and planar cell polarity in planarians. Proc Natl Acad Sci U S A 108:2813–2818
12. Rompolas P, Patel-King RS, King SM (2010) An outer arm dynein conformational switch is required for metachronal synchrony of motile cilia in Planaria. Mol Biol Cell 21:3669–3679
13. Newmark P, Reddien P, Cebria F, Sanchez-Alvarado A (2003) Ingestion of bacterially expressed double-stranded RNA inhibits gene expression in planaria. Proc Natl Acad Sci U S A 100:11861–11865
14. Dawar B (1973) A combined relaxing agent and fixative for Triclads (Planarians). Biotech Histochem 48:93–94
15. Patel-King RS, Gilberti RM, Hom EFY, King SM (2013) WD60/FAP163 is a dynein intermediate chain required for retrograde intraflagellar transport in cilia. Mol Biol Cell 24:2668–2677

Index

Peter Satir and Søren Tvorup Christensen (eds.), *Cilia: Methods and Protocols*, Methods in Molecular Biology, vol. 1454, DOI 10.1007/978-1-4939-3789-9, © Springer Science+Business Media New York 2016

D

E

Printed in the United States
By Bookmasters